Sociological Odyssey

Contemporary Readings in Sociology

PATRICIA A. ADLER
University of Colorado

PETER ADLER
University of Denver

Australia • Canada • Mexico • Singapore • Spain • United Kingdom • United States

Publisher: Eve Howard
Assistant Editor: Dee Dee Zobian
Editorial Assistant: Stephanie Monzon
Marketing Manager: Matthew Wright
Project Editor: Jerilyn Emori
Print Buyer: Karen Hunt
Permissions Editor: Robert Kauser
Production Service: Andrea Bednar/ Shepherd, Inc.
Copy Editor: Francine Banwarth
Illustrator: Jim Daggett
Cover Designer: Bill Stanton / Stanton Design
Text and Cover Printer: Webcom Limited
Compositor: Shepherd, Inc.

Printed in Canada
8 9 05 04

For permission to use material from this text, contact us by
Web: http://www.thomsonrights.com
Fax: 1-800-730-2215
Phone: 1-800-730-2214

Wadsworth/Thomson Learning
10 Davis Drive
Belmont, CA 94002-3098
USA

For more information about our products, contact us:
Thomson Learning Academic Resource Center
1-800-423-0563
http://www.wadsworth.com

International Headquarters
Thomson Learning
International Division
290 Harbor Drive, 2nd Floor
Stamford, CT 06902-7477
USA

UK/Europe/Middle East/South Africa
Thomson Learning
Berkshire House
168-173 High Holborn
London WC1V 7AA
United Kingdom

Asia
Thomson Learning
60 Albert Street, #15-01
Albert Complex
Singapore 189969

Canada
Nelson Thomson Learning
1120 Birchmount Road
Toronto, Ontario M1K 5G4
Canada

Library of Congress Cataloging-in-Publication Data
Sociological odyssey : contemporary readings in sociology / [edited by] Patricia A. Adler, Peter Adler.
p. cm.
ISBN 0–534–57053–4
1. Sociology. I. Adler, Patricia A. II. Adler, Peter, 1951–

HM585 .S6 2001
301—dc21

00–033449

This book is printed on acid-free recycled paper.

To Our Proteges

Kathy, Angela, Laurie, Dana, Joanna, Jen, Katy, Adina, and Kevin

May all teachers be blessed with students such as these

Contents

Preface

. . . what a long strange trip it's been . . .

So sang the Grateful Dead in the 1970s, when, in retrospect, their careers were only in an intermediate stage. The same refrain can be sung for sociology. This discipline, barely a century old (sociology is a relatively young science—only about 100 years old in Europe and slightly younger than that in America, a mere adolescent by intellectual standards), has witnessed the enormous social changes of the twentieth century. These changes include unsurpassed prosperity, a Depression that wreaked havoc with people's lives, several wars of varying lengths and questionable motivations, numerous natural disasters, an explosion in the production of knowledge, the rise of new media forms such as television, a grassroots student revolution that virtually altered the social patterns of society, so-called mind expanding drugs such as LSD, communication upheavals caused by innovations such as the Internet and cell phones, and myriad other inventions, events, and ideas that have shaped the world in which we now live. And sociology, the science of society, has been there to witness the impact of these profound transformations. Arguably, no other academic discipline is better suited to assess, analyze, evaluate, and predict what these changes have wrought.

You are about to begin an odyssey, a journey into a new realm of thinking that, if taken seriously, can change the way you view the world. The "sociological imagination," as the mid-century sociologist C. Wright Mills referred to it, can transform one's perspective on reality so that even the most mundane, seemingly trivial aspects of everyday life take on added significance. Further, the big issues that face us, such as social inequality, political corruption, crime, and environmental decay, can also be looked at with fresh eyes to lend increasing clarity to why these exist and how we can improve conditions. Like the visionary filmmaker Stanley Kubrick predicted decades ago, the early twenty-first century will be a space odyssey into domains we could not even have dreamed about a scant decade ago.

We are pleased to be your guides in these travels. Over the course of the next several months, with the assistance of your professor, we will explore parts of society that may either be very familiar to you or seem very far away. In either case, we hope to enlighten you about the social world that surrounds you. For the most part, the selections we have chosen are "experience-near" for you, such as on the change of identity you experienced from high school to college, the popularity of children in cliques, student participation in classrooms, public bathroom behavior, high school reunions, intermarriage, the Internet, and the behavior of some fraternity members. Other topics may not be as integral to your everyday lives, but relevant nevertheless. These include people on welfare, women with sexually transmitted diseases, skinheads, upper-level drug dealers and smugglers, illicit drug use by pharmacists, gang behavior, and the work of impoverished new immigrants. We hope that at the end of this journey you will have a better understanding of the social dynamics of society and the complexities that make up your world.

ACKNOWLEDGMENTS

This book came to full fruition during the hectic months of the Fall, 1999 semester. For years we had toyed with the idea of creating an exciting, new introductory reader for students to enjoy, but it wasn't until Eve Howard, stalwart acquisitions editor for Wadsworth, flew to Boulder to twist our arms that we finally agreed that the time was ripe. Throughout this process, Eve has been a strong supporter, listening to our worries, fielding our concerns, and doing more than we thought humanly possible to push the project through in record time. To her, we give our sincerest and heartfelt thanks.

We have also had the support of two departments at the University of Colorado and University of Denver, respectively, that have given us the foundation to pursue a project such as this. Our colleagues have been most open to hearing about our various selections, suggesting possible readings, and understanding the value of creating such a text. Most particularly, Paul Colomy (University of Denver), provided critical feedback at a time when we were unsure of the precise direction we wanted to take. His intellectual guidance will forever be appreciated. Our Chairs, Dennis Mileti and Nancy Reichman, gave us the intellectual and emotional support to pursue this endeavor. Other friends and colleagues, such as Dean Birkenkamp, Dan Cress, Gary Alan Fine, Chuck Gallmeier, Bob Granfield, Rik Scarce, and D. Angus Vail, gave us some ideas that were used in making our choices. Our children, Jori (now a superb sociologist in her own right) and Brye, offer us a perspective to evaluate what a college student wants to hear, as well as a model for what parents want to see in their progeny. No thank-you would be complete without the inclusion of Dorene Miller, Administrative Assistant at the University of Denver, who time and again has bailed us out. She unselfishly does the tough work and lends the type of support without which projects such as these could not flourish. Finally, we are fortunate to have a cadre of comrades, our proteges, who have also given us advice and encouragement along the way. Kathy Fox, Angela Yancik, Laurie Scarborough Voss, Dana Jones Hubbard, Joanna Gregson Higginson, Jennifer Lois, Katy Irwin, Adina Nack, and Kevin David Vryan are all now full-fledged collaborators with us in the joint sociological enterprise we share. We lovingly dedicate this book to them.

A team of sociological reviewers were also an important cog in assuring that the selections in this book had the most relevance to students' lives:

David Boden
Lake Forest College

John Bridges
Kutztown University

Kevin Early
Oakland University

Patti Guiffre
Southwest Texas State University

Jane Johnson
Southwest Texas State University

David Maines
Oakland University

Alvar Nieves
Wheaton College

We are grateful for the quick turnaround they gave in reviewing the book, as well as the thoughtful comments about the numerous ways that introductory sociology can be taught.

At Wadsworth Publishing Company, in addition to the steadfast work of Eve Howard, mentioned previously, there have been a number of individuals who have assisted us. Most especially, Dee Dee Zobian, assistant to Eve Howard, did more of the daily work than anyone else to assure the timeliness of the project. Dee Dee treated us as if we were her only project, when in reality she was simultaneously juggling a dozen or more ventures. Others, such as Jerilyn Emori, Bob Kauser, and Andrea Bednar made sure that the final product was produced with the professionalism that has become the hallmark of Wadsworth's operation. This list would not be complete without mentioning others in the publishing industry who either directly or indirectly helped us. Notably, Serina Beauparlant, formerly of Wadsworth but now holding court at Mayfield, was literally the first person to try to cajole us into doing this book. No other person deserves more credit than Serina for giving birth to the embryo that has culminated in this final product. Dean Birkenkamp (Rowman and Littlefield), Mitch Allen (Alta Mira), and Martha Heller (formerly of Rutgers University Press) are all editors with whom we've worked who remind us of the scholarly and intellectual fortunes, rather than merely economic ones, that are borne out through our books. They represent the best that publishing has to offer.

To our students, past, present, and future, and to our readers, we thank you for your patience, diligence, and understanding. We hope that we have been able to convey the sociological perspective in such a way as to make you as excited about applying it to your lives as it is for us to teach it to you.

About the Editors

Patricia A. Adler (Ph.D., University of California, San Diego) is Professor of Sociology at the University of Colorado. In 1999, she was named as the Outstanding Teacher in the faculty of Arts and Sciences. She has written and taught in the areas of deviance, drugs in society, sociology of gender, and the sociology of children. A second edition of her book *Wheeling and Dealing* (Columbia University Press), a study of upper-level drug traffickers, was released in 1993.

Peter Adler (Ph.D., University of California, San Diego) is Professor of Sociology at the University of Denver, where he served as Chair from 1987 to 1993. In 1998, he was named the University Lecturer, an award that represents outstanding achievement in scholarship and research. His research interests include social psychology, qualitative methods, and the sociology of sport and leisure.

Together the Adlers edited the *Journal of Contemporary Ethnography* and were the founding editors of *Sociological Studies of Child Development*. They are also editors of another Wadsworth book, *Constructions of Deviance,* a collection of readings on deviant behavior, now in its third edition. Among their other books are *Backboards and Blackboards,* a participant-observation study of college athletes that was published by Columbia University Press in 1991, and *Peer Power,* a study of the culture of elementary schoolchildren that was published by Rutgers University Press in 1998. Currently, they are studying the subculture of resort workers.

Patti and Peter have been writing together for thirty years, ever since they were undergraduates at Washington University in St. Louis. They have two children: a son, Brye, who recently graduated from Boulder High School and attends Emory University, and a daughter, Jori, a sociology major who recently graduated from Emory University. Their dog, Lanai, accompanies them to their classes.

If you have any questions for them, they can be reached at adler@spot.colorado.edu

About the Contributors

Emily Stier Adler is Professor of Sociology at Rhode Island College with areas of specialization in the sociology of gender, the family, and research methods. She is the coauthor, with Roger Clark, of *How It's Done: An Invitation to Social Research* (Wadsworth, 1999). Her current research projects include a study of college faculty members' views of teaching and the classroom experience and on family interaction patterns after retirement.

Elijah Anderson is the Charles and William L. Day Professor of the Social Sciences, Professor of Sociology, and Director of the Philadelphia Ethnography Project at the University of Pennsylvania. An expert on the sociology of black America, he is the author of the highly regarded work, *A Place on the Corner,* numerous articles on the black experience, and the recently published *Code of the Streets.* For his ethnographic study, *Streetwise,* he was honored with the Robert E. Park Award by the American Sociological Association.

Andrew Billingsley is Professor in the Department of Sociology, The Institute for Families in Society, and the African American Studies Program at the University of South Carolina. He received his Ph.D. in Social Policy and Social Research at the Florence Heller School of Brandeis University. His books include *Mighty Like a River: The Black Church and Social Reform* (Oxford University Press, 1999), *Climbing Jacob's Ladder* (Simon and Schuster, 1993), *Black Families in White America* (Simon and Schuster, 1968), and *Children of the Storm* (1972).

A. Ayres Boswell is the Director of the Intake Department for the Community Mental Health Foundation in Doylestown, Pennsylvania. She does evaluations, assessments, and makes recommendations for treatment as well as doing psychotherapy with clients.

William B. Brown is an Associate Professor of Sociology at the University of Michigan-Flint. He is the coauthor of *Youth Gangs in American Society* (Wadsworth, second edition, 2000). His current research centers on human rights violations of the state against the poor in America's inner cities, with hope that continued illumination of these violations will someday result in the adoption of social justice principles.

Alan Bryman is Professor of Social Research in the Department of Social Sciences, Loughborough University, England. His main research interests lie in research methodology, leadership studies, organizational analysis, and theme parks. He is author or coauthor of several books, including *Quantity and Quality in Social Research* (Routledge, 1988), *Charisma and Leadership in Organizations* (Sage, 1992), *Disney and His Worlds* (Routledge, 1995), and *Social Scientists* (Routledge, 1999). He is currently writing a book on social research methods (Oxford University Press, forthcoming) and *The Disneyization of Society* (Sage, forthcoming).

Spencer E. Cahill is Associate Director of Interdisciplinary Studies and Sociology at the University of South Florida. He has published articles on a range of topics including gender identity acquisition, public life, childhood, disability, and professional socialization. His current research examines adolescent culture and relations through notes and other personal documents that adolescents wrote and exchanged among themselves.

Cleopatra Howard Caldwell is an Assistant Professor in the Department of Health Behavior and Health Education, School of Public Health and Co-Associate Director of the Program for Research on Black Americans at the University of Michigan. She has published in the areas of help-seeking behaviors and informal social supports among African Americans, the black church as a social service institution, and race-related socialization and academic achievement among African American youth.

Joel Charon earned his Ph.D. in Sociology at the University of Minnesota. He has worked in the Department of Sociology and Criminal Justice at Moorhead State University, which he has chaired since 1993. He has written *Symbolic Interactionism* (seventh edition, Prentice Hall), *The Meaning of Sociology* (sixth edition, Prentice Hall), and *Ten Questions* (fourth edition, Wadsworth). He loves teaching, but has decided to devote the rest of his professional life to research and writing, beginning in 2001.

Randall Collins is Professor of Sociology at the University of Pennsylvania. His books include *Conflict Sociology* (1975), *The Credential Society* (1979), *Weberian Sociological Theory* (1986), *The Sociology of Philosophies* (1998), and *Macro-History: Essays in Sociology of the Long Run* (1999).

Paul Colomy is Professor of Sociology at the University of Denver. He is editor of *The Dynamics of Social Systems, Functionalist Sociology,* and *Neofunctionalist Sociology,* and has contributed articles to *Sociological Theory, Social Problems,* and *Sociological Forum.* His current work examines the origins and transformation of juvenile justice.

Laura Craig is research officer with the Canadian Broadcasting Corporation (CBC). At the CBC, Laura provides consultation on audience, demographic, and media trends. Of particular interest to her is the impact of the Internet on traditional media. Laura has a masters degree in Sociology from the University of Calgary, where she specialized in youth culture.

Dean A. Dabney is an Assistant Professor in the Department of Criminal Justice at Georgia State University. His research agenda is principally focused on advancing the understanding of forms of deviant and/or criminal behavior that occur in organizational settings. Namely, he has studied theft and substance abuse behaviors occurring among practicing pharmacists and nurses, incompetence among prison doctors, as well as shoplifting and employee theft in the retail industry.

Jay Demerath is Professor of Sociology at the University of Massachusetts, Amherst, and immediate past President of the Society for the Scientific Study of Religion. He is most recently the senior editor of *Sacred Companies: Organizational Aspects of Religion and Religious Aspects of Organizations* (Oxford University Press, 1998). The coauthor of *A Bridging of Faiths: Religion and Politics in a New England City* (Princeton University Press, 1992), he has taken the same topic from the local to the global in his forthcoming 14-nation comparative book, *Crossing the Gods: Religion, Violence, Politics, and the State Across the World* (Rutgers University Press).

G. William Domhoff is a Research Professor at the University of California, Santa Cruz, where he teaches a course on social psychology and power. He is most recently author of *State Autonomy or Class Dominance* (Aldine de Gruyter, 1996) and coauthor with Richard L. Zweigenhaft of *Diversity in the Power Elite: Have Women and Minorities Reached the Top?* (Yale University Press, 1998). Four of his earlier books, including the one from which the selection in this anthology is taken, appeared on the recent list of top 50 best sellers in sociology since 1945. He is currently studying the likelihood that a new liberal-environment-labor coalition can transform the Democratic Party due to the fact that the Voting Rights Act of 1964 made it possible for African American voters in the South to force conservative Southern whites into the Republican Party.

Paula Foster is the Clinical Director of Children's Services, a program of May Behavioral Health, Inc. In addition, she has a private practice specializing in trauma resolution, sexual offending, and family systems.

John H. Gagnon is Distinguished Professor of Sociology Emeritus at the State University of New York at Stony Brook. He received his undergraduate and graduate degrees from the University of Chicago. He is the author or coauthor of such books as *Sex Offenders* (1965), *Sexual Conduct* (1973), *Human Sexualities* (1977), and *The Social Organization of Sexuality* (1994). In addition, he has been the coeditor of a number of books, most recently *Conceiving Sexuality* (1995) and *Encounter with AIDS* (1997), as well as the author of many scientific articles. He is now Senior Scientist at the Center for Health and Policy Research.

Cecilia Garza is Chair of the Department of Psychology and Sociology at Texas A&M International University in Laredo, Texas, where she received the Excellence in Teaching, Scholarship, and Service Award and Scholar of the Year in 1996–97. She received her Ph.D. from Texas A&M University in 1993. Her research interests are in immigration issues, medical sociology, and inequality.

Paul S. Gray is Associate Professor in the Department of Sociology at Boston College. He received his A. B. in Politics from Princeton, an M.A. in Education from Stanford, and the Ph.D. in Sociology from Yale. Some of his research interests are: social change, sociology of education, and corporate-community relations. Dr. Gray is the author of *Unions and Leaders in Ghana* and coauthor of *The Research Craft,* a methods text. Currently, he does business consulting with the Center for Corporate Community Relations.

Sharon Hays is an Associate Professor of Sociology and Gender Studies at the University of Virginia. She is the author of *The Cultural Contradictions of Motherhood* (Yale University Press, 1996), an interview-based exploration of the cultural tensions between the ideology of "good" mothering and the ethos of a rationalized market society. Her current research, *Inside Welfare* (Oxford University Press, forthcoming), is an ethnographic study of the relationship between welfare reform and "family values." She also does research on family history, feminist theory, and issues of identity and cultural recognition. Her work has been published in *Sociological Theory, Journal of Marriage and the Family,* and *Media, Culture, and Society.*

Richard Hollinger is an Associate Professor at the University of Florida. He holds joint appointments in the Department of Sociology and the Center for Studies in Criminology and Law. Professor Hollinger is the author of numerous journal articles and three books: *Theft by Employees* (with J. P. Clark, 1983), *Dishonesty in the Workplace* (London House Press, 1989), and most recently, *Crime, Deviance, and the Computer* (Dartmouth Press, 1997). His principal research interests involve occupationally related crime and deviance. In addition to further work on the topic of drug abusing pharmacists, he is currently writing a paper explaining the causes of employee dishonesty.

Lynda Lytle Holmstrom is Professor of Sociology at Boston College and former chairperson of the department. Her books include: *The Two-Career Family* (Schenkman, 1972); *The Victims of Rape: Institutional Reactions,* with Ann Wolbert Burgess (Wiley, 1978; Transaction, 1983); and *Mixed Blessings: Intensive Care for Newborns,* with Jeanne Harley Guillemin (Oxford University Press, 1986). Her present research, with David A. Karp and Paul S. Gray, is on family dynamics and the college application process.

David Karp is Professor of Sociology at Boston College. His earlier books on cities, everyday life, aging, and depression commonly reflect an enduring interest in how people invest their daily worlds with meaning. His book entitled *Speaking of Sadness: Depression, Disconnection, and the Meaning of Illness* was the 1996 recipient of the Charles Horton Cooley Award. A new book, *Committed to Care: Mental Illness, Family Ties, and Moral Responsibility,* will be published by Oxford University Press in 2001.

Ilsoo Kim is currently affiliated with Mercy College in New York as an adjunct professor. His book on Koreans in New York is titled *New Urban Immigrants: The Korean Community in New York* (Princeton University Press, 1981). He is currently working on ethnographic research on Asian communities in New York.

Jonathan Kozol received a B.A. from Harvard University. His books include *Death at an Early Age, Rachel and Her Children,* and *Savage Inequalities.*

Edward O. Laumann is the George Herbert Mead Distinguished Service Professor in the Department of Sociology at the University of Chicago, Chair of the Department of Sociology, and Director of the Ogburn Stouffer Center for

Population and Social Organization. Previously he has been editor of the *American Journal of Sociology,* Dean of the Social Sciences Division, and Provost at the University. Among his books are *Bonds of Pluralism, The Organizational State* (with David Knoke), *Chicago Lawyers* (with John P. Heinz), *The Hollow Core* (with John P. Heinz, Robert Nelson, and Robert Salisbury), and *Designing for Technological Change* (with Gerry Nadler). Two volumes on human sexuality were published in 1994: *The Social Organization of Sexuality* and *Sex in America.* Laumann's research interests include social stratification, the sociology of the professions, occupations and formal organizations, social network analysis, the analysis of elite groups and national policy making, and the sociology of human sexuality.

Debra McPhee is Assistant Professor at Barry University School of Social Work in Miami, Florida, where she teaches both social welfare policy and social work practice. Dr. McPhee holds a B.A. from St. Mary's University in Nova Scotia, an M.S. from Columbia University, and a Ph.D. from the University of Toronto. She has international professional social work experience having practiced in the United States, The Netherlands, and Canada. Academic and research interests include child welfare, social welfare reform, and international health care policy. She has publications in the areas of child welfare, health care policy, and the social construction of social problems.

Robert T. Michael is the Eliakim Hastings Moore Distinguished Professor and Dean of the Harris Graduate School of Public Policy at the University of Chicago. From 1984 to 1989 he served as Director of the National Opinion Research Center (NORC).

Stuart Michaels served as Project Manager of the National Health and Social Life Survey (NHSLS).

Melissa A. Milkie is Assistant Professor of Sociology at the University of Maryland. Her research interests centers on questions regarding the interplay of culture, groups, and the self. Specifically, she focuses on continuity and change in cultural ideals related to gender and ethnicity, how individuals in various group contexts create meaning within those cultural boundaries, and some of the consequences of cultural beliefs for well-being. Currently, she is working on a project examining the relationship between cultural ideals of fathering and the actual practices of men in families across generations. Her work has been published in journals such as *Social Psychology Quarterly, American Sociological Review,* and *Journal of Marriage and the Family.*

C. Wright Mills received his Ph.D. in sociology from the University of Wisconsin and was Professor of Sociology at Columbia University. His best-known works are *White Collar* (1951), *The Power Elite* (1956), and *The Sociological Imagination* (1959).

Horace Miner earned his Ph.D. in Social Anthropology at the University of Chicago. He taught at the University of Michigan in the Department of Anthropology. His books include *The Primitive City in Timbuctoo, St. Denis: A French Canadian Parish,* and *The City in Modern Africa.*

Adina Nack is a Ph.D. candidate in sociology at the University of Colorado. Her teaching and research interests include medical sociology, feminist theory, ethnographic methods, and social psychology. She is currently studying self-conception and stigma management in adults living with chronic, non-HIV sexually transmitted diseases.

Steven Nock is Professor of Sociology at the University of Virginia. His most recent book is *Marriage in Men's Lives.* He is now conducting a five-year evaluation of the legal innovation known as Covenant Marriage in Louisiana.

Michael Novak is a theologian, author, and former U.S. ambassador. He currently holds the George Frederick Jewett Chair in Religion and Public Policy at the American Enterprise Institute in Washington, DC, where he is Director of Social and Political Studies. In 1994, he received the Templeton Prize for Progress in Religion (a million-dollar purse awarded at Buckingham Palace) and delivered the Templeton address in Westminster Abbey. Mr. Novak has written numerous books in the philosophy and theology of culture, some of which include *The Open Church* (1964), *Belief and Unbelief* (1965, 1994), *The Rise of Unmeltable Ethnics* (1972, 1994), *Moral Clarity in the Nuclear Age* (1983), *The Joy of Sports* (1976, 1994), and with his daughter, Jana, *Tell Me Why: A Father Answers His Daughter's Questions about God* (1998).

Irene Padavic is Associate Professor of Sociology at Florida State University. With Karin Brewster she has examined changes in ideology about gender (*Journal of Marriage and the Family,* 2000), and together they are currently working on a project about changes in child care arrangements over the last 20 years. With Mindy Stombler (*Social Problems,* 1997) and with Alexandra Berkowitz (*Journal of Contemporary Ethnography,* 1999) she has explored differences between African American and white women's groups on college campuses. Most of her research, however, has centered on gender and work, including several articles about segregation written independently or in collaboration with Barbara Reskin, and articles about sexual harassment and women veterans' earnings.

Joel Perlmann is Senior Scholar, Jerome Levy Economics Institute of Bard College and Levy Institute Research Professor at the College. He received his Ph.D. in history and sociology at Harvard University. Among his publications are *Ethnic Differences* (Cambridge University Press), *Women's Work? American Schoolteachers 1650–1920* (with Robert Margo, forthcoming, University of Chicago Press), and (with Roger Waldinger), "Immigrants Past and Present: A Reconsideration," in *The Handbook of International Migration.* He is at work on the history of American ethnic intermarriage; a comparison of contemporary second-generation economic prospects with those of second generations from the last great wave of American immigration; and a demographic history of the Russian Jews on the eve of their emigrations to the West.

Robert Putnam is the Peter and Isabel Malkin Professor of Public Policy at Harvard University. He is the author of many books, including *Bowling Alone* (1999), *Making Democracy Work* (1993), *Double-Edged Diplomacy* (1993), *Hanging Together* (1984), *Bureaucrats and Politicians in Western Democracies* (1981), *Comparative Study of Political Elites* (1976), and *Beliefs in Politicians* (1973). A recipient of numerous scholarly honors and a consultant to various government and international organizations, he is a Fellow of the American Academy of Arts and Sciences.

Mark R. Rank received his Ph.D. in sociology from the University of Wisconsin. He is currently Associate Professor in the George Warren Brown School of Social Work at Washington University in St. Louis. Rank's research agenda has focused on the areas of welfare recipients, poverty, social inequality, and family dynamics.

Barbara Reskin is Professor of Sociology at Harvard University. Her research focuses on sex, race, and ethnic inequality in jobs. Her most recent books are *The Realities of Affirmative Action in Employment* (1998), *Women and Men at Work* (with Irene Padavic, 1994), *Job Queues, Gender Queues: Explaining Women's Inroads into Male Occupations* (with Patricia Roos, 1990). She is currently studying the impact of employers' personnel practices and government actions on levels of employment discrimination by sex and race.

Barbara J. Risman is Professor of Sociology at North Carolina State University, where she also directs the graduate program for the Women and Gender Studies Program. She is currently coeditor of *Contemporary Sociology.* Currently, her research is on women's activism in post-Soviet Russia.

George Ritzer is best known for *The McDonaldization of Society,* but he has authored other works like it, including *Enchanting a Disenchanting World.* He is cofounding editor of *The Journal of Consumer Culture.* The vast majority of his work is in the area of social theory and metatheory. Sage (England) will soon publish two volumes of his collected works, one volume focusing on theory and the other on consumption. He has been Professor of Sociology at the University of Maryland since 1974.

Andrew Shapiro is a writer and legal scholar with a particular interest in the Internet. He is the author of *The Control Revolution: How the Internet is Putting Individuals in Charge and Changing the World We Know.* He is also cofounder of Kind.com and a senior advisor to the Markle Foundation.

Edward Shapiro is Professor of History at Seton Hall University, where he has taught since 1969. Professor Shapiro earned his Ph.D. from Harvard University. He is the author of over two hundred books, reviews, and essays, many of which are on American Jewish history. His book, *A Time for Healing: American Jews since World War II* (1992), was nominated for the Pulitzer Prize in history. A collection of his essays titled, *We Are Many: Essays on American Jewish History and Culture,* is forthcoming with Syracuse University Press.

Ruth Sidel is Professor of Sociology at Hunter College of the City University of New York. Professor Sidel has studied the role of women and provision of human services in urban areas in the United States and in several other countries. Her books include *Women and Child Care in China, Women and Children Last, On Her Own: Growing Up in the Shadow of the American Dream, Battling Bias,* and *Keeping Women and Children Last.*

Ken Silverstein is a writer based in Washington, DC. His work has appeared in *The Nation, Harper's, Mother Jones, Sierra,* among others. He is the author of *Washington on $10 Million a Day,* about beltway lobbying, and of the upcoming *Private Warriors,* which looks at the current activities of a group of Cold War-era arms dealers, generals, and covert operators.

Joan Z. Spade is Associate Professor of Sociology at Lehigh University. She is the author of several articles in the area of gender, work, and family and also has written on tracking in education. She is currently working on a study of the governance of higher education.

Deborah Tannen is on the linguistics department faculty at Georgetown University, where she is one of only three who hold the distinguished rank of University Professor. Among her scholarly books are *Talking Voices* (Cambridge University Press), *Gender and Discourse* (Oxford University Press), and *Conversational Style* (Ablex). Among general audiences Tannen is best known as the author of *You Just Don't Understand: Women and Men in Conversation* (Ballantine, 1990), which was on the *New York Times* best-seller list for nearly four years; *That's Not What I Meant!: How Conversational Style Makes or Breaks Relationships* (Ballantine, 1986); *Talking from 9 to 5: Women and Men in the Workplace: Language, Sex, and Power* (Avon, 1994); and *The Argument Culture: Stopping America's War of Words* (Ballantine, 1999).

Vered Vinitzky-Seroussi is Ph.D. lecturer at the Department of Sociology and Anthropology and Institute of Criminology, Hebrew University of Jerusalem, Israel. She has written *After Pomp and Circumstance: High School Reunion as an Autobiographical Occasion* (University of Chicago Press, 1998). Currently, she is working on the commemoration of a difficult past using the assassination of Yitzhak Rabin as a case study.

Howard Waitzkin is Professor and Director at the Division of Community Medicine, Department of Family and Community Medicine, at the University of New Mexico. Since receiving his doctorate in sociology and a degree in medicine from Harvard University, he has worked with several community-based health projects and has written numerous articles and books on health policy, community medicine, and patient-doctor communication, including *The Politics of Medical Encounters* and *The Exploitation of Illness in Capitalist Society.*

Alan Wolfe is Professor of Political Science and Director of the Center for Religion and American Public Life at Boston College. He is the author or editor of more than ten books, including *Marginalized in the Middle* (University of Chicago Press, 1997) and *One Nation, After All* (Viking Penguin, 1998). His current research interests revolve around a project, financed by the Lilly Endowment and the Institute for Civil Society, which deals with the question of moral freedom.

William C. Yoels is Professor of Sociology and previously directed the Ph.D. program in Medical Sociology at the University of Alabama-Birmingham. He has coauthored *Being Urban* (Praeger), *Sociology in Everyday Life* (Waveland), and *Experiencing the Life Cycle* (Charles C. Thomas Publishers). In addition, he has published numerous articles in journals, such as *Symbolic Interaction, Journal of Contemporary Ethnography, Archives of Physical Medicine and Rehabilitation,* and *Sociology of Health and Illness.*

Kevin Young currently is affiliated with Loughborough University in the United Kingdom. His research and teaching interests include youth, sport, and cultural theory. He is vice president of the International Sociology of Sport Association.

Robert Zussman teaches sociology at the University of Massachusetts-Amherst. He is the author of *Mechanics of the Middle Class: Work and Politics among American Technical Workers* and *Intensive Care: Medical Ethics and the Medical Profession.* He is the editor of *Qualitative Sociology* and is working on a history of the social structure of autobiographical accounts.

PART I

THE SOCIOLOGICAL VISION

One of its early founders, August Comte, in developing his new approach to the study of human groups, dubbed sociology the "queen" of all the social sciences. By this he meant that sociology represented the most fully developed social scientific perspective, at the apex of all the others since, in many respects, it combines elements of the other disciplines. For instance, it builds on the understandings generated by economics into financial life and other aspects of commercial exchange; by political science into our government and official policies; by psychology into mental and cognitive processes; by anthropology into different cultures; by history into the large-scale trends of the past; and by communication into the ways we convey thoughts and information to each other, either face-to-face or through the mass media. Sociology has also served as the "parent" of several other "spin-off" disciplines, including criminology and criminal justice, the study of crime and our social institutions that deal with it (police, courts, jails); gerontology, the study of aging and the life course; women's studies, the examination of sex and gender; ethnic studies, the analysis of race and ethnicity; urban studies, the inquiry into life in cities; social work, an applied field that helps people navigate their way through our social institutions, and several more that are still developing. Even marketing, a discipline in the field of business, is heavily influenced by sociology, by a need to know how people will act, what they will like or dislike, what they will want and buy.

Positioned to both succeed its predecessors and lead its offspring, sociology's depiction, by Comte, as the "royalty" of the social sciences may have some truth. In its wide-ranging focus, one can imagine a sociology of almost anything, from the most minuscule worlds of how individuals pause to let others take a turn in a conversation, to middle-level scenes and subcultures such as hip-hop or auto thieves, to macro arenas such as the decline or evolution of the family, or the global flow of markets and workers. In its substance, there can be a sociology of such diverse, idiosyncratic group behaviors as surfing, stamp collecting, religious fundamentalism, and dog breeding to large-scale trends relating to fertility rates, ethnic intermarriage, and international health care systems.

Yet, ironically, despite its omnipresence, sociology is in some ways still a "hidden" discipline. Quite possibly, you never even heard of it until you opened up your college catalogue and saw it listed there. You've, no doubt, heard of political science at least annually in our voting polls, of psychology in conversations about neuroses and disorders, of communication in the news and the mass media, of anthropology in the *National Geographic* or in films of foreign cultures, and of economics in the way the stock market fluctuates and the government regulates it. That may be because sociology consists of an approach, a *perspective* for studying the world rather than a specific empirical focus within it. Sociology's subject matter focuses on the *trends and patterns* into which human behavior forms and the (sometimes invisible) *social forces* that mold it into these shapes. These social forces may be large-scale social institutions or policies, the *norms* (behavioral guidelines) and *values* (shared goals, desires, beliefs) circulating within our groups, or specific, face-to-face encounters people have with concrete others who influence them. Students of sociology have the greatest opportunity to see their subject matter in the everyday world around them of any discipline, including the sciences, the humanities, and the social sciences. This sometimes leads people to think that sociology is nothing more than the commonsensical, that which we experience in our daily lives. We think that the readings presented in this book will show you that, quite to the contrary, sociology proves that "things are not always what they appear to be."

In this book we want to take you on a **Sociological Odyssey** through some of the fascinating and important revelations that this discipline has to offer. We want to show you what sociology has discovered about the everyday world and the way you can use this perspective to keep discovering new insights into the patterns and connections of our world. Contemporary society is evolving and changing rapidly, and it is our goal to fill this book with fresh selections that keep you up-to-date with current developments that will have the most meaning and relevance to your everyday lives.

In this first part of the book we introduce you to the "vision" of sociology, the unique perspective that differentiates it from your taken-for-granted realities. We begin by offering some writings that spell out the key elements of the sociological perspective. First, Randall Collins introduces us to the field of sociology by defining its domain of study. He shows how we can all recognize the subject matter and abstract forces of sociology in our routine, everyday experiences, and shows how these can inspire intellectual curiosity and satisfaction. Next, C. Wright Mills talks about the importance of recognizing the connection between concrete, everyday behavior and the larger, invisible forces of history and social structure. He suggests that a greater understanding and appreciation of the world can be gained by developing the "sociological imagination" and learning to recognize the invisible forces that surround us. Joel Charon then differentiates between a stereotype and a sociologically scientific generalization, pointing out the dangers and flaws of the former and the careful construction of the latter. All of these selections are designed to alert you to this new way of sociological thinking that your professor is teaching you.

In the next section, our authors discuss the nature of theory and methods in sociology. Paul Colomy offers an explanation of the three main **Theoretical Perspectives** (functionalism, conflict theory, and symbolic interactionism) that form the foundation of sociology. These can be applied to help understand the larger functioning of society as a whole or the smaller workings of its sub-sections. Theories can help explain the "big picture" of how parts of society work separately or in interrelation, as well as the way people are guided to act within various social situations and contexts.

Macro theories (functionalism and conflict) focus on the broad features of society. Sociologists who use this approach analyze such things as social institutions and social class to see how large entities are related to each other. They believe that in order to understand human behavior, we need to see how social structure both restricts and guides social behavior. *Social structure* is the framework of society that has already been laid out before you were born; it is the patterns of society, such as the relationships between men and women, education and religion, or ethnicity and politics. Critical to determining your behavior are such central sociological *demographic variables* as your ethnicity, religion, age, gender, occupational and marital status, or socioeconomic status. Most sociologists believe that the difference between people is not based on biological differences (nature), but on people's place in the social structure and what they learn there (nurture).

Micro theories (symbolic interactionism) place the focus of study on social interaction. Here, we look at what people do when they come together and interact in a face-to-face environment. Micro sociologists are more interested in forms of interaction, observing how people talk, gesture, and make sense of their everyday worlds.

Rather than looking at the broad structures of society, they focus on the small details of social life, how people negotiate with each other, and how people's perceptions and interpretations of their world are influenced by each other.

Part I concludes with illustrations of the three major methodological approaches in sociology (experimental design, survey research, field research). Sociologists need to study the social world in order to learn about it, construct theories to explain how it works, and make social policies to help make it better. As social scientists, we are tied to many of the canons of science and the scientific method. However, ours is not a world of merely atoms and molecules. We study a distinctly social phenomenon: human beings. Studying the social world requires us to create methods that can capture the fullness of human behavior.

Human life is so complex, there are so many variables to consider, that it is nearly impossible to study people with exact precision. Yet human behavior is also filled with patterns and regularities. We behave in fashions predictable enough for people to roughly anticipate what others will think and do. All of us are "raw" sociologists; every time we enter a scene we take into account numerous variables about others at the same time as we, ourselves, broadcast out our own information. At a party, for instance, we make assessments about who looks interesting, who we want to talk to, or who we should stand or sit next to in a room. We classify and categorize people by their age, gender, race, class, fashion, appearance, actions, and expressions. We form expectations about the kind of people they are and how they might behave. If you are at a residential college, for instance, your first few days on campus were probably spent "checking out" various people to see who might likely become your friends. We also tend to act "normatively," following rules of behavior that operate under the surface of society, in such things as our seating patterns, traffic patterns, dress codes, and our rules of decorum and demeanor. In all of these respects, we think sociologically. It is our hope that with this book (and course), you may be able to hone these skills further.

As social scientists, sociologists have the problem of trying to study human behavior, even though humans are both unpredictable and highly patterned. Sociological methodology is the means by which social scientists solve this dilemma. The three methods available offer us various systematic approaches to studying different aspects of society and its people. They offer us ways to be rigorous and controlled, broad and precise, or empathetic and deep, in studying people, so that we can invest our findings with some accuracy, validity, and reliability.

In the selections that follow, we present reflections on sociology's three major **Methods.** Emily Stier Adler and Paula Foster begin by discussing the logic and intent of one of the earliest methods used, the experiment. They carefully guide us through the logic of experimental design and its implementation, showing how a precise program of focus, matching, and control can help social scientists to isolate the influence of specific sociological variables on other factors. Their study examines the impact of a literature-based education on students' moral values. Then, Edward Laumann and his coauthors lay out the logic that they used in designing the most extensive, contemporary survey of American sexual behavior completed. They explain some of the reasons behind the way they organized and carried out their sampling, questionnaire design, and interviewing. Finally, in the selection on field research, one of us, Patti, describes the way we infiltrated a group of upper-level drug dealers and smugglers, managing to gain their trust and become peripheral members of the group. Fraught with excitement as well as danger, this research employed participant-observation methods to find out about the highly illegal and hidden behavior of drug traffickers, and to understand the complex contradictions and motivations rippling throughout their lives and their subculture. Participant-observation offers researchers insight into people's deep, inner thoughts and feelings about what they do, shows how people and their situations change over time, and reveals the effects of multiple causal factors working together in the natural setting, the one where people actually live.

1

The Sociological Eye

RANDALL COLLINS

Collins attempts to define the core and fascination of sociology in this passionate essay on the sociological perspective. He carves out the subject matter of sociology and shows how it is uniquely evident for examination in the routine world of everyday life. This immediate accessibility is the cause, he suggests, for not only people's initial attraction to sociology, as they are curiously drawn to examine and understand the circumstances that surround them, but their continuing interest in sociology as well, as we can all develop deeper and more sophisticated analyses of society as we learn more and consider social conditions and connections more thoroughly. As you read this book and take this course, we challenge you to apply the ideas you learn to both your everyday experiences and your broader thoughts about what makes society work. Then ask yourself the question Collins posits: is sociology not the most relevant of all perspectives to understanding and living in the world around us?

Does sociology have a core? Yes, but it is not an eternal essence; not a set of texts or ideas, but an activity.

This is not the same as saying the discipline of sociology will always exist. Sociology became a self-conscious community only in the mid-1800s, about five generations ago, and has been an academic discipline for four generations or less. Disciplines go in and out of existence. The very concept of disciplinary specialization as we know it was created in the Napoleonic period at the time of reorganization of the French Academies, as Johan Heilbron has shown in *The Rise of Social Theory* (1995). There is no guarantee that any particular discipline will remain fixed. Biology, a discipline first recognized by Auguste Comte, has repeatedly shifted its boundaries, combining with physics and chemistry, or spinning off genetics and ecology, making up a shifting array of new fields. Discoveries do not respect administrative boundary lines. Major advances in research or theory tend to pull followers after them, who institutionalize themselves in turn for a while in some organizational form, if only until the next big round of discovery.

From "The Sociological Eye and Its Blinders." *Contemporary Sociology.* (Jan. 1988) pp. 2–3. Reprinted by permission of the American Sociological Assn.

In much the same way, sociologists keep forming hybrid communities on their borders, for example, with economics, literary theory, or computer science. In recent decades, hybrid disciplines have split off from, overlapped with, or encroached upon sociology as criminal justice, ethnic studies, gender studies, management, science and technology studies (i.e., what was once "sociology of science"), and no doubt more to come. There is nothing to lament in this. A glance at the history of long-term intellectual networks, and of academic organizations, shows that branching and recombining are central to what drives intellectual innovation. (The pattern of such long-term networks is documented in my *The Sociology of Philosophies* [1998].)

Sociology, like everything else, is a product of particular historical conditions. But I also believe we have hit upon a distinctive intellectual activity. Its appeal is strong enough to keep it alive, whatever its name will be in the future and whatever happens to the surrounding institutional forms. The lure of this activity is what drew many of us into sociology. One becomes hooked on being a sociologist. The activity is this: It is looking at the world around us, the immediate world you and I live in, through the sociological eye.

There is a sociology of everything. You can turn on your sociological eye no matter where you are or what you are doing. Stuck in a boring committee meeting (for that matter, a sociology department meeting), you can check the pattern of who is sitting next to whom, who gets the floor, who makes eye contact, and what is the rhythm of laughter (forced or spontaneous) or of pompous speechmaking. Walking down the street, or out for a run, you can scan the class and ethnic pattern of the neighborhood, look for lines of age segregation, or for little pockets of solidarity. Waiting for a medical appointment, you can read the professions and the bureaucracy instead of old copies of *National Geographic.* Caught in a traffic jam, you can study the correlation of car models with bumper stickers or with the types of music blaring from radios. There is literally nothing you can't see in a fresh way if you turn your sociological eye to it. Being a sociologist means never having to be bored.

But doesn't every discipline have its special angle on all of reality? Couldn't a physicist see the laws of motion everywhere, or an economist think of supply curves of whatever happens in everyday life? I still think sociology is uniquely appealing in this respect. What physicists or chemists can see in everyday life is no doubt rather banal for them, and most of their discoveries in recent centuries have been made by esoteric laboratory equipment. Fields like economics, it is true, could probably impose an application of some of their theories upon a great many things. But for virtually all disciplines, the immediate world is a sideshow. For sociologists, it is our arena of discovery, and the source at which we renew our energies and our enthusiasm. . . .

All of us who are turned on by sociology, who love doing what we do, have the sociological eye. It is this that gives us new theoretical ideas and makes alive the theories that we carry from the past. The world a sociologist can see is not bounded by the immediate microsituation. Reading the newspaper, whether the business section or the personal ads, is for us like an astronomer training his or her telescope on the sky. Where the ordinary reader is pulled into the journalistic mode, reading the news through one or another political bias or schema of popular melodrama, the sociological eye sees suggestions of social movements mobilizing or winding down, indications of class domination or conflict, or perhaps the organizational process whereby just this kind of story ended up in print, defined as news. For us, novels depict the boundaries of status groups and the saga of social mobility, just as detective stories show us about backstages. Whatever we read with the sociological eye becomes a clue to the larger patterns of society, here or in the past. The same goes for the future: Today's sociologists are not just caught up in the fad of the Internet; they are already beginning to look at it as another frontier for sociological discovery.

I want to claim, in short, that all kinds of sociologists, microethnographers and statisticians, historical comparativists and theorists alike, have the sociological eye. I think that virtually all of the most productive sociologists among us do. We all went through a gestalt switch in our way of looking at the world, sometime early in our careers, that was the key moment in our initiation into sociology. Turn on the sociological and go look at something. Don't take someone else's word for what there is to see, or some common cliché (even a current trendy one), above all not a media-hype version of what is there; go and see it yourself. Make it observationally strange, as if you'd never seen it before. The energy comes back. In that way, I suspect, sociologists are probably more energized by their subject matter than practitioners of virtually any other discipline.

REFERENCES

Collins, Randall. 1998. *The Sociology of Philosophies.* Cambridge, MA: Harvard University Press.

Heilbron, Johan. 1995. *The Rise of Social Theory.* Minneapolis: University of Minnesota Press.

2

The Promise of Sociology

C. WRIGHT MILLS

In a classic statement, Mills points out the significance of the social context framing people and their actions. We sometimes overlook the role of larger historical and institutional factors in affecting our situations, failing to recognize the connection between the tangible, micro level of how people act in everyday life and the invisible forces containing social trends and structures. Learning to recognize these forces is the challenge Mills poses to inquiring minds. He calls this an enriching yet frightening opportunity, for in discovering the role of the broader sweeps of history that guide and constrain us, we see ourselves as small parts of the larger microcosm. With this call to enlightenment we open this collection of sociological essays, ones that will help you develop your own "sociological imagination" and enter the intellectual domain of grand, exciting ideas. Hopefully, after taking this course, you will share in his enthusiasm about the unique perspective that sociology offers us. How is the sociological vision of society different from the commonsensical one? What makes it unique?

Nowadays men often feel that their private lives are a series of traps. They sense that within their everyday worlds, they cannot overcome their troubles, and in this feeling, they are often quite correct: What ordinary men are directly aware of and what they try to do are bounded by the private orbits in which they live; their visions and their powers are limited to the close-up scenes of job, family, neighborhood; in other milieux, they move vicariously and remain spectators. And the more aware they become, however vaguely, of ambitions and of threats which transcend their immediate locales, the more trapped they seem to feel.

Underlying this sense of being trapped are seemingly impersonal changes in the very structure of continent-wide societies. The facts of contemporary history are also facts about the success and the failure of individual men and women. When a society is industrialized, a peasant becomes a worker; a feudal lord is liquidated or becomes a businessman. When classes rise or fall, a man is employed or unemployed; when the rate of investment goes up or down, a man takes new heart or goes broke. When wars happen, an insurance salesman becomes a rocket launcher; a store clerk, a radar man; a wife lives alone; a child grows up without a father. Neither the life of an individual nor the history of a society can be understood without understanding both.

This introductory piece was written in 1959 before writers became sensitive to sexist language. Please excuse the use of the word "men" in this selection, where more appropriately the word "humans" should appear.

Yet men do not usually define the troubles they endure in terms of historical change and institutional contradiction. The well-being they enjoy, they do not usually impute to the big ups and downs of the societies in which they live. Seldom aware of the intricate connection between the patterns of their own lives and the course of world history, ordinary men do not usually know what this connection means for the kinds of men they are becoming and for the kinds of history-making in which they might take part. They do not possess the quality of mind essential to grasp the interplay of man and society, or biography and history, of self and world. They cannot cope with their personal troubles in such ways as to control the structural transformations that usually lie behind them.

Surely it is no wonder. In what period have so many men been so totally exposed at so fast a pace to such earthquakes of change? That Americans have not known such catastrophic changes as have the men and women of other societies is due to historical facts that are now quickly becoming "merely history." The history that now affects every man is world history. Within this scene and this period, in the course of a single generation, one-sixth of mankind is transformed from all that is feudal and backward into all that is modern, advanced, and fearful. Political colonies are freed; new and less visible forms of imperialism installed. Revolutions occur; men feel the intimate grip of new kinds of authority. Totalitarian societies rise, and are smashed to bits—or succeed fabulously. After two centuries of ascendancy, capitalism is shown up as only one way to make society into an industrial apparatus. After two centuries of hope, even formal democracy is restricted to a quite small portion of mankind. Everywhere in the underdeveloped world, ancient ways of life are broken up and vague expectations become urgent demands. Everywhere in the overdeveloped world, the means of authority and of violence become total in scope and bureaucratic in form. Humanity itself now lies before us, the super-nation at either pole concentrating its most coordinated and massive efforts upon the preparation of World War Three.

The very shaping of history now outpaces the ability of men to orient themselves in accordance with cherished values. And which values? Even when they do not panic, men often sense that older ways of feeling and thinking have collapsed and that newer beginnings are ambiguous to the point of moral stasis. Is it any wonder that ordinary men feel they cannot cope with the larger worlds with which they are so suddenly confronted? That they cannot understand the meaning of their epoch for their own lives? That—in defense of selfhood—they become morally insensible, trying to remain altogether private men? Is it any wonder that they come to be possessed by a sense of the trap?

It is not only information that they need—in this Age of Fact, information often dominates their attention and overwhelms their capacities to assimilate it. It is not only the skills of reason that they need—although their struggles to acquire these often exhaust their limited moral energy.

What they need, and what they feel they need, is a quality of mind that will help them to use information and to develop reason in order to achieve lucid summations of what is going on in the world and of what may be happening within themselves. It is this quality, I am going to contend, that journalists and scholars, artists and publics, scientists and editors are coming to expect of what may be called the sociological imagination.

The sociological imagination enables its possessor to understand the larger historical scene in terms of its meaning for the inner life and the external career of a variety of individuals. It enables him to take into account how individuals, in the welter of their daily experience, often become falsely conscious of their social positions. Within that welter, the framework of modern society is sought, and within that framework the psychologies of a variety of men and women are formulated. By such means the personal uneasiness of individuals is focused upon explicit troubles and the indifference of publics is transformed into involvement with public issues.

The first fruit of this imagination—and the first lesson of the social science that embodies it—is the idea that the individual can understand his own experience and gauge his own fate only by locating himself within his period, that he can know his own chances in life by becoming aware of those of all individuals in his circumstances. In many ways it is a terrible lesson; in many ways a magnificent one. We do not know the limits of men's capacities for supreme effort or willing degradation, for agony or glee, for pleasurable brutality or the sweetness of reason. But in our time we have come to know that the limits of "human nature" are frighteningly broad. We have come to know that every individual lives, from one generation to the next, in some society; that he lives out a biography, and that he lives it out within some historical sequence. By the fact of his living he contributes, however minutely, to the shaping of this society and to the course of its history, even as he is made by society and by its historical push and shove.

The sociological imagination enables us to grasp history and biography and the relations between the two within society. That is its task and its promise. To recognize this task and this promise is the mark of the classic social analyst. It is characteristic of Herbert Spencer—turgid, polysyllabic, comprehensive; of E. A. Ross—graceful, muckraking, upright; of August Comte and Emile Durkheim; of the intricate and subtle Karl Mannheim. It is the quality of all that is intellectually excellent in Karl Marx; it is the clue to Thorstein Veblen's brilliant and ironic insight, to Joseph Schumpeter's many-sided constructions of reality; it is the basis of the psychological sweep of W. E. H. Lecky no less than of the profundity and clarity of Max Weber. And it is the signal of what is best in contemporary studies of man and society.

No social study that does not come back to the problems of biography, of history, and of their intersections within a society has completed its intellectual journey. Whatever the specific problems of the classic social analysts, however limited or however broad the features of social reality they have examined, those who have been imaginatively aware of the promise of their work have consistently asked three sorts of questions:

1. What is the structure of this particular society as a whole? What are its essential components, and how are they related to one another? How does it differ from other varieties of social order? Within it, what is the meaning of any particular feature for its continuance and for its change?

2. Where does this society stand in human history? What are the mechanics by which it is changing? What is its place within and its meaning for the development of humanity as a whole? How does any particular feature we are examining affect, and how is it affected by, the historical period in which it moves? And this period—what are its essential features? How does it differ from other periods? What are its characteristic ways of history-marking?
3. What varieties of men and women now prevail in this society and in this period? And what varieties are coming to prevail? In what ways are they selected and formed, liberated and repressed, made sensitive and blunted? What kinds of "human nature" are revealed in the conduct and character we observe in this society in this period? And what is the meaning for "human nature" of each and every feature of the society we are examining?

Whether the point of interest is a great power state or a minor literary mood, a family, a prison, a creed—these are the kinds of questions the best social analysts have asked. They are the intellectual pivots of classic studies of man in society—and they are questions inevitably raised by any mind possessing the sociological imagination. For that imagination is the capacity to shift from one perspective to another—from the political to the psychological; from examination of a single family to comparative assessment of the national budgets of the world; from the theological school to the military establishment; from considerations of an oil industry to studies of contemporary poetry. It is the capacity to range from the most impersonal and remote transformations to the most intimate features of the human self—and to see the relations between the two. Back of its use there is always the urge to know the social and historical meaning of the individual in the society and in the period in which he has his quality and his being.

That, in brief, is why it is by means of the sociological imagination that men now hope to grasp what is going on in the world, and to understand what is happening in themselves as minute points of the intersections of biography and history within society. In large part, contemporary man's self-conscious view of himself as at least an outsider, if not a permanent stranger, rests upon an absorbed realization of social relativity and of the transformative power of history. The sociological imagination is the most fruitful form of this self-consciousness. By its use men whose mentalities have swept only a series of limited orbits often come to feel as if suddenly awakened in a house with which they had only supposed themselves to be familiar. Correctly or incorrectly, they often come to feel that they can now provide themselves with adequate summations, cohesive assessments, comprehensive orientations. Older decisions that once appeared sound now seem to them products of a mind unaccountably dense. Their capacity for astonishment is made lively again. They acquire a new way of thinking, they experience a transvaluation of values: In a word, by their reflection and by their sensibility, they realize the cultural meaning of the social sciences.

Perhaps the most fruitful distinction with which the sociological imagination works is between "the personal troubles of milieu" and "the public issues of social structure." This distinction is an essential tool of the sociological imagination and a feature of all classic work in social science.

Troubles occur within the character of the individual and within the range of his immediate relations with others; they have to do with his self and with those limited areas of social life of which he is directly and personally aware. Accordingly, the statement and the resolution of troubles properly lie within the individual as a biographical entity and within the scope of this immediate milieu—the social setting that is directly open to his personal experience and to some extent his willful activity. A trouble is a private matter: Values cherished by an individual are felt by him to be threatened.

Issues have to do with matters that transcend these local environments of the individual and the range of his inner life. They have to do with the organization of many such milieux into the institutions of an historical society as a whole, with the ways in which various milieux overlap and interpenetrate to form the larger structure of social and historical life. An issue is a public matter: Some value cherished by publics is felt to be threatened. Often there is a debate about what that value really is and about what it is that really threatens it. This debate is often without focus if only because it is the very nature of an issue, unlike even widespread trouble, that it cannot very well be defined in terms of the immediate and everyday environments of ordinary men. An issue, in fact, often involves a crisis in institutional arrangements, and often too it involves what Marxists call "contradictions" or "antagonisms."

In these terms, consider unemployment. When, in a city of 100,000, only one man is unemployed, that is his personal trouble, and for its relief we properly look to the character of the man, his skills, and his immediate opportunities. But when in a nation of 50 million employees, 15 million men are unemployed, that is an issue, and we may not hope to find its solution within the range of opportunities open to any one individual. The very structure of opportunities has collapsed. Both the correct statement of the problem and the range of possible solutions require us to consider the economic and political institutions of the society, and not merely the personal situation and character of a scatter of individuals.

Consider war. The personal problem of war, when it occurs, may be how to survive it or how to die in it with honor; how to make money out of it; how to climb into the higher safety of the military apparatus; or how to contribute to the war's termination. In short, according to one's values, to find a set of milieux and within it to survive the war or make one's death in it meaningful. But the structural issues of war have to do with its causes; with what types of men it throws up into command; with its effects upon economic and political, family and religious institutions, with the unorganized irresponsibility of a world of nation-states.

Consider marriage. Inside a marriage a man and a woman may experience personal troubles, but when the divorce rate during the first four years of marriage is 250 out of every 1,000 attempts, this is an indication of a structural issue having to do with the institutions of marriage and the family and other institutions that bear upon them.

Or consider the metropolis—the horrible, beautiful, ugly, magnificent sprawl of the great city. For many upper-class people, the personal solution to "the problem of the city" is to have an apartment with private garage under it in the heart of the city and, forty miles out, a house by Henry Hill, garden by Garrett Eckbo, on a hundred acres of private land. In these two controlled environments—with a small staff at each end and a private helicopter connection—most people could solve many of the problems of personal milieux caused by the facts of the city. But all this, however splendid, does not solve the public issues that the structural fact of the city poses. What should be done with this wonderful monstrosity? Break it up into scattered units, combining residence and work? Refurbish it as it stands? Or, after evacuation, dynamite it and build new cities according to new plans in new places? What should those plans be? And who is to decide and to accomplish whatever choice is made? These are structural issues; to confront them and to solve them requires us to consider political and economic issues that affect innumerable milieu.

Insofar as an economy is so arranged that slumps occur, the problem of unemployment becomes incapable of personal solution. Insofar as war is inherent in the nation-state system and in the uneven industrialization of the world, the ordinary individual in his restricted milieu will be powerless—with or without psychiatric aid—to solve the troubles this system or lack of system imposes upon him. Insofar as the family as an institution turns women into darling little slaves and men into their chief providers and unweaned dependents, the problem of a satisfactory marriage remains incapable of purely private solution. Insofar as the overdeveloped megalopolis and the overdeveloped automobile are built-in features of the overdeveloped society, the issues of urban living will not be solved by personal ingenuity and private wealth.

What we experience in various and specific milieux, I have noted, is often caused by structural changes. Accordingly, to understand the changes of many personal milieux we are required to look beyond them. And the number and variety of such structural changes increase as the institutions within which we live become more embracing and more intricately connected with one another. To be aware of the idea of social structure and to use it with sensibility is to be capable of tracing such linkages among a great variety of milieux. To be able to do that is to possess the sociological imagination. . . .

3

Should We Generalize about People?

JOEL CHARON

Charon's description of sociology highlights its generalizing potential, discussing how categorizations and extrapolations are made accurately. While generalization is an important aspect of everyday human life, people often mix stereotypes in with their empirical categorizations. Charon sensitively distinguishes between stereotyping and generalizing, showing how the inaccurate and potentially harmful characteristics of the former are replaced by the measured and carefully empirically constructed latter. In so doing, he offers you a guide for how to assess useful and damaging categories, and shows the careful work undergirding sociological analysis. Is generalizing, as a goal for sociology, good or bad? What is the difference between social scientists' generalizations and laypeople's stereotypes?

CATEGORIES AND GENERALIZATIONS

The Importance of Categories and Generalizations to Human Beings

Sociology is a social science, and therefore it makes generalizations about people and their social life. "The top positions in the economic and political structures are far more likely to be filled by men than by women." "The wealthier the individual, the more likely he or she will vote Republican." "In the United States the likelihood of living in poverty is greater among the African-American population than among whites." "American society is segregated." "Like other industrial societies, American society has a class system in which more than three-fourths of the population end up in approximately the same social class as what they were at birth."

But such generalizations often give me a lot of trouble. I know that the sociologist must learn about people and generalize about them, but I ask myself: "Are such generalizations worthwhile? Shouldn't we simply study and treat people as individuals?" An English professor at my university was noted for explaining to his class that "you should not generalize about people—that's the same as stereotyping and everyone knows that educated people are not supposed to stereotype. Everyone is an individual." (Ironically, this is *itself* a generalization about people.)

However, the more I examine the situation, the more I realize that all human beings categorize and generalize. They do it every day in almost every situation they enter, and they almost always do it when it comes to other people. In fact, we have no choice in the matter. "Glass breaks and can be dangerous." We have learned what "glass" is, what "danger" means, and what "breaking" is. These are all categories we apply to situations when we enter so that we can understand how to act. We generalize from our past. "Human beings who have a cold are contagious, and, unless we want to catch a cold, we should not get close to them." We are here generalizing about "those with colds," "how people catch colds," and "how we should act around those with colds." In fact, every noun and verb we use is a generalization that acts as a guide for us. The reality is that we are unable to escape generalizing about our environment. That is one aspect of our essence as human beings. This is what language does to us. Sometimes our generalizations are fairly accurate; sometimes they are unfounded. However, we do in fact generalize: all of us, almost all the time! The question that introduces this chapter is a foolish one. *Should we generalize about people* is not a useful question simply because we have no choice. A much better question is:

How can we develop accurate generalizations about people?

The whole purpose of social science is to achieve accurate categorizations and generalizations about human beings. Indeed, the purpose of almost all academic pursuits involves learning, understanding, and developing accurate categories and generalizations.

For a moment let us consider other animals. Most are prepared by instinct or simple conditioning to respond in a certain way to a certain stimulus in their environment. So, for example, when a minnow swims in the presence of a hungry fish, then that particular minnow is immediately responded to and eaten. The fish is able to distinguish that type of stimulus from other stimuli, and so whenever something identical to it or close to it appears, the fish responds. The minnow is a concrete object that can be immediately sensed (seen, smelled, heard, touched), so within a certain range the fish is able to easily include objects that look like minnows and to exclude those that do not. Of course, occasionally a lure with a hook is purposely used to fool the fish, and a slight mistake in perception ends the fish's life.

Human beings are different from the fish and other animals because we have *words for objects and events* in the environment, and this allows us to *understand* that environment and not just respond to it. With words we are able to make many more distinctions, and we are able to apply knowledge from one situation to the next far more easily. We are far less dependent on immediate physical stimuli. So, for example, we come to learn what fish, turtles, and whales are, as well as what minnows, worms, lures, and boats are. We read and learn what qualities all fish have, how fish differ from whales, and what differences fish have from one another. We learn how to catch fish, and we are able to apply what we learn to some fish but not other fish. We begin to understand the actions of all fish—walleyes, big walleyes, big female walleyes. Some of us decide to study pain, and we try to determine if all fish feel pain, if some do, or if all do not. Humans do not then simply respond to the environment, but they label that environment, study and understand that environment, develop categories and subcategories for objects in that environment, and constantly try to generalize from what they learn in specific situations about those categories. Through understanding a category we are able to see important and subtle similarities and distinctions that are not available to animals who do not categorize and generalize with words.

Generalizing allows us to walk into situations and apply knowledge learned elsewhere to understanding objects there. When we enter a classroom we know what a teacher is, and we label the person at the front of the room as a teacher. We know from past experience that teachers give grades, usually know more than we do about things we are about to learn in that classroom, have more formal education than we do, and usually resort to testing us to see if we learned something they regard as important. We might have also learned that teachers are usually kind (or mean), sensitive (or not sensitive), authoritarian (or democratic); or we might have had so many diverse experiences with teachers that whether a specific teacher is any of these things will depend on that specific individual. If we do finally decide that a given teacher is in fact authoritarian, then we will now see an "authoritarian teacher," and we will now apply what we know about such teachers from our past.

This is a remarkable ability. We are able to figure out how to act in situations we enter because we understand many of the objects we encounter there by applying relevant knowledge about them that we learned in the past. This allows us to intelligently act in a wide diversity of situations, some of which are not even close to what we have already experienced. If we are open-minded and reflective, we can even evaluate how good or how poor our generalizations are, and we can alter what we know as we move from situation to situation.

The problem for almost all of us, however, is that many of our generalizations are not carefully arrived at or accurate, and it is sometimes difficult for us to recognize this and change them. Too often our generalizations actually stand in the way of our understanding, especially when we generalize about human beings.

To better understand what human beings do and how that sometimes gets us into trouble, let us look more closely at what "categories" and "generalizations" are.

The Meaning of Categorization

Human beings categorize their environment; that is, we isolate a chunk out of our environment, distinguish that chunk from all other parts of the environment, give it a name, and associate certain ideas with it. Our chunks—or categories—arise in interaction; they are socially created. We discuss our environment, and we categorize it with the words we take on in our social life: "living things," "animals," "reptiles," "snakes," "poisonous snakes," "rattlers." A category is created, and once we understand it, we are able to compare objects in situations we encounter to that category. The number of distinctions we are able to make in our environment increases manyfold. It is not only nouns that represent categories (*men, boys*) but also verbs (*run, walk, fall*), adverbs (*slow, fast*), and adjectives (*weak, strong, intelligent, married*). Much of our learning is simply aimed at understanding what various categories mean, and this involves understanding the qualities that make up those categories and the ideas associated with them. . . .

The Meaning of Generalization

A *category* is an isolated part of our environment that we notice. We generalize about that category by observing specific instances of objects included in it and by isolating common qualities that seem to characterize those included in that category, including other yet unobserved members we might observe in the future. We watch birds build a nest, and we assume that all birds build nests out of sticks (including birds other than the robins and sparrows we observed). We continue to observe and note instances where birds use materials other than sticks, and then we learn that some birds do not build nests but dig them out. More often, our generalizations are a mixture of observation and learning from others: we learn that wealthy people often drive Mercedes, and that police officers usually carry guns. On the basis of generalizing about a category, we are able to predict future events where that category comes into play. When we see a wealthy person, we expect to see a Mercedes (or something that we learn is comparable); and when we see a police officer, we expect to see a gun. That is what a generalization is.

A generalization describes the category. It is a statement that characterizes objects within the category and defines similarities and differences with other categories. "This is what an educated person is!" (in contrast to an uneducated person). "This is what wealthy people do to help ensure that privilege is passed down to their children." This is what U.S. presidents have in common." "This is what Catholic people believe in." . . .

THE STEREOTYPE

When it comes to people, generalization is very difficult to do well. The principal reason for this is that we are judgmental, and too often it is much easier for us to generalize for the purpose of evaluating (condemning or praising) others than for the purpose of understanding them. When we do this we fall into the practice of *stereotyping.*

A stereotype is a certain kind of categorization. It is a category and a set of generalizations characterized by the following qualities:

1. *A stereotype is judgmental.* It is not characterized by an attempt to understand, but by an attempt to condemn or praise the category. It makes a value judgment, and it has a strong emotional flavor. Instead of simple description of differences, there is a moral evaluation of those differences. People are judged good or bad because of the category.
2. *A stereotype tends to be an absolute category.* That is, there is a sharp distinction made between those inside and those outside the category. There is little recognition that the category is merely a guide to understanding and that, in reality, there will be many individuals within a category who are exceptions to any generalization.
3. *The stereotype tends to be a category that overshadows all others in the mind of the observer.* All other categories to which the individual belongs tend to be ignored. A stereotype treats the human being as simple and unidimensional, belonging to only one category of consequence. In fact, we are all part of a large number of categories.
4. *A stereotype does not change with new evidence.* When one accepts a stereotype, the category and the ideas associated with it are rigidly accepted, and the individual who holds it is unwilling to alter it. The stereotype, once accepted, becomes a filter through which evidence is accepted or rejected.
5. *The stereotype is not created carefully in the first place.* It is either learned culturally and simply accepted by the individual or created through uncritical acceptance of a few concrete personal experiences.
6. *The stereotype does not encourage a search for understanding why human beings are different from one another.* Instead of seeking to understand the cause as to why a certain quality is more in evidence in a particular category of people, a stereotype aims at exaggerating and judging differences. There is often an underlying assumption that "this is the way these people are," it is part of their "essence," and there seems to be little reason to try to understand the cause of differences any further than this.

Stereotypes are highly oversimplified, exaggerated views of reality. They are especially attractive to people who are judgmental of others and who are quick to condemn people who are different from them. They have been used to justify ethnic discrimination, war, and systematic murder of whole categories of people. Far from arising out of careful and systematic analysis, stereotypes arise out of hearsay and culture, and instead of aiding our understanding of the human being, they always stand in the way of accurate understanding. . . .

SOCIAL SCIENCE: A REACTION TO STEREOTYPES

Creating categories about people and generalizing intelligently is very difficult to do unless we work hard at it. A big part of a university education is to uncover and critically evaluate stereotypes in order to obtain a better understanding of reality. Each discipline in its own way attempts to teach the student to be more careful about categorizing and generalizing.

Because this book focuses on the perspective of sociology and social science, I would like to show how social science tries to rid us of stereotypes through the careful development of accurate categories and generalizations about human beings. Social science is a highly disciplined process of investigation whose purpose is to question many of our uncritically accepted stereotypes and generalizations. Social science does not always succeed. There are many instances of inaccuracies and even stereotyping that have resulted from poor science or from scientists simply not being sensitive to their own biases. It is important, however, to recognize that even though scientists make mistakes in their attempts to describe reality accurately, the whole thrust and spirit of social science is to control personal bias, to uncover unfounded assumptions about people, and to understand reality as objectively as possible. Here are some of the ways that social science (as it is supposed to work) aims at creating accurate categories and generalizations about human beings:

1. *Social science tries hard not to be judgmental about categories of people.* We recognize that generalizations and categories must not condemn or praise but must simply be guides to understanding. To stereotype is to emphasize qualities in others that we dislike or to emphasize qualities in others that are similar to our own that we like. . . .
2. *Categories and generalizations in social science are rarely—if ever—absolute.* Social scientists begin with the assumption that it is difficult to generalize about people and that every time we do exceptions are likely, and often a large number. By definition, all atheists do not believe in God, but there is absolutely nothing else we can say about all atheists. However, we might contend that atheists tend to be more educated (but there are many exceptions to this), male rather than female (but there are many exceptions to this), and raised by atheist parents (but there are many exceptions to this). We can tease out generalizations about atheists from carefully studying them, but we will never find a quality that all of them have other than their belief that there is no God. This goes for every category of people we try to understand: those who commit suicide, those who abuse drugs, those who commit violent acts against children, serial killers, and students who do not finish college. We can generalize, but we must be careful, and we must assume

exceptions within every category we create. The scientific generalization is treated as a probability rather than an absolute. . . .

3. *Categories in social science are not assumed to be all-important for understanding the individual.* A stereotype is itself an assumption that a certain category necessarily dominates an individual's life. We might meet a young African-American single male artist. The role of each of these categories may or may not be important to the individual. For some individuals, being male or single or an artist will be most influential; for others it will be being African American. For those of us who stereotype by race, it will almost always be African American. . . .

4. *Social science tries to create categories and generalizations through carefully gathered evidence.* Stereotypes tend to be cultural; that is, they are taught by people around us who have generalized based on what they have simply accepted from others or what they have learned through personal experience (which is usually extremely limited in scope, unsystematic, subject to personal and social biases, and uncritically observed). Science tries hard to encourage accurate generalizations through making explicit how generalizations must be arrived at. . . .

5. *Generalizations in social science are tentative and subject to change* because evidence is constantly being examined. Stereotypes, on the other hand, are unconditionally held. Once held, a stereotype causes the individual to select out evidence that only reaffirms that stereotype. A stereotype resists change. When we believe that whites have superior abilities to nonwhites, we tend to notice only those individuals who support our stereotype. If we believe that politicians are selfish bureaucrats, we tend to forget all those political leaders who are unselfish and who get things done. (Note: the category "politician" gives away that one is stereotyping rather than simply generalizing, because politician has come to mean someone who is not worth our respect.) Because the purpose of a stereotype is to condemn or praise a category of people, it becomes difficult to evaluate evidence. The stereotype is embedded in the mind of the observer, it takes on an emotional flavor, and evidence that might contradict it is almost impossible to accept.

   A generalization in social science about a category of people is subject to change as soon as new evidence is discovered. The final truth about people is never assumed to have been found. The generalization is always taken as a tentative guide to understanding rather than a quality that is etched in stone. . . .

6. *Scientists do not categorize as an end in itself.* Instead, scientists categorize because they seek a certain kind of generalization: they seek to understand cause. In social science that means we seek to know *why* a category of people tends to have a certain quality. We generalize about categories of people *to better understand what causes* the existence of qualities that belong to a given category. We seek to understand the cause of schizophrenia, but we can only do this after we understand what characterizes those people who are schizophrenic. . . .

Real generalization in science therefore is to uncover why certain qualities make up a category, and why they are less in evidence in other categories. *Why* is there increasing individualism among Americans? *Why* do some people graduate college and not others? *Why* are women absent from the top political and economic positions in American life? *Why* is there an increase in the number of people who are experiencing downward social mobility in the United States? *Why* is there a rising suicide rate among young people? In every one of these cases we find a category, we describe those who make up that category, and we attempt to generalize as to why a certain quality exists in that category. To judge? No. True of everyone in the category? No. The only category of importance? No. A fixed category that clearly and absolutely distinguishes between one group and another? No. A generalization that we can regard as true without reservation? No. . . .

Finally, we should examine once more the question that we started with: Should we generalize about people? This is not an easy question to answer.

We must begin our answer by admitting that we all take whatever we know about categories of people and apply it to situations we encounter. When we see that the individual is a child or an elderly person, male or female, single or married, a professor or a physician, wealthy or poor, kind or insensitive, that information guides us in our actions. If we are careful, we will recognize that our view of the other must be tentative, that the individual may in fact be an exception in our category, and that we must be ready to change whatever we think as we get to know that person as a unique individual. In fact the category we use may end up being unimportant for understanding this particular person.

In a society where we cry out for individual recognition, few of us will admit we want others to place us into categories and generalize. "Do not categorize me. I'm an individual!" Yet, if we are honest, we will recognize that those who do not know us will be forced to categorize us. It actually is not too bad if the category is a positive one. If we apply for a job we want the employer to categorize us as dependable, hardworking, knowledgeable, intelligent, and so on. We will even try to control how we present ourselves in situations so we can influence the other to place us in favorable categories: I'm cool, intelligent, sensitive, athletically talented, educated. The doctor may try to let people know "I am a physician" so that they will think highly of him or her as an individual. The individual who announces himself as a boxer is telling us that he is tough, the rock musician is telling us that she is talented, the minister that he or she is caring—in many such cases it does not seem so bad if we are being categorized. For almost all of us, however, it is the *negative* categorization that we wish to avoid. And this makes good sense: no one wants to be put into a category and negatively judged without having a chance to prove himself or herself as an individual.

But no matter how we might feel about others categorizing us and applying what they know to understanding us as a member of that category, the fact is that, except for those we know well, human beings can only be understood if we categorize and generalize. If we do this carefully, we can understand much about them, but if we are sloppy, we sacrifice understanding and end up making irrationally based value judgments about people before we have an opportunity to know them as individuals. . . .

If we have to generalize, let's try to be careful. Stereotyping does not serve our own interests well because it blocks understanding; nor does it help those we stereotype.

4

Three Sociological Perspectives

PAUL COLOMY

In this chapter, Colomy overviews the three perspectives most often associated with sociology: functionalism, conflict theory, and symbolic interactionism. Each of these three theories have contributed a great deal to our understanding of human behavior and group life. The first theory, symbolic interactionism, is a micro theory of society, focusing on social interaction and how people act in face-to-face meetings with one another. The last two, functionalism and conflict, are considered to be macro theories of society, looking at how the social structure operates to determine people's behavior. Each theory has its strengths and weaknesses as an explanatory scheme, but taken together, you will see the power of sociological reasoning. From this article, can you differentiate the major foundational ideas that they represent? Are there other ideas that you have to account for social behavior that are not represented by these three approaches?

When conducting research, sociologists typically draw on one or more perspectives. Sociological perspectives provide very general ways of conceptualizing the social world and its basic elements. A perspective consists of a set of fairly abstract assumptions about the nature of human action and the character of social organization. Each perspective can be likened to a spotlight that brightly illuminates select aspects of behavior and social relations while leaving other areas shrouded in darkness. Because a single perspective supplies only a partial or one-sided view, a comprehensive understanding of social life requires becoming familiar with several different perspectives.

Sociology contains a large number of distinct perspectives, and they can be divided into two broad categories: micro and macro. In very general terms, micro perspectives are oriented toward small time and small space, while macro perspectives are oriented toward big time and big space (Collins 1981). That is, micro perspectives are usually concerned with the conduct of individuals and small groups as it unfolds in relatively small spatial contexts and over short durations of time. Macro perspectives, on the other hand, focus on larger entities—not indi-

Reprinted by permission of the author.

viduals and small groups, but institutions, entire societies, and even the global system—and on how these entities emerge, maintain themselves, and change over decades, centuries, and millennia. The following section outlines one micro perspective (symbolic interactionism) and two macro perspectives (functionalism and the conflict approach).

SYMBOLIC INTERACTIONISM

Symbolic interactionism's intellectual roots reside in pragmatism, a philosophical tradition developed by such prominent, early twentieth-century American thinkers as John Dewey, William James, George Herbert Mead, and Charles Peirce. The sociological implications of pragmatism were articulated by several innovative sociologists, including Robert Park, W. I. Thomas, Herbert Blumer, Everett Hughes, and Erving Goffman, who taught or studied at the University of Chicago between 1910 and 1960. Because it originated at the University of Chicago, symbolic interactionism is sometimes referred to as the Chicago School.

Symbolic interactionism is based on five core ideas. First, it assumes that human beings act in terms of the meanings they assign to objects in their environment. (Interactionists define the term object very broadly to include material things, events, symbols, actions, and other people and groups.) Using slightly different terminology to make the same point, interactionists maintain that people's conduct is powerfully influenced by their definition of the situation. This assumption can be clarified by contrasting it to a rudimentary model of social action advanced by a psychological perspective known as behaviorism. The behaviorist approach characterizes conduct as a response to objective stimuli, and suggests that human behavior resembles a series of stimulus-response chains:

stimulus ——————> response.

Rejecting the notion that individuals respond directly to an objective stimulus, interactionists insist that people interpret, or assign meanings to, the stimulus before they act:

stimulus ————> interpretation ————> response.

Athletes' reactions to coaches' criticisms, for instance, depends largely on whether they interpret that criticism as a constructive attempt to improve their play or as a malicious attack on their character.

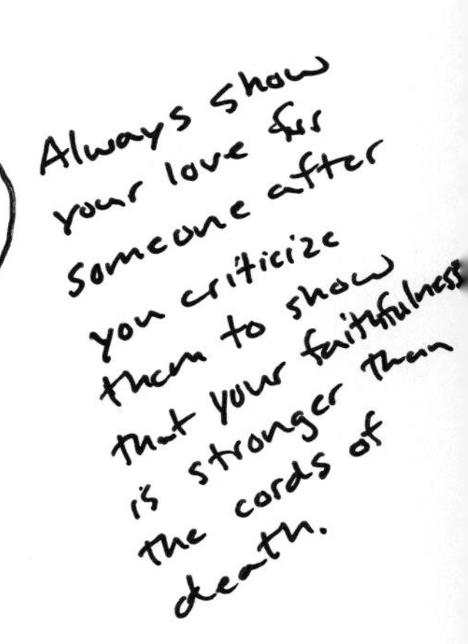

Even when a definition of the situation is demonstrably false, it can still exert a powerful effect on behavior. As W. I. Thomas once said, "A situation defined as real is real in its consequences." Many adults, for example, perceive Halloween as filled with potential danger, and believe that their young children are vulnerable to sadistic strangers dispensing drug-tainted candy or apples laced with razor blades. The belief that such acts of Halloween sadism are widespread is, in fact, an urban legend with virtually no factual basis (Best and Horiuchi 1985). Nevertheless, millions of parents are convinced that the threat is genuine and, acting in terms of their definition of the situation, continue to inspect their children's treats for signs of tampering.

Symbolic interactionism's second assumption asserts that social action typically involves making a series of adjustments and readjustments as an individual's interpretation of the situation changes. Interactionists reject the notion that behavior is the unmediated product of a variable or cause. Instead, they view action as something that is continually being built up, modified, re-directed, and transformed (Blumer 1969). People's initial definition of the situation is always subject to change, and as they redefine the situation their conduct changes accordingly. Effective teachers, for example, routinely interpret students' comments, facial expressions, and other gestures to determine whether the subject matter is being communicated clearly. They rely on this feedback to define and redefine the unfolding classroom situation and to make corresponding adjustments in their presentations. When students look confused, they may introduce a familiar example; if students' attention should wander, the instructors may call on them; and if students are visibly upset, they may ask to meet privately with them after class.

Third, interactionists assume that the meanings imputed to an object are socially constructed (Berger and Luckmann 1966). Meanings do not, in other words, simply reflect a quality or essence built into the very nature of an object. Other than its size and color, the cloth used to make handkerchiefs is virtually identical to that used to produce American flags. Though handkerchiefs and flags are sewn from the same physical material, the meanings attached to these two objects differ in dramatic ways. Rather than being intrinsic to an object, then, meanings are attributed to it by individuals, groups, and communities.

Elaborating this logic, some interactionists treat the self as an object whose meanings are socially constructed. In other words, the kind of person you assume yourself to be, and that others take you to be, mirror the meanings that individuals and groups have assigned to you. If, from a young age, family members, friends, and teachers have said you were "brilliant" and have acted toward you in a manner consistent with that characterization, then one assumption you are likely to make about yourself—one meaning that you are likely to assign to yourself—is that you are a highly intelligent person.

Fourth, symbolic interactionism holds that in modern, heterogeneous societies, different groups often assign divergent meanings to the "same" object. Contemporary societies contain a wide variety of groups (e.g., occupational, religious, age-based, racial and ethnic groups). Since group members interact and communicate frequently with one another, they tend to develop a common "universe of discourse" (Mead 1934) or shared meanings about the objects comprising their social world. Not surprisingly, discrepancies are likely to arise between the distinctive systems of meanings devised by different groups. Parents and adolescents, for instance, commonly attach opposing meanings to curfews, underage drinking, and body piercing.

Discrepant meanings can be a significant source of social conflict, with rival groups mobilizing to insure that their definition of the situation is officially acknowledged and enforced by the larger society. A classic interactionist study discovered that Prohibition was largely a battle over the meanings assigned to drinking (Gusfield 1963). In the early twentieth century, small-town, middle-class, WASPS (white, Anglo-Saxon, Protestants) regarded drinking as sinful, while the working-class and largely Catholic immigrants from Southeastern Europe who settled in the nation's largest cities viewed drinking as an integral part of everyday life. More adept at organizing and lobbying politicians, the small-town Protestants succeeded in inscribing their interpretation into law: the Eighteenth Amendment, approved by Congress in 1919, outlawed (at least for a time) the manufacture and sale of alcoholic liquors.

Fifth, established meanings are always subject to transformation, and interactionists maintain that the emergence and diffusion of novel definitions of reality are a critically important feature of social change. At any given time, the meanings attached to some objects and practices are so entrenched that they appear natural and beyond question. Behavior that deviates sharply from these prescribed meanings is regarded as threatening, immoral, and even a little crazy. Interactionists examine how social movements, broad cultural shifts, and/or deviant individuals and groups sometimes challenge long-standing meanings and replace them with alternative conceptions of reality. From an interactionist perspective, one of the most significant consequences of the feminist movement is its redefinition of what it means to be a woman. A generation or two ago, it was simply assumed, particularly by members of the middle class and by women as well as men, that a woman's "proper place" was in the home taking care of her household, husband, and children. In the late 1960s and throughout the 1970s, feminists questioned this assumption and ushered in a strikingly different conception of women, one that affirmed a woman's right to work outside the home and to be treated as an equal, in every respect, to her male colleagues. Today, feminism's once novel and radical definition of reality has been institutionalized, and the earlier view, which at the time was widely accepted and regarded as obvious and commonsensical, has now been redefined as an arbitrary infringement on women's freedom.

FUNCTIONALISM

Functionalism is a macro perspective that examines the creation, maintenance, and alteration of durable social practices, institutions, and entire societies. Emile Durkheim, a great French sociologist who published several provocative books between 1890 and 1915, is often regarded as the classic founder of functionalism. This approach was articulated most forcefully, however, during the twenty-five years between 1945 and 1970 by a group of American sociologists, most of whom were trained at Harvard or Columbia. Key figures in this group include Robert Bellah, Robert Merton, Wilbert Moore, Talcott Parsons, Neil Smelser, and Robin Williams.

Functionalism assumes, first, that societies can be likened to problem-solving entities. If a society is to persist, functionalists argue, it must address a large (but not infinite) number of problems in a reasonably satisfactory way. (Functionalists sometimes refer to these problems as requirements, functions, prerequisites, or functional prerequisites.) An enduring society must, for example, socialize its youngest members, distribute food and other goods and services, and devise mechanisms to control deviance and contain conflict. If a society does not satisfactorily address these (and other) problems, it will experience considerable strain, and if its failure to address these problems continues, it will collapse.

Extending the metaphor that likens societies to problem-solving entities, functionalism portrays persisting practices and institutions as answers or solutions to the kinds of problems mentioned above. Customary practices and institutions are established to meet the problems every society must confront. Families and schools, for instance, are institutions that arise to answer the problem of socializing and educating the young; free markets, on the other hand, are created to address the problem of producing and distributing goods and services, while police and prisons are mechanisms for controlling deviance and containing conflict.

Second, functionalists assume that during the course of human history societies have developed many different answers to basic needs. This assumption can be termed the principle of institutional alternatives. In traditional societies, for instance, the extended family had sole responsibility for supporting dependents, whether they be very young or very old. In many modern societies, however, insurance policies, pensions, social security, and welfare programs share responsibility for the problem of caring for dependents. From a functionalist perspective, these programs are institutional alternatives to the extended family.

The principle of institutional alternatives implies that any single functional prerequisite can be met in many different ways. Many functionalists argue, however, that in an attempt to address prerequisites more efficiently and more effectively, modern societies have increasingly replaced multi-functional institutions with more specialized ones (Parsons 1977). Two hundred years ago, the family was a multi-functional institution in that it assumed primary responsibility for many different tasks, such as economic production, procreation, socialization, care for the infirm elderly, and social control. Today, however, many of these problems have been delegated to specialized institutions. Economic production, for instance, is no longer addressed by the family but by business enterprises located (for the most part) outside the home, while the control of deviance is a problem for specialized social control agents like the police and criminal courts. The family, too, has become a more specialized institution, one whose primary tasks include procreation, socialization of the very young, and emotional support for family members.

Third, functionalism presumes that the particular practices and institutions that arise in response to one problem have crucial repercussions for the practices and institutions devised to address other problems. A society, in other words, can be viewed as a system of practices and institutions. The notion that persisting practices or institutions are part of a larger system has led some functionalists to develop a distinctive protocol for studying the inter-relations between different parts of social systems. Referred to as functional analysis (Merton 1968), this method examines the effects a practice or institution has on other institutions and on the larger society. These effects or consequences assume four principal forms. Manifest functions refer to the consequences or objectives an institution explicitly attempts to achieve. Universities, for instance, are designed to impart knowledge and skills that will enable students to become productive workers and thoughtful citizens. Latent functions, on the other hand, identify effects that typically go unnoticed by the general public and frequently appear unrelated or starkly incompatible with an institution's (or the larger society's) explicit objectives. For example, while many citizens routinely denounce crime, functionalists argue that it—or, more specifically, the condemnation crime provokes—has the important latent function of clearly defining and affirming a community's normative boundaries (i.e., its sense of right and wrong). Prisons, too, have a latent function: by serving as "schools of crime," they insure that many ex-prisoners will commit new crimes which, in turn, will elicit still more community outrage and additional affirmation of a society's moral code.

In addition to being either manifest or latent, the functions served by a practice or institution can be positive or negative. Positive functions are evident when an institution facilitates the operation of other institutions and/or contributes to the overall stability and effectiveness of the larger society. In this vein, Davis and Moore (1945) hypothesize that attaching unequal financial and social rewards to different occupations has the positive effect of attracting the most talented and qualified individuals to a society's most "functionally important" positions. Negative consequences, which are sometimes called dysfunctions, occur when a practice or institution impedes the operation of other institutions and/or produces instability. For example, the "soft money" donated by corporations and other large contributors to political campaigns fosters the perception that contemporary politics is corrupt and convinces many citizens that their votes "don't count." That perception, moreover, is partially responsible for shockingly low rates of voter turnout.

Fourth, functionalism suggests that in contemporary societies containing scores of specialized institutions and hundreds of heterogenous sub-groups, societal integration is a recurring but manageable problem. In modern social orders, societal integration is achieved in two primary ways. First, specialized integrative institutions and processes—e.g., religious ceremonies, athletic contests, media events, and nationally celebrated holidays—heighten cohesion among people who otherwise share little in common. Second, consensus or agreement on such core values as individualism, freedom, achievement, and equal opportunity also serves to integrate complex societies. Incorporated into different institutions and internalized by individuals (during the course of their early socialization), shared values enable the diverse components of a large, differentiated society to co-exist and bond rather than dissolve into chronic chaos or a "war of all against all." According to one prominent functionalist, the relative stability of American society over the last two hundred years (the Civil War being a glaring exception) is largely attributable to the continuing consensus on the values of achievement and liberty (Lipset 1979).

Fifth, functionalism asserts that deviance and conflict arise from social strains, or contradictions within an institution or between institutions. That is, the primary source of contention and crime are inconsistencies inherent in the social system itself. In an influential essay, Robert Merton (1938) contends that in American society everyone, regardless of his or her station in life, is encouraged to pursue the American Dream. At the same time, however, the institutional means (e.g., a quality education and well-connected friends or acquaintances on the job market) for attaining success are not equally distributed: middle- and upper-class people are, in general, much more likely to have access to these institutional means than are working-class people. Confronted with this contradiction between a cultural goal (i.e., success) and the institutional means (e.g., a quality education) to achieve it, some individuals and groups will turn to crime (e.g., selling drugs). Note that in Merton's terms, crime often involves using "innovative," illegal means to realize a cultural goal prized both by criminals and law-abiding citizens. Under certain conditions, the same contradiction between cultural goals and institutional means can prompt widespread rebellion, with various groups replacing established cultural goals and the standard, institutional means with radically different values and means.

THE CONFLICT PERSPECTIVE

Like functionalism, the conflict perspective is a macrosociological approach that examines the emergence, persistence, and transformation of long-standing practices, institutions, and societies. Karl Marx, whose work first appeared in the mid-nineteenth century, is usually credited with crystallizing the key principles of this approach. Max Weber, an early twentieth-century German sociologist, is also recognized as a founding figure of conflict sociology. Leading contemporary conflict theorists include William Chambliss, Randall Collins, Ralf Dahrendorf, William Domhoff, and C. Wright Mills.

The conflict perspective rejects the functionalist notion that societies can be accurately portrayed as problem-solving entities. It also disputes the complementary idea that long-standing practices and institutions represent reasonably satisfactory answers to problems and as such contribute to a society's general welfare. Conflict sociologists embrace a very different orienting assumption: societies are arenas in which groups with fundamentally antagonistic interests struggle against one another. Different theorists within this tradition differ in terms of which particular groups and struggles they emphasize. Marx, for instance, highlights the conflicts between social classes, while Weber focuses on competing status groups (e.g., racial, ethnic, religious, age-based, etc.) and Dahrendorf and Collins draw attention to the battles between those who have authority and those subject to that authority. Despite disagreements about which groups and struggles are most important, all conflict sociologists believe that the interests that divide groups (whether classes, status groups, and so on) are built into the very fabric of a social order; these opposing interests are not readily negotiated, compromised, or resolved, nor can they be wished away or papered-over.

Second, conflicts among classes, status groups, and between those exercising authority and those subject to it supply the energy and the motivation for constructing and maintaining (as well as challenging and transforming) practices and institutions. Platt's (1977) well-known study of the origins of the juvenile court, for example, contends that this institution was created (in 1899) by social and economic elites and was employed to target and control the children of working-class immigrants residing in large cities. Conflict sociologists argue that today the nation's newly constructed maximum security prisons are, in practice, reserved predominantly for young, minority males raised in inner-city areas where good jobs are scarce (Chambliss 1999). On the other hand, white-collar, middle-class criminals, if they receive a prison sentence at all, are rarely housed in these types of facilities.

Third, the conflict perspective characterizes on-going practices and institutions as structures of domination that promote the interests of a relatively powerful, superordinate group while subverting the interests of relatively powerless, subordinate groups, even though the latter are usually much larger, numerically, than the powerful elites. Consequently, this perspective's orienting question is: which group's interests are served by a specific practice or institution? Kozol's (1991) investigation of how public schools are funded found that schools located in well-to-do suburban areas receive substantially more support than inner-city schools, which often lack textbooks, desks, and even serviceable plumbing. Far from enabling students from economically disadvantaged backgrounds to compete fairly on a level playing field, the current school system simply reflects and reproduces existing class inequalities.

Fourth, the conflict perspective reconceptualizes what functionalism terms values as ideologies. The primary purpose of an ideology is to protect and promote the distinctive interests of a particular class (or status or authority group). This legitimating purpose is best served when the ideology is presented in universal terms; when its ideas are stated as if they apply to everyone equally. According to the conflict approach, achievement and equal opportunity are most accurately viewed not as widely shared values but as a dominant ideology that operates to preserve (and reproduce) existing systems of inequality. In essence, the ideology of achievement and equal opportunity asserts that individuals and groups with great wealth, prestige, and power are rightfully entitled to these rewards because they have sacrificed, worked hard, and/or displayed exceptional talent. This ideology also explains why many people have few or none of these rewards: they are lazy, unwilling to make the sacrifices necessary for success, and/or lack the requisite talent. This ideology justifies the unequal distribution of social rewards by referring to individuals' character and moral virtues (or lack thereof). At the same time, it draws attention away from the structural inequities that largely explain why members of some groups are much more likely to "succeed" than are members of other groups.

Fifth, the conflict approach holds that significant social change usually reflects the efforts of groups mobilizing to advance their collective interests, often at the expense of other groups' interests. In this regard, proponents of conflict sociology question the functionalist claim that the substantial financial rewards enjoyed by physicians are due to the fact that medical doctors perform tasks that are, objectively, of great functional importance to society. Physicians' impressive incomes are more persuasively explained, conflict sociologists contend, with the observation that the American medical profession has established, in effect, a monopoly on the provision of health care (Starr 1982). Prior to 1850, this monopoly did not exist, and physicians were poorly paid and given little esteem. After the Civil War, however, doctors began to organize in earnest, and by the late nineteenth century they secured legislation prohibiting other groups and individuals from providing health care. The exclusion of competitors paved the way for a remarkable surge in doctors' income and a parallel rise in their prestige.

CONCLUSION

Each of the readings in this book employ one or more of the perspectives outlined above. The authors of these readings, it must be acknowledged, are not always fully explicit about which perspective (or combination of perspectives) they have used. Nevertheless, the perspectives they draw on, implicitly or explicitly, powerfully inform how they formulate the empirical problem under investigation and the type of explanation they devise to account for it. By keeping the basic assumptions of symbolic interactionism, functionalism, and the conflict perspective in mind, you will acquire a deeper and more complete understanding of the chapters that follow.

REFERENCES

Berger, Peter and Thomas Luckmann. 1966. *The Social Construction of Reality.* New York: Doubleday.

Best, Joel and Gerald Horiuchi. 1985. "The Razor Blade in the Apple: The Social Construction of Urban Legends." *Social Problems* 32: 488–499.

Blumer, Herbert. 1969. *Symbolic Interactionism.* Englewood Cliffs: Prentice-Hall.

Chambliss, William J. 1999. *Power, Politics, and Crime.* Boulder: Westview Press.

Collins, Randall. 1981. "On the Microfoundations of Macrosociology." *American Journal of Sociology* 86: 984–1014.

Davis, Kingsley and Wilbert E. Moore. 1945. "Some Principles of Stratification." *American Sociological Review* 10: 242–249.

Gusfield, Joseph R. 1963. *Symbolic Crusade.* Urbana: University of Illinois Press.

Kozol, Jonathan. 1991. *Savage Inequalities.* New York: HarperCollins.

Lipset, Seymour M. 1979. *The First New Nation.* New York: W. W. Norton.

Mead, George H. 1934. *Mind, Self and Society.* Edited by Charles W. Morris. Chicago: University of Chicago Press.

Merton, Robert K. 1938. "Social Structure and Anomie." *American Sociological Review* 3: 672–682.

_____. 1968. *Social Theory and Social Structure.* Enlarged edition. New York: Free Press.

Parsons, Talcott. 1977. *The Evolution of Societies.* Edited and with an Introduction by Jackson Toby. Englewood Cliffs: Prentice-Hall.

Platt, Anthony M. 1977. *The Child Savers.* 2nd edition. Chicago: University of Chicago Press.

Starr, Paul. 1982. *The Social Transformation of American Medicine.* Cambridge: Harvard University Press.

5

Anatomy of an Experiment: Values through Literature

EMILY STIER ADLER
AND PAULA J. FOSTER

The fundamental design and logic of experimental research are clearly and carefully laid out in this simple, yet complete study. Experimental research, long a mainstay of basic and medical science, is used in sociological research to search out specific, isolated causes and effects. The principle of the design is used to focus a clear relationship between the independent variable that is the action, and the measure of the dependent variable that occurs prior to and after its occurrence. Unlike other methods that deal with multiple factors that serve as compound influences, the experiment is highly focused and controlled. Precise matching of the two groups and measurement of the dependent variable enables researchers to specify, with mathematical quantification, the strength and direction of the outcomes. In thinking about this experiment you may want to ask yourself what you think of the strengths and weaknesses of this method. How well are abstract concepts such as "values" able to be measured? Do you think that the way they assessed their dependent variable was effective? Thinking beyond this experiment, have there ever been times when you have conducted informal experiments in life, testing what people thought or how some things influenced others? What are the specific values that we can glean from a method that targets the effect of one factor in isolation, without the contaminating influences of others?

The intentional teaching of morals and values was common in American public education for almost three hundred years and had widespread support until the 1930s. After decades of a "hands off" approach to the teaching of values, there have been calls for schools to return to explicit moral or character education in recent years (Kilpatrick, 1992; Wynne and Ryan, 1993).

Reprinted by permission of Libra Publishures.

One proposed curriculum revision suggests the use of literature to "help youngsters grow in courage, charity, justice and other virtues" (Kilpatrick, 1992: 268).

At the same time, advocacy of literature-based approaches has increased in educational psychology and social work practice. The literature-based approach, sometimes called bibliotherapy, has been championed for numerous uses—a way to teach children to relate moral principles to real life (Dana and Lynch-Brown, 1991), an effective deterrent to substance abuse (Bump, 1990; Pardeck, 1991), a guide to self-understanding and improved self-concept (Calhoun, 1987; Hebert, 1991; Lenkowsky and Lenkowsky 1978; Miller, 1993), and as a way to help children deal with parental divorce (Early, 1993). The advocacy has occurred despite the caution that the approach has been awarded a "scientific respectability it does not have," and that the application of the technique "far outstrips the tight validating studies supporting its use" (Riordan, 1991: 306). In fact, our review of the literature found no studies of bibliotherapy or other literature-based approaches to teaching attitudes, values, or character traits that used an experimental design to evaluate the effectiveness of the method. It is this evaluation which our study seeks to provide.

THE HYPOTHESIS

We were interested in determining the impact of a literature-based curriculum on student values. Constructing a causal hypothesis, we designed a specific curriculum for use as our independent variable and identified a specific value, the value of caring for others, as the dependent variable. We hypothesized that the use of a literature-based curriculum would increase students' belief in the value of caring for others. We went into the field—the real world of students, classrooms, and teachers—in order to test our hypothesis.

THE SAMPLE

A middle school in an urban community with a fairly homogeneous population of approximately 40,000 (94 percent were non-Hispanic and white, 3 percent Hispanic, 2 percent Asian, and 1 percent African American) was selected because administrators were willing to allow a new curricular approach to be implemented on a small scale. The school principal and the teachers of one large seventh grade class agreed to work with us. We did not select the community, school, or students randomly and recognize that our results cannot be generalized beyond the students in our sample.

The class was team-taught, with two teachers, 57 students, and a room which could be divided into two rooms with a movable partition. The class structure varied between the teachers teaching all 57 students in one space, the teachers working with many small groups, and two separate classrooms, each with a teacher. The students were an average of 13.1 years old when the study began. There were 31 girls and 26 boys; 50 students were white and 7 were Asian.

Using the characteristics of gender, race, age, and academic ability, the teachers matched the students into pairs and created two approximately equal groups. One of the groups was randomly selected to be the experimental group while the other became the control group.

THE INDEPENDENT VARIABLE: THE READING PROJECT

The independent variable was the ten-week reading project developed for this study. In consultation with the teachers, we selected three books with appropriate themes and designed class exercises and discussions to accompany them.[1] The theme of each book is the importance of caring for others. The main characters make choices and ultimately decide to care for others, even at the expense of achieving personal goals. In all three books, the main characters are adolescents, two are female and one is male, two are white and one is Asian.[2] Several times each week for ten weeks, the students in the experimental group participated in classroom discussions and exercises designed to reinforce the theme of caring for others. During the same class time, the students in the control group worked with the other teacher in the other half of the divided classroom space, reading and discussing the regular seventh grade literature (books with animal main characters, such as *Call of the Wild*).

THE DEPENDENT VARIABLE: CARING FOR OTHERS

The dependent variable, support for the value of caring for others, is defined conceptually as the extent to which students "see and respond to need" and advocate "taking care of the world by sustaining the web of relationship so that no one is left alone" (Gilligan, 1982: 62). The extent to which the students supported this value was measured both before and after the reading project by a set of three essays on caring for others. Students in both groups completed the essays during the two weeks before and the two weeks after the ten-week reading project. The teachers used class time and asked the students to write the essays during the language arts class. The students were told that there were no "right" answers and were asked to express their opinions or make up a story without worrying about what they were "supposed to say." Student essays were identified by number. Due to absenteeism, some students did not complete all six essays.

After considering several additional essay questions, we used three questions to measure caring attitudes:

1. John and Bill both have jobs in a movie theater. The manager is deciding on the work schedule for the weekend. Both John and Bill want Friday night off. John wants to spend the evening visiting his grandmother who is sick and in the hospital, and Bill has plans to go out with his friends. The manager

tells them that he needs one of them to work on Friday night, but that they can either decide between themselves who will work or he'll flip a coin to decide. What do you think should happen and why? (GRANDMOTHER)

2. Susan reads in the local newspaper about a family that lost all their belongings when there was a fire in their home. She reads that they didn't have much insurance to replace their belongings. She decides to help out by sending them her whole allowance of $5.00 a week for the next three weeks. Describe how you feel about Susan's behavior. (ALLOWANCE)
3. Sometimes people have small families. Do you think that friends can be like families? In what ways? In what ways are they different from families? Write about your opinions. (FRIENDS)

Each essay was read for content by both researchers and coded into one of three categories. For GRANDMOTHER and ALLOWANCE, each was coded from most supportive for caring for others to least supportive of caring for others. For GRANDMOTHER, answers that defined John's and Bill's needs as equivalent were defined as least supportive of caring (for example, "Bill and John should split the hours in half that Friday," or "The manager should flip a coin since it's fair and quicker"). Answers that gave some priority to John's visiting his grandmother were coded as moderately supportive of caring (for example, "I think Bill should work because he should understand that John really needs to see his grandmother this week. Next week John will work and let Bill have the time off"). Essays that identified John's desire to visit his grandmother as having more priority than Bill's spending time with friends were classified as most supportive of the value of caring. For example, "If Bill has any decency, then he should work instead of John. John seeing his sick grandmother is more important than going out with friends. Friends last forever, but grandparents don't."

Inter-coder reliability was 95 percent for GRANDMOTHER. On this essay, 63 percent of the sample (57 percent of the experimental group and 68 percent of the control group) was classified as most supportive of caring for others at the pretest.

Essays in response to the ALLOWANCE question were judged to be least supportive of the value of caring if the student disagreed with Susan's behavior. Examples include the student who said, "I wouldn't do that. Susan's five dollars wouldn't help very much anyway," the one who said, "I don't care. I think she was nice, but stupid to do that," and the response, "Susan's behavior is not logical. If she read the newspaper I read, then she would be broke. I doubt $15.00 will help the poor family anyway." Essays that were partially enthusiastic about Susan's behavior or felt she would get some sort of reward for it were coded as being somewhat supportive of caring. The student who said "She is very kind, but I don't think she should give all her money away. She should get a group of people and start a collection for the family so Susan won't have to give all her money" was placed in this category. If the student felt that Susan had done the right thing and expected nothing in return, the answers were classified as most supportive of caring. Illustrative of this response was the student who said "I think Susan is doing a great thing because most people wouldn't care. I also think she has a big heart to give up her allowance for three weeks. That is a really great thing to do."

Inter-coder reliability was 95 percent for this essay. Seventy-five percent of the sample (78 percent of the experimental group and 71 percent of the control group) was classified as most supportive of caring before the introduction of the literature project.

For the essay FRIENDS, student answers were coded into three categories: friends and families seen as very different from each other, friends and families seen as somewhat similar, and friends seen as serving the same emotional and support functions as families do, even if the people live in different households. An example of the first category is the essay that said, "I don't think friends are like family. Friends are people to hang around with and have fun with. You can have fun with family but it's different. I wouldn't consider my friends my family." An example of the second category, seeing some similarities between the two units, is the student who wrote, "In some ways friends can be like family, sometimes you can trust them and they can care for you. In other ways, they aren't like family because they can't know you as well as family members do." Illustrative of the essays that were coded as seeing the units as serving the same emotional and support functions are these two: "I think friends can be like family. You can like them as well as family and they can like you in return. You can trust them even if you haven't grown up with them" and "I do think friends can be like families. They can do special things for you, or just show you they care. The only way they are different is that they are not blood related. There could be other differences, but it doesn't matter as long as they care for you."

The inter-coder reliability for this essay was 94 percent. At the pretest, 59 percent of the sample (50 percent of the experimental group and 71 percent of the control group) thought that friends can provide the same sorts of emotional and social supports as families.

FINDINGS

To determine the effect of the reading project, the essays each student wrote before and after completing one of the ten-week curriculums were compared. Based on the two essays, each student was classified as becoming more supportive of caring for others, less supportive, or showing no change in values. Most students in the experimental and control groups were consistent in their values at both time 1 and time 2 and showed no change in support for the value of caring. For the students who did change, the results for two of the essays were in the hypothesized direction.

There was a statistically significant difference between the two groups for the FRIENDS essay. Thirty percent of the experimental group and none of the control group became more supportive of the belief that friends can be like family in providing emotional and social support to people. For GRANDMOTHER, the results approached statistical significance, with 20 percent of the control group and none of the experimental group becoming less supportive of caring for others after the reading project, and slightly more (3 percent) of the experimental group becoming more caring. For ALLOWANCE, the results, which were not statistically significant, were nevertheless not supportive of the hypothesis. Twenty-three percent of the control group and only 8 percent of the experimental group became more supportive of caring values after the completion of the reading project. Analyses controlling for gender are very similar to the results with no control variable and also demonstrate some support for the hypothesis that a literature-based curriculum can have an impact on caring values.

DISCUSSION AND CONCLUSION

The strongest support for the hypothesis that a literature-based curriculum can influence values is the finding that the experimental group became significantly more committed to the view that friends can fulfill many of the same emotional and social support functions as families. The data for one of the other two essays also showed changes in the hypothesized direction. The experimental group remained more supportive of the caring value for the GRANDMOTHER essay, providing weak support for the hypothesis. However, for the ALLOWANCE essay, the control group became more supportive of caring than the experimental group, a finding that does not support the effectiveness of the reading project.

Several issues, including methodological ones, are important for interpreting the results of the research. The specific indicators used to measure the value of caring can be questioned. The large percentage of the student essays at time 1 which advocated caring values could be a source of concern. Honesty might be an issue as students could have written the essays that they thought their teachers wanted, rather than sharing their own beliefs. Another concern is variability. With such a large proportion of students being categorized as "most caring" at time 1, only a relatively small increase in caring was possible. A final methodological concern is the amount of missing data. With a large number of students not completing the essays both times due to absences, as many as sixteen cases were omitted from each analysis.

Even with the methodological weaknesses, one interpretation of the findings is that there were changes in caring values as a result of reading and discussing selected literature. It is clear that any changes were small, perhaps because values are not quickly or easily affected by a modest change in the curriculum. We can assume that these students, like other American children and teens, watch an average of three hours of television per day (*Time,* 1995: 74), engage in many out-of-school activities, and have other curricula materials that present values explicitly and implicitly. Therefore, it should not be surprising that the ten-week reading project in this study had, at best, a modest effect. Teaching values and developing character is a complex and time-consuming task. If the value system presented in popular culture and the rest of the school curriculum does not include a great deal of support for positive values and pro-social behavior, substantial character or moral education with small, discrete projects may not be feasible. Schools interested in character education should consider a multiyear curriculum starting in the early grades with frequent reinforcement. The curriculum could include the reading and discussion of books with specific values as well as community service projects, the development of rules within the school, the inclusion of families in curriculum development, and other techniques. While small reading projects may be able to have some effect on some values, only when part of a larger framework will they be able to provide the kind of socialization that the proponents of moral and character education advocate.

THE READING PROJECT AS AN EXPERIMENT

The reading project asked whether an independent variable had *caused* a change in or *had an effect* on the dependent variable. As such, it was an explanatory study with a causal hypothesis. The decisions about research strategies, such as sample selection, measurement of variables, and collecting data more than once, were influenced by the explanatory purpose of the research. Analysis of the reading project points to some of the benefits and disadvantages inherent in the use of the controlled experiment, a study design that is very useful for testing causal hypotheses.

To test the hypothesis that participating in the reading project would affect "caring," it was important to determine changes in the dependent variable. For this reason, student support for caring was measured two times—before and after the implementation of the reading project. In experiments, the terms for the before and after measurements of the dependent variable are the **pre-test** and the **post-test.**

In this study, the independent variable—whether a student participated in the literature-based curriculum focused on caring or the regular curriculum—was not "measured." Instead the literature-based curriculum was "introduced" to the experimental group and the control group received the regular seventh grade reading curriculum. The ability to control the timing of the independent variable and to determine who is exposed to each condition are central elements in the experimental design.

In experiments, the control group is exposed to *all* the influences that the experimental group is exposed to *except* for the **stimulus** (the experimental condition of the independent variable). The researcher tries to have the two groups treated exactly alike, except that instead of a stimulus, the control group receives no treatment, an alternative treatment, or **placebo** (a simulated treatment that is supposed to appear authentic). For maximum internal validity, the experimental design tries to eliminate systematic differences between the experimental and control group beyond introducing the independent variable. Remember that **internal validity** is concerned with factors that affect the internal links between the independent and dependent variables, specifically factors that support alternative explanations for variations in the dependent variable. In experiments, internal validity focuses on whether the independent variable or other factors, such as prior differences between the experimental and control groups, are explanations for the observed differences in the dependent variable.

If, as in the reading project, the data analysis reveals at least some association between the independent and dependent variables, then the advantages of the experimental design become evident. An experiment offers evidence of time-order because the dependent variable is measured both *before* and *after* the independent variable is introduced, and *only one group* is exposed to the stimulus in the interval between the measurements. In addition, the method of selecting the experimental and control groups tends to minimize the chances of preexisting systematic differences between the two groups.

In the reading project study, two of the three indicators of the dependent variable showed small changes in the hypothesized direction. In addition, the time-order of the variables is known. However, demonstration of change in the dependent variable and a time-order in which the independent variable precedes changes in the dependent variable offer *necessary,* but not *sufficient* support for a cause and effect relationship. It is possible that another, unobserved factor brought about the change in the dependent variable. In the reading project, for example, perhaps one or more significant events occurred in the community, school, or classroom during the weeks that students were reading and discussing the books. It could be that new staff members or students arrived at the school in those weeks. Or maybe a new television show, game, or movie became very popular during the experimental time period. We might be concerned that these or other factors influenced student caring.

Concern about the effects of other factors is the rationale for one of the essential design elements in the controlled experiment—the use of two groups, an experimental and a control group. In the focal research, although the students in the class might have been exposed to new students, staff, games, movies, or television programs, we know that only the experimental group was directly exposed to the stimulus. With two groups and the ability to control who is exposed to a stimulus, researchers using an experimental design can be fairly confident of their ability to discern effects of the independent variable.

There are several possible challenges to the assumption that there were no systematic differences between the two groups. It is possible, for example, that the two teachers were different in a way that could have affected the results—perhaps one was a better teacher, more caring or more enthusiastic. Another possibility is that the kinds of exercises used by the reading project encouraged closer student and teacher interaction than did the traditional class work. It could also be that the two groups were unequal in some important way to begin with even though the teachers matched them on age, race, gender, and academic ability. For instance, we don't know if there were more students in the experimental group who were avid readers or if there were more in the control group who had recently moved to the community. Using matching to select experimental and control groups in this and other studies has clear limitations. In hindsight, both Emily and her coauthor believe that an alternative method of selecting the two groups might have been better for internal validity.

NOTES

1. Activities for students in the experimental group were based on methods presented in *Webbing With Literature* (Bromley, 1991). In "webbing," a core concept is selected and emphasized by using "webs" or lines connecting the concept to information. In the experiment, the concept of caring was the one highlighted in the discussion of each book. Class exercises included writing favorite quotes and feelings in a journal, making a group collage that expressed the theme and the feelings in each novel, and webbing how the characters in the books were connected to each other (by marriage, birth, friendship, etc.).

2. *Friends Are Like That* by Patricia Hermes (New York: Scholastic Inc., 1984) is the story of Tracy, an eighth grader who weighs popularity against supporting her best friend, Kelly, a nonconformist and a social pariah. Tempted to become part of the school's "popular" crowd, Tracy instead chooses friendship. *Red Cap* by G. Clifton Wisler (New York: Lodestar Books, 1991) is the true story of Ransom Powell, who in 1862, at age thirteen, joined the Union Army and was captured with his company and sent to Camp Sumter, the Confederate prison at Andersonville. Under terrible conditions, the prisoners in Ransom's company help and support each other in ways that a family might. Ransom survived imprisonment because of this care. *The Clay Marble* by Minfong Ho (New York: Farrar Straus Giroux, 1991) is set in Cambodia in the early 1980s. Twelve-year-old Dara and her family are among the thousands who are forced to flee from their villages to a refugee camp on the border of Cambodia and Thailand. Her family meets and becomes emotionally attached to another family. Even though some family members die, the two units are as one at the story's end.

REFERENCES

Bromley, K. 1991. *Webbing with literature: A practical guide.* Boston: Allyn & Bacon.

Bump, J. 1990. Innovative bibliotherapy approaches to substance abuse education. *Arts in Psychotherapy* 17: 355–362.

Calhoun, G. 1987. Enhancing self-perception through bibliotherapy. *Adolescence* 22 (Winter): 939–943.

Dana, N. Fichtman, and C. Lynch-Brown. 1991. Moral development of the gifted: Making a case for children's literature. *Roeper Review* 14 (Sept.): 13–16.

Early, B. P. 1993. The healing magic of myth: Allegorical tales and the treatment of children of divorce. *Child and Adolescent Social Work Journal* 10 (2) (April): 97–106.

Gilligan, C. 1982. *In a different voice.* Cambridge, MA: Harvard University Press.

Hebert, T. P. 1991. Meeting the affective needs of bright boys through bibliotherapy. *Roeper Review* 13 (June): 207–212.

Kilpatrick, W. 1992. *Why Johnny can't tell right from wrong.* New York: Simon & Schuster.

Lenkowsky, B. E., and R. S. Lenkowsky. 1978. Bibliotherapy for the LD adolescent. *Academic Therapy* 14 (1978): 179–185.

Miller, D. 1993. The Literature Project: Using literature to improve the self-concept of at-risk adolescent females. *Journal of Reading* 36 (March): 442–448.

Pardeck, J. 1991. Using books to prevent and treat adolescent chemical dependency. *Adolescence* 26 (Spring): 201–208.

Riordan, R. J. 1991. "Bibliotherapy Revisited." *Psychological Reports* 68: 306.

Time. 1995. Special Report: The State of the Union. Jan. 30: 74.

Wynne, E. A., and K. Ryan. 1993. *Reclaiming our schools.* New York: Macmillan.

METHODS:
Survey Research

6

Survey of Sexual Behavior of Americans

EDWARD O. LAUMANN,
JOHN H. GAGNON,
ROBERT T. MICHAEL,
AND STUART MICHAELS

In the most wide-ranging, contemporary study of American sexual behavior, Laumann et al. discuss their approach to conducting this survey. After the survey was released, it was critically acclaimed, but many other social scientists and laypeople question how accurate the results can be given the highly secretive and sensitive nature of the topic. The authors of this article lay out in clear language the methodological theories that led them to conduct each step of the process and how they acted carefully to minimize distortion and enhance generalizability. This research shows us the way sociologists generate correlational connections between various types of demographics, attitudes, and behavior. Thus, the authors feel that, despite problems, survey research is a viable tool for understanding even the most delicate aspects of the human condition. Where do you stand on this issue? Is survey research a useful way to learn about people's sexual inclinations? What are some of the problems you foresee with this study? Nonetheless, why do you think that these researchers were able to carry out such a study with such critical acclaim from many members of the scientific community?

Most people with whom we talked when we first broached the idea of a national survey of sexual behavior were skeptical that it could be done. Scientists and laypeople alike had similar reactions: "Nobody will agree to participate in such a study." "Nobody will answer questions like these, and, even if they do, they won't tell the truth." "People don't know enough about sexual practices as they relate to disease transmission or even to pleasure or physical and emotional satisfaction to be able to answer questions accurately." It

From *The Sociological Organization of Sexuality,* pp. 35–73, by Laumann, E.O., et al.
Reprinted by permission of the author and publisher.

would be dishonest to say that we did not share these and other concerns. But our experiences over the past seven years, rooted in extensive pilot work, focus-group discussions, and the fielding of the survey itself, resolved these doubts, fully vindicating our growing conviction that a national survey could be conducted according to high standards of scientific rigor and replicability. . . .

The society in which we live treats sex and everything related to sex in a most ambiguous and ambivalent fashion. Sex is at once highly fascinating, attractive, and, for many at certain stages in their lives, preoccupying, but it can also be frightening, disturbing, or guilt inducing. For many, sex is considered to be an extremely private matter, to be discussed only with one's closest friends or intimates, if at all. And, certainly for most if not all of us, there are elements of our sexual lives never acknowledged to others, reserved for our own personal fantasies and self-contemplation. It is thus hardly surprising that the proposal to study sex scientifically, or any other way for that matter, elicits confounding and confusing reactions. Mass advertising, for example, unremittingly inundates the public with explicit and implicit sexual messages, eroticizing products and using sex to sell. At the same time, participants in political discourse are incredibly squeamish when handling sexual themes, as exemplified in the curious combination of horror and fascination displayed in the public discourse about Long Dong Silver and pubic hairs on pop cans during the Senate hearings in September 1991 on the appointment of Clarence Thomas to the Supreme Court. We suspect, in fact, that with respect to discourse on sexuality there is a major discontinuity between the sensibilities of politicians and other self-appointed guardians of the moral order and those of the public at large, who, on the whole, display few hang-ups in discussing sexual issues in appropriately structured circumstances. This book is a testament to that proposition.

The fact remains that, until quite recently, scientific research on sexuality has been taboo and therefore to be avoided or at best marginalized. While there is a visible tradition of (in)famous sex research, what is, in fact, most striking is how little prior research exists on sexuality in the general population. Aside from the research on adolescence, premarital sex, and problems attendant to sex such as fertility, most research attention seems to have been directed toward those believed to be abnormal, deviant, criminal, perverted, rare, or unusual, toward sexual pathology, dysfunction, and sexually transmitted disease—the label used typically reflecting the way in which the behavior or condition in question is to be regarded. "Normal sex" was somehow off limits, perhaps because it was considered too ordinary, trivial, and self-evident to deserve attention. To be fair, then, we cannot blame the public and the politicians entirely for the lack of sustained work on sexuality at large—it also reflects the prejudices and understandings of researchers about what are "interesting" scientific questions. There has simply been a dearth of mainstream scientific thinking and speculation about sexual issues. We have repeatedly encountered this relative lack of systematic thinking about sexuality to guide us in interpreting and understanding the many findings reported in this book.

. . . In order to understand the results of our survey, the National Health and Social Life Survey (NHSLS), one must understand how these results were generated. To construct a questionnaire and field a large-scale survey, many research design decisions must be made. To understand the decisions made, one needs to understand the multiple purposes that underlie this research project. Research design is never just a theoretical exercise. It is a set of practical solutions to a multitude of problems and considerations that are chosen under the constraints of limited resources of money, time, and prior knowledge.

SAMPLE DESIGN

The sample design for the NHSLS is the most straightforward element of our methodology because nothing about probability sampling is specific to or changes in a survey of sexual behavior. . . .

Probability sampling, that is, sampling where every member of a clearly specified population has a known probability of selection—what by commentators often somewhat inaccurately call random sampling—is the sine qua non of modern survey research (see Kish 1965, the classic text on the subject). There is no other scientifically acceptable way to construct a representative sample and thereby to be able to generalize from the actual sample on which data are collected to the population that that sample is designed to represent. Probability sampling as practiced in survey research is a highly developed practical application of statistical theory to the problem of selecting a sample. Not only does this type of sampling avoid the problems of bias introduced by the researcher or by subject self-selection bias that come from more casual techniques, but it also allows one to quantify the variability in the estimates derived from the sample. . . .

Sample Size

How large should the sample be? There is real confusion about the importance of sample size. In general, for the case of a probability sample, the bigger the sample, the higher the precision of its estimates.[1] This precision is usually measured in terms of the amount of sampling error accruing to the statistics calculated from the sample. The most common version of this is the statement that estimated proportions (e.g., the proportion of likely voters planning to vote for a particular candidate) in national political polls are estimated as being within ± 2 or 3 percent of the overall population figure. The amount of this discrepancy is inversely related to the size of the sample: the larger the sample, the smaller the likely error in the estimates. This is not, however, a simple linear relation. Instead, in general, as the sample size increases, the precision of the estimates derived from the sample increases by the square root of the sample size. For example, if we quadruple the sample size, we improve the estimate only by a factor of two. That is, if the original sample has a sampling error of ± 10 percent, then the quadrupled sample size will have an error of ± 5 percent.

In order to determine how large a sample size for a given study should be, one must first decide how precise the estimates to be derived need to be. To illustrate this reasoning process, let us take one of the simplest and most commonly used statistics in survey research, the proportion. Many of the most important results reported in this book are proportions. For example, what proportion of the population had more than five sex partners in the last year? What proportion engaged in anal intercourse? With condoms? Estimates based on our sample will differ from the true proportion in the population because of sampling error (i.e., the random fluctuations in our estimates that are due to the fact that they are based on samples rather than on complete enumerations or censuses). If one drew repeated samples using the same methodology, each would produce a slightly different estimate. If one looks at the distribution of these *estimates,* it turns out that they will be normally distributed (i.e., will follow the famous bell-shaped curve known as the Gaussian or normal distribution) and centered around the true proportion in the population. The larger the sample size, the tighter the distribution of estimates will be.

We began with an area probability sample, which is a sample of households, that is, of addresses, not names. Rather than approach a household by knocking on the door without advance warning, we followed NORC's standard practice of sending an advance letter, hand addressed by the interviewer, about a week before the interviewer expected to visit the address. In this case, the letter was signed by the principal investigator, Robert Michael, who was identified as the dean of the Irving B. Harris Graduate School of Public Policy Studies of the University of Chicago. The letter briefly explained the purpose of the survey as helping "doctors, teachers, and counselors better understand and prevent the spread of diseases like AIDS and better understand the nature and extent of harmful and of healthy sexual behavior in our country." The intent was to convince the potential respondent that this was a legitimate scientific study addressing personal and potentially sensitive topics for a socially useful purpose. AIDS was the original impetus for the research, and it certainly seemed to provide a timely justification for the study. But any general purpose approach has drawbacks. One problem that the interviewers frequently encountered was potential respondents who did not think that AIDS affected them and therefore that information about their sex lives would be of little use.

Gaining respondents' cooperation requires mastery of a broad spectrum of techniques that successful interviewers develop with experience, guidance from the research team, and careful field supervision. This project required extensive training before entering the field. While interviewers are generally trained to be neutral toward topics covered in the interview, this was especially important when discussing sex, a topic that seems particularly likely to elicit emotionally freighted sensitivities both in the respondents and in the interviewers. Interviewers needed to be fully persuaded about the legitimacy and importance of the research. Toward this end, almost a full day of training was devoted to presentations and discussions with the principal investigators in addition to the extensive advance study materials to read and comprehend. Sample answers to frequently asked questions by skeptical respondents and brainstorming about strategies to convert reluctant respondents were part of the training exercises. A set of endorsement letters from prominent local and national notables and refusal conversion letters were also provided to interviewers. A hotline to the research office at the University of Chicago was set up to allow potential respondents to call in with their concerns. Concerns ranged from those about the legitimacy of the survey, most fearing that it was a commercial ploy to sell them something, to fears that the interviewers were interested in robbing them. Ironically, the fact that the interviewer initially did not know the name of the respondent (all he or she knew was the address) often led to behavior by the interviewer that appeared suspicious to the respondent. For example, asking neighbors for the name of the family in the selected household and/or questions about when the potential respondent was likely to be home induced worries that had to be assuaged. Another major concern was confidentiality—respondents wanted to know how they had come to be selected and how their answers were going to be kept anonymous.

Mode of Administration: Face-to-face, Telephone, or Self-Administered

Perhaps the most fundamental design decision, one that distinguishes this study from many others, concerned how the interview itself was to be conducted. In survey research, this is usually called the *mode* of interviewing or of questionnaire administration. We chose face-to-face interviewing, the most costly mode, as the primary vehicle for data collection in the NHSLS. What follows is the reasoning behind this decision.

A number of recent sex surveys have been conducted over the telephone, . . . The principal advantage of the telephone survey is its much lower cost. Its major disadvantages are the length and complexity of a questionnaire that can be realistically administered over the telephone and problems of sampling and sample control. . . . The NHSLS, cut to its absolute minimum length, averaged about ninety minutes. Extensive field experience suggests an upper limit of about forty-five minutes for phone interviews of a cross-sectional survey of the population at large. Another disadvantage of phone surveys is that it is more difficult to find people at home by phone and, even once contact has been made, to get them to participate. . . . One further consideration in evaluating the phone as a mode of interviewing is its unknown effect on the quality of responses. Are people more likely to answer questions honestly and candidly or to dissemble on the telephone as opposed to face-to-face? Nobody knows for sure.

The other major mode of interviewing is through self-administered forms distributed either face-to-face or through the mail.[2] When the survey is conducted by mail, the questions must be self-explanatory, and much prodding is typically required to obtain an acceptable response rate. . . . This procedure has been shown to produce somewhat higher rates of reporting socially undesirable behaviors, such as engaging in criminal acts and substance abuse. We adopted the mixed-mode strategy to a limited extent by using four short, self-administered forms, totaling nine pages altogether, as part of our interview. When filled out, these forms were placed in a "privacy envelope" by the respondent so that the interviewer never saw the answers that were given to these questions. . . .

The fundamental disadvantage of self-administered forms is that the questions must be much simpler in form and language than those that an interviewer can ask. Complex skip patterns must be avoided. Even the simplest skip patterns are usually incorrectly filled out by some respondents on self-administered forms. One has much less control over whether (and therefore much less confidence that) respondents have read and understood the questions on a self-administered form. The NHSLS questionnaire (discussed below) was based on the idea that questions about sexual behavior must be framed as much as possible in the specific contexts of particular patterns and occasions. We found that it is impossible to do this using self-administered questions that are easily and fully comprehensible to people of modest educational attainments.

To summarize, we decided to use face-to-face interviewing as our primary mode of administration of the NHSLS for two principal reasons: it was most likely to yield a substantially higher response rate for a more inclusive cross section of the population at large, and it would permit more complex and detailed questions to be asked. While by far the most expensive approach, such a strategy provides a solid benchmark against which other modes of interviewing can and should be judged. The main unresolved question is whether another mode has an edge over face-to-face interviewing when highly sensitive questions likely to be upsetting or threatening to the respondent are being asked. As a partial control and test of this question, we have asked a number of sensitive questions in both formats so that an individual's responses can be systematically compared. . . . Suffice it to say at this point that there is a stunning consistency in the responses secured by the different modes of administration.

Recruiting and Training Interviewers

We firmly believed that it was very important to recruit and train interviewers for this study very carefully. In particular, we worried that interviewers who were in any way uncomfortable with the topic of sexuality would not do a good job and would adversely affect the quality of the interview. We thus took special steps in recruiting interviewers to make it clear what the survey was about, even showing them especially sensitive sample questions. We also assured potential recruits that there would be no repercussions should they not want to work on this study; that is, refusal to participate would not affect their future employment with NORC. None of these steps seemed to hinder the recruitment effort. In general, interviewers like challenging studies. Any survey that is not run of the mill and promises to be of current public relevance is regarded as a good and exciting assignment—one to pursue enthusiastically. In short, we had plenty of interviewers eager to work on this study. Of course, a few interviewers did decline to participate because of the subject matter.

THE QUESTIONNAIRE

The questionnaire itself is probably the most important element of the study design. It determines the content and quality of the information gathered for analysis. Unlike issues related to sample design, the construction of a questionnaire is driven less by technical precepts and more by the concepts and ideas motivating the research. It demands even more art than applied sampling design requires. . . .

The problem that we faced in writing the questionnaire was figuring out how best to ask people about their sex lives. There are two issues here that should be highlighted. One is conceptual, having to do with how to define sex, and the second has to do with the level or kind of language to be used in the interview.

Very early in the design of a national sexual behavior survey, in line with our goal of not reducing this research to a simple behavioral risk inventory, we faced the issue of where to draw the boundaries in defining the behavioral domain that would be encompassed by the concept of sex. This was particularly crucial in defining sexual activity that would lead to the enumeration of a set of sex partners. There are a number of activities that commonly serve as markers for sex and the status of sex partner, especially intercourse and orgasm. While we certainly wanted to include these events and their extent in given relationships and events, we also felt that using them to define and ask about sexual activity might exclude transactions or partners that should be included. Since the common meaning and uses of the term *intercourse* involve the idea of the intromission of a penis, intercourse in that sense as a defining act would at the very least exclude a sexual relationship between two women. There are also many events that we would call sexual that may not involve orgasm on the part of either or both partners.

Another major issue is what sort of language is appropriate in asking questions about sex. It seemed obvious that one should avoid highly technical language because it is unlikely to be understood by many people. One tempting alternative is to use colloquial language and even slang since that is the only language that some people ever use in discussing sexual matters. There is even some evidence that one can improve reporting somewhat by allowing respondents to select their own preferred terminology (Blair et al. 1977; Bradburn et al. 1978; Bradburn and Sudman 1983). Slang and other forms of colloquial speech, however, are likely to be problematic in several ways. First, the use of slang can produce a tone in the interview that is counterproductive because it downplays the distinctiveness of the interviewing situation itself. An essential goal in survey interviewing, especially on sensitive topics like sex, is to create a neutral, nonjudgmental, and confiding atmosphere and to maintain a certain professional distance between the interviewer and the respondent. A key advantage that the interviewer has in initiating a topic for discussion is being a stranger or an outsider who is highly unlikely to come in contact with the respondent again. It is not intended that a longer-term bond between the interviewer and the respondent be formed, whether as an advice giver or a counselor or as a potential sex partner.[3]

The second major shortcoming of slang is that it is highly variable across class and education levels, ages, regions, and other social groupings. It changes meanings rapidly and is often imprecise. Our solution was to seek the simplest possible language—standard English—that was neither colloquial nor highly technical. For example, we chose to use the term *oral sex* rather than the slang *blow job* and *eating pussy* or the precise technical but unfamiliar terms *fellatio* and *cunnilingus.* Whenever possible, we provided definitions when terms were first introduced in a questionnaire—that is, we tried to train our respondents to speak about sex in our terms. Many terms that seemed clear to us may not, of course, be universally understood; for example, terms like *vaginal* or *heterosexual* are not understood very well by substantial portions of the population. Coming up with simple and direct speech was quite a challenge because most of the people working on the questionnaire were highly educated, with strong inclinations toward the circumlocutions and indirections of middle-class discourse on sexual themes. Detailed reactions from field interviewers and managers and extensive pilot testing with a broad cross section of recruited subjects helped minimize these language problems.

ON PRIVACY, CONFIDENTIALITY, AND SECURITY

Issues of respondent confidentiality are at the very heart of survey research. The willingness of respondents to report their views and experiences fully and honestly depends on the rationale offered for why the study is important and on the assurance that the information provided will be treated as confidential. We offered respondents a strong rationale for the study, our interviewers made great efforts to conduct the interview in a manner that protected respondents' privacy, and we went to great lengths to honor the assurances that the information would be treated confidentially. The subject matter of the NHSLS makes the issues of confidentiality especially salient and problematic because there are so many easily imagined ways in which information voluntarily disclosed in an interview might be useful to interested parties in civil and criminal cases involving wrongful harm, divorce proceedings, criminal behavior, or similar matters.

NOTES

1. This proposition, however, is not true when speaking of nonrandom samples. The original Kinsey research was based on large samples. As noted earlier, surveys reported in magazines are often based on very large numbers of questionnaires. But, since these were not representative probability samples, there is no necessary relation between the increase in the sample size and how well the sample estimates population parameters. In general, nonprobability samples describe only the sample drawn and cannot be generalized to any larger population.

2. We ruled out the idea of a mail survey because its response rate is likely to be very much lower than any other mode of interviewing (see Bradburn, Sudman, et al. 1978).

3. Interviewers are not there to give information or to correct misinformation. But such information is often requested in the course of an interview. Interviewers are given training in how to avoid answering such questions (other than clarification of the meaning of particular questions). They are not themselves experts on the topics raised and often do not know the correct

answers to questions. For this reason, and also in case emotionally freighted issues for the respondent were raised during the interview process, we provided interviewers with a list of toll-free phone numbers for a variety of professional sex- and health-related referral services (e.g., the National AIDS Hotline, an STD hotline, the National Child Abuse Hotline, a domestic violence hotline, and the phone number of a national rape and sexual assault organization able to provide local referrals).

REFERENCES

Blair, Ellen, Seymour Sudman, Norman M. Bradburn, and Carol Stacking. 1977. "How to Ask Questions About Drinking and Sex: Response Effects in Measuring Consumer Behavior." *Journal of Marketing Research* 14: 316–321.

Bradburn, Norman M., and Seymour Sudman. 1983. *Asking Questions: A Practical Guide to Questionnaire Design.* San Francisco: Jossey-Bass.

Bradbum, Norman M., Seymour Sudman, Ed Blair, and Carol Stacking. 1978. "Question Threat and Response Bias." *Public Opinion Quarterly* 42: 221–234.

Kish, Leslie. 1965. *Survey Sampling.* New York: Wiley

METHODS:
Field Research

7

Researching Dealers and Smugglers

PATRICIA A. ADLER

Based on a study we conducted in the 1970s, Patti shows the difficulties, dangers, and rich payoffs that can come from the depth investigations associated with ethnographic research in this participant-observation study of upper-level drug dealers and smugglers. She traces the natural history of the research process, from the way we entered the hidden world of drug traffickers and became a part of this community, to making close friendships and spending years gaining a richer understanding of participants' complex and often contradictory motivations. She discusses some of the research decisions we were forced to make, some of the tightropes we had to tread, and some of the practical and ethical issues that arose from this research. Our work illustrates the potential of field research to yield highly accurate understandings of people's behavior and the way this is lodged within the norms and values of distinct subcultures. Can participant-observation methodology be used to study all kinds of group behavior? What problems would you imagine that you might encounter? Why is participant-observation considered to be the only method that one could use in certain situations? Now that you have read about all three methods, you might want to figure out which method is best at answering specific types of questions in sociology. Why are some methods applicable to certain problems whereas other methods would not do as good a job of addressing the central concerns of a particular study?

I strongly believe that investigative field research (Douglas 1976), with emphasis on direct personal observation, interaction, and experience, is the only way to acquire accurate knowledge about deviant behavior. Investigative techniques are especially necessary for studying groups such as drug dealers and smugglers because the highly illegal nature of their occupation makes them secretive, deceitful, mistrustful, and paranoid. To insulate themselves from the straight world, they construct multiple false fronts, offer lies and misinformation, and withdraw into their group. In fact, detailed, scientific information about upper-level drug dealers and smugglers is lacking precisely because of the difficulty sociological researchers have had in penetrating into their midst. As a result, the only way I could possibly get close enough to these individuals to discover what they were doing and to understand their world from their perspectives (Blumer 1969) was to take a membership role in the setting. While my different values and goals precluded my becoming converted to complete membership in the subculture, and my fears presented my ever becoming "actively" involved in their trafficking activities, I was able to assume a "peripheral" membership role (Adler and Adler 1987). I became a member of the dealers' and smugglers' social world and participated in their daily activities on that basis. In this chapter, I discuss how I gained access to this group, established research relations with members, and how personally involved I became in their activities.

GETTING IN

When I moved to Southwest County [California] in the summer of 1974, I had no idea that I would soon be swept up in a subculture of vast drug trafficking and unending partying, mixed with occasional cloak-and-dagger subterfuge. I had moved to California with my husband, Peter, to attend graduate school in sociology. We rented a condominium townhouse near the beach and started taking classes in the fall. We had always felt that socializing exclusively with academicians left us nowhere to escape from our work, so we tried to meet people in the nearby community. One of the first friends we made was our closest neighbor, a fellow in his late twenties with a tall, hulking frame and gentle expression. Dave, as he introduced himself, was always dressed rather casually, if not sloppily, in T-shirts and jeans. He spent most of his time hanging out or walking on the beach with a variety of friends who visited his house, and taking care of his two young boys, who lived alternately with him and his estranged wife. He also went out of town a lot. We started spending much of our free time over at his house, talking, playing board games late into the night, and smoking marijuana together. We were glad to find someone from whom we could buy marijuana in this new place, since we did not know too many people. He also began treating us to a fairly regular supply of cocaine, which was a thrill because this was a drug we could rarely afford on our student budgets. We noticed right away, however, that there was something unusual about his use and knowledge of drugs: while he always had a plentiful supply and was fairly expert about marijuana and cocaine, when we tried to buy a small bag of marijuana from him he had little idea of the going price. This incongruity piqued our curiosity and raised suspicion. We wondered if he might be dealing in larger quantities. Keeping our suspicions to ourselves, we began observing Dave's activities a little more closely. Most of his friends were in their late twenties and early thirties and, judging by their lifestyles and automobiles, rather wealthy. They came and left his house at all hours, occasionally extending their parties through the night and the next day into the following night. Yet throughout this time we never saw Dave or any of his friends engage in any activity that resembled a legitimate job. In most places this might have evoked community suspicion, but few of the people we encountered in Southwest County seemed to hold traditionally structured jobs. Dave, in fact, had no visible means of financial support. When we asked him what he did for a living, he said something vague about being a real estate speculator, and we let it go at that. We never voiced our suspicions directly since he chose not to broach the subject with us.

We did discuss the subject with our mentor, Jack Douglas, however. He was excited by the prospect that we might be living among a group of big dealers, and urged us to follow our instincts and develop leads into the group. He knew that the local area was rife with drug trafficking, since he had begun a life history case study of two drug dealers with another graduate student several years previously. That earlier study was aborted when the graduate student quit school, but Jack still had many hours of taped interviews he had conducted with them, as well as an interview that he had done with an undergraduate student who had known the two dealers independently, to serve as a cross-check on their accounts. He therefore encouraged us to become friendlier with Dave and his friends. We decided that if anything did develop out of our observations of Dave, it might make a nice paper for a field methods class or independent study.

Our interests and background made us well suited to study drug dealing. First, we had already done research in the field of drugs. As undergraduates at Washington University we had participated in a nationally funded project on urban heroin use (see Cummins et al. 1972). Our role in the study involved using fieldwork techniques to investigate the extent of heroin use and distribution in St. Louis. In talking with heroin users, dealers, and rehabilitation personnel, we acquired a base of knowledge about the drug world and the subculture of drug trafficking. Second, we had a generally open view toward soft drug use, considering moderate consumption of marijuana and cocaine to be generally nondeviant. This outlook was partially etched by our 1960s-formed attitudes, as we had first been introduced to drug use in an environment of communal friendship, sharing, and counterculture ideology. It also partially reflected the widespread acceptance accorded to marijuana and cocaine use in the surrounding local culture. Third, our age (mid-twenties at the start of the study) and general appearance gave us compatibility with most of the people we were observing.

We thus watched Dave and continued to develop our friendship with him. We also watched his friends and got to know a few of his more regular visitors. We continued to build friendly relations by doing, quite naturally, what Becker (1963), Polsky (1969), and Douglas (1972) had advocated for the early stages of field research: we gave them a chance to know us and form judgments about our trustworthiness by jointly pursuing those interests and activities which we had in common.

Then one day something happened which forced a breakthrough in the research. Dave had two guys visiting him from out of town and, after snorting quite a bit of cocaine, they turned their conversation to a trip they had just made from Mexico, where they piloted a load of marijuana back across the border in a small plane. Dave made a few efforts to shift the conversation to another subject, telling them to "button their lips," but they apparently thought that he was joking. They thought that anybody as close to Dave as we seemed to be undoubtedly knew the nature of his business. They made further allusions to his involvement in the operation and discussed the outcome of the sale. We could feel the wave of tension and awkwardness from Dave when this conversation began, as he looked toward us to see if we understood the implications of what was being said, but then he just shrugged it off as done. Later, after the two guys left, he discussed with us what happened. He admitted to us that he was a member of a smuggling crew and a major marijuana dealer on the side. He said that he knew he could trust us, but that it was his practice to say as little as possible to outsiders about his activities. This inadvertent slip, and Dave's subsequent opening up, were highly significant in forging our entry into Southwest County's drug world. From then on he was open in discussing the nature of his dealing and smuggling activities with us.

He was, it turned out, a member of a smuggling crew that was importing a ton of marijuana weekly and 40 kilos of cocaine every few months. During that first winter and spring, we observed Dave at work and also got to know the other members of his crew, including Ben, the smuggler himself. Ben was also very tall and broad shouldered, but his long black hair, now flecked with gray, bespoke his earlier membership in the hippie subculture. A large physical stature, we observed, was common to most of the male participants involved in this drug community. The women also had a unifying physical trait: they were extremely attractive and stylishly dressed. This included Dave's ex-wife, Jean, with whom he reconciled during the spring. We therefore became friendly with Jean and through her met a number of women ("dope chicks") who hung around the dealers and smugglers. As we continued to gain the friendship of Dave and Jean's associates we were progressively admitted into their inner circle and apprised of each person's dealing or smuggling role.

Once we realized the scope of Ben's and his associates' activities, we saw the enormous research potential in studying them. This scene was different from any analysis of drug trafficking that we had read in the sociological literature because of the amounts they were dealing and the fact that they were importing it themselves. We decided that, if it was at all possible, we would capitalize on this situation, to "opportunistically" (Riemer 1977) take advantage of our prior expertise and of the knowledge, entrée, and rapport we had already developed with several key people in this setting. We therefore discussed the idea of doing a study of the general subculture with Dave and several of his closest friends (now becoming our friends). We assured them of the anonymity, confidentiality, and innocuousness of our work. They were happy to reciprocate our friendship by being of help to our professional careers. In fact, they basked in the subsequent attention we gave their lives.

We began by turning first Dave, then others, into key informants and collecting their life histories in detail. We conducted a series of taped, depth interviews with an unstructured, open-ended format. We questioned them about such topics as their backgrounds, their recruitment into the occupation, the stages of their dealing careers, their relations with others, their motivations, their lifestyle, and their general impressions about the community as a whole.

We continued to do taped interviews with key informants for the next six years until 1980, when we moved away from the area. After that, we occasionally did follow-up interviews when we returned for vacation visits. These later interviews focused on recording the continuing unfolding of events and included detailed probing into specific conceptual areas, such as dealing networks, types of dealers, secrecy, trust, paranoia, reputation, the law, occupational mobility, and occupational stratification. The number of taped interviews we did with each key informant varied, ranging between 10 and 30 hours of discussion.

Our relationship with Dave and the others thus took on an added dimension—the research relationship. As Douglas (1976), Henslin (1972), and Wax (1952) have noted, research relationships involve some form of mutual exchange. In our case, we offered everything that friendship could entail. We did routine favors for them in the course of our everyday lives, offered them insights and advice about their lives from the perspective of our more respectable position, wrote letters on their behalf to the authorities when they got in trouble, testified as character witnesses at their non drug-related trials, and loaned them money when they were down and out. When Dave was arrested and brought to trial for check-kiting, we helped Jean organize his defense and raise the money to pay his fines. We spelled her in taking care of the children so that she could work on his behalf. When he was eventually sent to the state prison we maintained close ties with her and discussed our mutual efforts to buoy Dave up and secure his release. We also visited him in jail. During Dave's incarceration, however, Jean was courted by an old boyfriend and gave up her reconciliation with Dave. This proved to be another significant turning point in our research because, desperate for money, Jean looked up Dave's old dealing connections and went into the business herself. She did not stay with these marijuana dealers and smugglers for long, but soon moved into the cocaine business. Over the next several years her experiences in the world of cocaine dealing brought us into contact with a different group of people. While these people knew Dave and his associates (this was very common in the Southwest County dealing and smuggling community), they did not deal with them directly. We were thus able to gain access to a much wider and more diverse range of subjects than we would have had she not branched out on her own.

Dave's eventual release from prison three months later brought our involvement in the research to an even deeper level. He was broke and had nowhere to go. When he showed up on our doorstep, we took him in. We offered to let him stay with us until he was back on his feet again and could afford a place of his own. He lived with us for seven months, intimately sharing his daily experiences with us. During this time we witnessed, firsthand, his transformation from a scared ex-con who would never break the law again to a hard-working legitimate employee who only dealt to get money for his children's Christmas presents, to a full-time dealer with no pretensions at legitimate work. Both his process of changing attitudes and the community's gradual reacceptance of him proved very revealing.

We socialized with Dave, Jean, and other members of Southwest County's dealing and smuggling community on a near-daily basis, especially during the first four years of the research (before we had a child). We worked in their legitimate businesses, vacationed together, attended their weddings, and cared for their children. Throughout their relationship with us, several participants became co-opted to the researcher's perspective[1] and actively sought out instances of behavior which filled holes in the conceptualizations we were developing. Dave, for one, became so intrigued by our conceptual dilemmas that he undertook a "natural experiment" entirely on his own, offering an unlimited supply of drugs to a lower-level dealer to see if he could work up to higher levels of dealing, and what factors would enhance or impinge upon his upward mobility.

In addition to helping us directly through their own experiences, our key informants aided us in widening our circle of contacts. For instance, they let us know when someone in whom we might be interested was planning on dropping by, vouching for our trustworthiness and reliability as friends who could be included in business conversations. Several times we were even awakened in the night by phone calls informing us that someone had dropped by for a visit, should we want to "casually" drop over too. We rubbed the sleep from our eyes, dressed, and walked or drove over, feeling like sleuths out of a television series. We thus were able to snowball, through the active efforts of our key informants,[2] an expanded study population. This was supplemented by our own efforts to cast a research net and befriend other dealers, moving from contact to contact slowly and carefully through the domino effect.

THE COVERT ROLE

The highly illegal nature of dealing in illicit drugs and dealers' and smugglers' general level of suspicion made the adoption of an overt research role highly sensitive and problematic. In discussing this issue with our key informants, they all agreed that we should be extremely discreet (for both our sakes and theirs). We carefully approached new individuals before we admitted that we were studying them. With many of these people, then, we took a covert posture in the research setting. As nonparticipants in the business activities which bound members together into the group, it was difficult to become fully accepted as peers. We therefore tried to establish some sort of peripheral, social membership in the general crowd, where we could be accepted as "wise" (Goffman 1963) individuals and granted a courtesy membership. This seemed an attainable goal, since we had begun our involvement by forming such relationships with our key informants. By being introduced to others in this wise rather than overt role, we were able to interact with people who would otherwise have shied away from us. Adopting a courtesy membership caused us to bear a courtesy stigma,[3] however, and we suffered since we, at times, had to disguise the nature of our research from both lay outsiders and academicians.

In our overt posture we showed interest in dealers' and smugglers' activities, encouraged them to talk about themselves (within limits, so as to avoid acting like narcs), and ran home to write field notes. This role offered us the advantage of gaining access to unapproachable people while avoiding researcher effects, but it prevented us from asking some necessary, probing questions and from tape recording conversations.[4] We therefore sought, at all times, to build toward a conversion to the overt role. We did this by working to develop their trust.

DEVELOPING TRUST

Like achieving entrée, the process of developing trust with members of unorganized deviant groups can be slow and difficult. In the absence of a formal structure separating members from outsiders, each individual must form his or her own judgment about whether new persons can be admitted to their confidence. No gatekeeper existed to smooth our path to being trusted, although our key informants acted in this role whenever they could by providing introductions and references. In addition, the unorganized nature of this group meant that we met people at different times and were constantly at different levels in our developing relationships with them. We were thus trusted more by some people than by others, in part because of their greater familiarity with us. But as Douglas (1976) has noted, just because someone knew us or even liked us did not automatically guarantee that they would trust us.

We actively tried to cultivate the trust of our respondents by tying them to us with favors. Small things, like offering the use of our phone, were followed with bigger favors, like offering the use of our car, and finally really meaningful favors, like offering the use of our home. Here we often trod a thin line, trying to ensure our personal safety while putting ourselves in enough of a risk position, along with our research subjects, so that they would trust us. While we were able to build a "web of trust" (Douglas 1976) with some members, we found that trust, in large part, was not a simple status to attain in the drug world. Johnson (1975) has pointed out that trust is not a one-time phenomenon, but an ongoing developmental process. From my experiences in this research I would add that it cannot be simply assumed to be a one-way process either, for it can be diminished, withdrawn, reinstated to varying degrees, and re-questioned at any point. Carey (1972) and Douglas (1972) have remarked on this waxing and waning process, but it was especially pronounced for us because our subjects used large amounts of cocaine over an extended period of time. This tended to make them alternately warm and cold to us. We thus lived through a series of ups and downs with the people we were trying to cultivate as research informants.

THE OVERT ROLE

After this initial covert phase, we began to feel that some new people trusted us. We tried to intuitively feel when the time was right to approach them and go overt. We used two means of approaching people to inform them that we were involved in a study of dealing and smuggling: direct and indirect. In some cases our key informants approached their friends or connections and, after vouching for our absolute trustworthiness, convinced these associates to talk to us. In other instances, we approached people directly, asking for their help with our project. We worked our way through a progression with these secondary contacts, first discussing the dealing scene overtly and later moving to taped life history interviews. Some people reacted well to us, but others responded skittishly, making appointments to do taped interviews only to break them as the day drew near, and going through fluctuating stages of being honest with us or putting up fronts about their dealing activities. This varied, for some, with their degree of active involvement in the business. During the times when they had quit dealing, they would tell us about their present and past activities, but when they became actively involved again, they would hide it from us.

This progression of covert to overt roles generated a number of tactical difficulties. The first was the problem of *coming on too fast* and blowing it. Early in the research we had a dealer's old lady (we thought) all set up for the direct approach. We knew many dealers in common and had discussed many things tangential to dealing with her without actually mentioning the subject. When we asked her to do a taped interview of her bohemian lifestyle, she agreed without hesitation. When the interview began, though, and she found out why we were interested in her, she balked, gave us a lot of incoherent jumble, and ended the session as quickly as possible. Even though she lived only three houses away we never saw her again. We tried to move more slowly after that.

A second problem involved simultaneously *juggling our overt and covert roles* with different people. This created the danger of getting our cover blown with people who did not know about our research (Henslin 1972). It was very confusing to separate the people who knew about our study from those who did not, especially in the minds of our informants. They would make occasional veiled references in front of people, especially when loosened by intoxicants, that made us extremely uncomfortable. We also frequently worried that our snooping would someday be mistaken for police tactics. Fortunately, this never happened.

CROSS-CHECKING

The hidden and conflictual nature of the drug dealing world made me feel the need for extreme certainty about the reliability of my data. I therefore based all my conclusions on independent sources and accounts that we carefully verified. First, we tested information against our own common sense and general knowledge of the scene. We adopted a hard-nosed attitude of suspicion, assuming people were up to more than they would originally admit. We kept our attention especially riveted on "reformed" dealers and smugglers who were living better than they could outwardly afford, and were thereby able to penetrate their public fronts.

Second, we checked out information against a variety of reliable sources. Our own observations of the scene formed a primary reliable source, since we were involved with many of the principals on a daily basis and knew exactly what they were doing. Having Dave live with us was a particular advantage because we could contrast his statements to us with what we could clearly see was happening. Even after he moved out, we knew him so well that we could generally tell when he was lying to us or, more commonly, fooling himself with optimistic dreams. We also observed other dealers' and smugglers' evasions and misperceptions about themselves and their activities. These usually occurred when they broke their own rules by selling to people they did not know, or when they commingled other people's money with their own. We also cross-checked our data against independent, alternative accounts. We were lucky, for this purpose, that Jean got reinvolved in the drug world. By interviewing her, we gained additional insight into Dave's past, his early dealing and smuggling activities, and his ongoing involvement from another person's perspective. Jean (and her connections) also talked to us about Dave's associates, thereby helping us to validate or disprove their statements. We even used this pincer effect to verify information about people we had never directly interviewed. This occurred, for instance, with the tapes that Jack Douglas gave us from his earlier study. After doing our first round of taped interviews with Dave, we discovered that he knew the dealers Jack had interviewed. We were excited by the prospect of finding out what had happened to these people and if their earlier stories checked out. We therefore sent Dave to do some investigative work. Through some mutual friends he got back in touch with them and found out what they had been doing for the past several years.

Finally, wherever possible, we checked out accounts against hard facts: newspaper and magazine reports; arrest records; material possessions; and visible evidence. Throughout the research, we used all these cross-checking measures to evaluate the veracity of new information and to prod our respondents to be more accurate (by abandoning both their lies and their self-deceptions).[5]

After about four years of near-daily participant observation, we began to diminish our involvement in the research. This occurred gradually, as first pregnancy and then a child hindered our ability to follow the scene as intensely and spontaneously as we had before. In addition, after having a child, we were less willing to incur as many risks as we had before; we no longer felt free to make decisions based solely on our own welfare. We thus pulled back from what many have referred to as the "difficult hours and dangerous situations" inevitably present in field research on deviants (see Becker 1963; Carey 1972; Douglas 1972). We did, however, actively maintain close ties with research informants (those with whom we had gone overt), seeing them regularly and periodically doing follow-up interviews.

PROBLEMS AND ISSUES

Reflecting on the research process, I have isolated a number of issues which I believe merit additional discussion. These are rooted in experiences which have the potential for greater generic applicability.

The first is the *effect of drugs on the data-gathering process.* Carey (1972) has elaborated on some of the problems he encountered when trying to interview respondents who used amphetamines, while Wax (1952, 1957) has mentioned the difficulty of trying to record field notes while drinking sake. I found that marijuana and cocaine had nearly opposite effects from each other. The latter helped the interview process, while the former hindered it. Our attempts to interview respondents who were stoned on marijuana were unproductive for a number of reasons. The primary obstacle was the effects of the drug. Often, people became confused, sleepy, or involved in eating to varying degrees. This distracted them from our purpose. At times, people even simulated overreactions to marijuana to hide behind the drug's supposed disorienting influence and thereby avoid divulging information. Cocaine, in contrast, proved to be a research aid. The drug's warming and sociable influence opened people up, diminished their inhibitions, and generally increased their enthusiasm for both the interview experience and us.

A second problem I encountered involved *assuming risks while doing research*. As I noted earlier, dangerous situations are often generic to research on deviant behavior. We were most afraid of the people we studied. As Carey (1972), Henslin (1972), and Whyte (1955) have stated, members of deviant groups can become hostile toward a researcher if they think that they are being treated wrongfully. This could have happened at any time from a simple occurrence, such as a misunderstanding, or from something more serious, such as our covert posture being exposed. Because of the inordinate amount of drugs they consumed, drug dealers and smugglers were particularly volatile, capable of becoming malicious toward each other or us with little warning. They were also likely to behave erratically owing to the great risks they faced from the police and other dealers. These factors made them moody, and they vacillated between trusting us and being suspicious of us.

At various times we also had to protect our research tapes. We encountered several threats to our collection of taped interviews from people who had granted us these interviews. This made us anxious, since we had taken great pains to acquire these tapes and felt strongly about maintaining confidences entrusted to us by our informants. When threatened, we became extremely frightened and shifted the tapes between different hiding places. We even ventured forth one rainy night with our tapes packed in a suitcase to meet a person who was uninvolved in the research at a secret rendezvous so that he could guard the tapes for us.

We were fearful, lastly, of the police. We often worried about local police or drug agents discovering the nature of our study and confiscating or subpoenaing our tapes and field notes. Sociologists have no privileged relationship with their subjects that would enable us legally to withhold evidence from the authorities should they subpoena it.[6] For this reason we studiously avoided any publicity about the research, even holding back on publishing articles in scholarly journals until we were nearly ready to move out of the setting. The closest we came to being publicly exposed as drug researchers came when a former sociology graduate student (turned dealer, we had heard from inside sources) was arrested at the scene of a cocaine deal. His lawyer wanted us to testify about the dangers of doing drug-related research, since he was using his research status as his defense. Fortunately, the crisis was averted when his lawyer succeeded in suppressing evidence and had the case dismissed before the trial was to have begun. Had we been exposed, however, our respondents would have acquired guilt by association through their friendship with us.

Our fear of the police went beyond our concern for protecting our research subjects, however. We risked the danger of arrest ourselves through our own violations of the law. Many sociologists (Becker 1963; Carey 1972; Polsky 1969; Whyte 1955) have remarked that field researchers studying deviance must inevitably break the law in order to acquire valid participant observation data. This occurs in its most innocuous form from having "guilty knowledge": information about crimes that are committed. Being aware of major dealing and smuggling operations made us an accessory to their commission, since we failed to notify the police. We broke the law, secondly, through our "guilty observations," by being present at the scene of a crime and witnessing its occurrence (see also Carey 1972). We knew it was possible to get caught in a bust involving others, yet buying and selling was so pervasive that to leave every time it occurred would have been unnatural and highly suspicious. Sometimes drug transactions even occurred in our home, especially when Dave was living there, but we finally had to put a stop to that because we could not handle the anxiety. Lastly, we broke the law through our "guilty actions," by taking part in illegal behavior ourselves. Although we never dealt drugs (we were too scared to be seriously tempted), we consumed drugs and possessed them in small quantities. Quite frankly, it would have been impossible for a nonuser to have gained access to this group to gather the data presented here. This was the minimum involvement necessary to obtain even the courtesy membership we achieved. Some kind of illegal action was also found to be a necessary or helpful component of the research by Becker (1963), Carey (1972), Johnson (1975), Polsky (1969), and Whyte (1955).

Another methodological issue arose from the *cultural clash between our research subjects and ourselves.* While other sociologists have alluded to these kinds of differences (Humphreys 1970; Whyte 1955), few have discussed how the research relationships affected them. Relationships with research subjects are unique because they involve a bond of intimacy between persons who might not ordinarily associate together, or who might otherwise be no more than casual friends. When fieldworkers undertake a major project, they commit themselves to maintaining a long-term relationship with the people they study. However, as researchers try to get depth involvement, they are apt to come across fundamental differences in character, values, and attitudes between their subjects and themselves. In our case, we were most strongly confronted by differences in present versus future orientations, a desire for risk versus security, and feelings of spontaneity versus self-discipline. These differences often caused us great frustration. We repeatedly saw dealers act irrationally, setting themselves up for failure. We wrestled with our desire to point out their patterns of foolhardy behavior and offer advice, feeling competing pulls between our detached, observer role which advised us not to influence the natural setting, and our involved, participant role which called for us to offer friendly help whenever possible.[7]

Each time these differences struck us anew, we gained deeper insights into our core, existential selves. We suspended our own taken-for-granted feelings and were able to reflect on our culturally formed attitudes, character, and life choices from the perspective of the other. When comparing how we might act in situations faced by our respondents, we realized where our deepest priorities lay. These revelations had the effect of changing our self-conceptions: whereas we, at one time, had thought of ourselves as what Rosenbaum (1981) has called "the hippest of non-addicts" (in this case nondealers), we were suddenly faced with being the straightest members of the crowd. Not only did we not deal, but we had a stable, long-lasting marriage and family life, and needed the security of a reliable monthly paycheck. Self-insights thus emerged as one of the unexpected outcomes of field research with members of a different cultural group.

The final issue I will discuss involved the various *ethical problems* which arose during this research. Many fieldworkers have encountered ethical dilemmas or pangs of guilt during the course of their research experiences (Carey 1972; Douglas 1976; Humphreys 1970; Johnson 1975; Klockars 1977, 1979; Rochford 1985). The researchers' role in the field makes this necessary because they can never fully align themselves with their subjects while maintaining their identity and personal commitment to the scientific community. Ethical dilemmas, then, are directly related to the amount of deception researchers use in gathering the data, and the degree to which they have accepted such acts as necessary and therefore neutralized them.

Throughout the research, we suffered from the burden of intimacies and confidences. Guarding secrets which had been told to us during taped interviews was not always easy or pleasant. Dealers occasionally revealed things about themselves or others that we had to pretend not to know when interacting with their close associates. This sometimes meant that we had to lie or build elaborate stories to cover for some people. Their fronts therefore became our fronts, and we had to weave our own web of deception to guard their performances. This became especially disturbing during the writing of the research report, as I was torn by conflicts between using details to enrich the data and glossing over description to guard confidences.[8]

Using the covert research role generated feelings of guilt, despite the fact that our key informants deemed it necessary, and thereby condoned it. Their own covert experiences were far more deeply entrenched than ours, being a part of their daily existence with non–drug world members. Despite the universal presence of covert behavior throughout the setting, we still felt a sense of betrayal every time we ran home to write research notes on observations we had made under the guise of innocent participants.

We also felt guilty about our efforts to manipulate people. While these were neither massive nor grave manipulations, they involved courting people to procure information about them. Our aggressively friendly postures were based on hidden ulterior motives: we did favors for people with the clear expectation that they could only pay us back with research assistance. Manipulation bothered us in two ways: immediately after it was done, and over the long run. At first, we felt awkward, phony, almost ashamed of ourselves, although we believed our rationalization that the end justified the means. Over the long run, though, our feelings were different. When friendship became intermingled with research goals, we feared that people would later look back on our actions and feel we were exploiting their friendship merely for the sake of our research project.

The last problem we encountered involved our feelings of whoring for data. At times, we felt that we were being exploited by others, that we were putting more into the relationship than they, that they were taking us for granted or using us. We felt that some people used a double standard in their relationship with us: they were allowed to lie to us, borrow money and not repay it, and take advantage of us, but we were at all times expected to behave honorably. This was undoubtedly an outgrowth of our initial research strategy where we did favors for people and expected little in return. But at times this led to our feeling bad. It made us feel like we were selling ourselves, our sincerity, and usually our true friendship, and not getting treated right in return.

CONCLUSIONS

The aggressive research strategy I employed was vital to this study. I could not just walk up to strangers and start hanging out with them as Liebow (1967) did, or be sponsored to a member of this group by a social service or reform organization as Whyte (1955) was, and expect to be accepted, let alone welcomed. Perhaps such a strategy might have worked with a group that had nothing to hide, but I doubt it. Our modern, pluralistic society is so filled with diverse subcultures whose interests compete or conflict with each other that each subculture has a set of knowledge which is reserved exclusively for insiders. In order to serve and prosper, they do not ordinarily show this side to just anyone. To obtain the kind of depth insight and information I needed, I had to become like the members in certain ways. They dealt only with people they knew and trusted, so I had to become known and trusted before I could reveal my true self and my research interests. Confronted with secrecy, danger, hidden alliances, misrepresentations, and unpredictable changes of intent, I had to use a delicate combination of overt and covert roles. Throughout, my deliberate cultivation of the norm of reciprocal exchange enabled me to trade my friendship for their knowledge, rather than waiting for the highly unlikely event that information would be delivered into my lap. I thus actively built a web of research contacts, used them to obtain highly sensitive data, and carefully checked them out to ensure validity.

Throughout this endeavor I profited greatly from the efforts of my husband, Peter, who served as an equal partner in this team field research project. It would have been impossible for me to examine this social world as an unattached female and not fall prey to sex role stereotyping which excluded women from business dealings. As a couple, our different genders allowed us to relate in different ways to both men and women (see Warren and Rasmussen 1977). We also protected each other when we entered the homes of dangerous characters, buoyed each other's initiative and courage, and kept the conversation going when one of us faltered. Conceptually, we helped each other keep a detached and analytical eye on the setting, provided multiperspectival insights, and corroborated, clarified, or (most revealingly) contradicted each other's observations and conclusions.

Finally, I feel strongly that to ensure accuracy, research on deviant groups must be conducted in the settings where it naturally occurs. As Polsky (1969: 115–16) has forcefully asserted:

> This means—there is no getting away from it—the study of career criminals *au natural,* in the field, the study of such criminals as they normally go about their work and play, the study of "uncaught" criminals and the study of others who in the past have been caught but are not caught at the time you

> study them. . . . Obviously we can no longer afford the convenient fiction that in studying criminals in their natural habitat, we would discover nothing really important that could not be discovered from criminals behind bars.

By studying criminals in their natural habitat I was able to see them in the full variability and complexity of their surrounding subculture, rather than within the artificial environment of a prison. I was thus able to learn about otherwise inaccessible dimensions of their lives, observing and analyzing firsthand the nature of their social organization, social stratification, lifestyle, and motivation.

NOTES

1. Gold (1958) discouraged this methodological strategy, cautioning against overly close friendship or intimacy with informants, lest they lose their ability to act as informants by becoming too much observers. Whyte (1955), in contrast, recommended the use of informants as research aides, not for helping in conceptualizing the data but for their assistance in locating data which support, contradict, or fill in the researcher's analysis of the setting.

2. See also Biernacki and Waldorf 1981; Douglas 1976; Henslin 1972; Hoffman 1980; McCall 1980; and West 1980 for discussions of "snowballing" through key informants.

3. See Kirby and Corzine 1981; Birenbaum 1970; and Henslin 1972 for more detailed discussion of the nature, problems, and strategies for dealing with courtesy stigma.

4. We never considered secret tapings because, aside from the ethical problems involved, it always struck us as too dangerous.

5. See Douglas (1976) for a more detailed account of these procedures.

6. A recent court decision, where a federal judge ruled that a sociologist did not have to turn over his field notes to a grand jury investigating a suspicious fire at a restaurant where he worked, indicates that this situation may be changing (Fried 1984).

7. See Henslin 1972 and Douglas 1972, 1976 for further discussions of this dilemma and various solutions to it.

8. In some cases I resolved this by altering my descriptions of people and their actions as well as their names so that other members of the dealing and smuggling community would not recognize them. In doing this, however, I had to keep a primary concern for maintaining the sociological integrity of my data so that the generic conclusions I drew from them would be accurate. In places, then, where my attempts to conceal people's identities from people who know them have been inadequate, I hope that I caused them no embarrassment. See also Polsky 1969; Rainwater and Pittman 1967; and Humphreys 1970 for discussions of this problem.

REFERENCES

Adler, Patricia A., and Peter Adler. 1987. *Membership Roles in Field Research.* Beverly Hills, CA: Sage.

Becker, Howard. 1963. *Outsiders.* New York: Free Press.

Biernacki, Patrick, and Dan Waldorf. 1981. "Snowball sampling." *Sociological Methods and Research* 10: 141–63.

Birenbaum, Arnold. 1970. "On managing a courtesy stigma." *Journal of Health and Social Behavior* 11: 196–206.

Blumer, Herbert. 1969. *Symbolic Interactionism.* Englewood Cliffs, NJ: Prentice-Hall.

Carey, James T. 1972. "Problems of access and risk in observing drug scenes." In

Jack D. Douglas, ed., *Research on Deviance,* pp. 71–92. New York: Random House.

Cummins, Marvin, et al. 1972. *Report of the Student Task Force on Heroin Use in Metropolitan Saint Louis.* Saint Louis: Washington University Social Science Institute.

Douglas, Jack D. 1972. "Observing deviance." In Jack D. Douglas, ed., *Research on Deviance,* pp. 3–34. New York: Random House.

_____. 1976. *Investigative Social Research.* Beverly Hills, CA: Sage.

Fried, Joseph P. 1984. "Judge protects waiter's notes on fire inquiry." *New York Times,* April 8: 47.

Goffman, Erving. 1963. *Stigma.* Englewood Cliffs, NJ: Prentice-Hall.

Gold, Raymond. 1958. "Roles in sociological field observations." *Social Forces* 36: 217–23.

Henslin, James M. 1972. "Studying deviance in four settings: research experiences with cabbies, suicidees, drug users and abortionees." In Jack D. Douglas, ed., *Research on Deviance,* pp. 35–70. New York: Random House.

Hoffman, Joan E. 1980. "Problems of access in the study of social elites and boards of directors." In William B. Shaffir, Robert A. Stebbins, and Allan Turowetz, eds., *Fieldwork Experience,* pp. 45–56. New York: St. Martin's.

Humphreys, Laud. 1970. *Tearoom Trade.* Chicago: Aldine.

Johnson, John M. 1975. *Doing Field Research.* New York: Free Press.

Kirby, Richard, and Jay Corzine. 1981. "The contagion of stigma." *Qualitative Sociology* 4: 3–20.

Klockars, Carl B. 1977. "Field ethics for the life history." In Robert Weppner ed., *Street Ethnography,* pp. 201–26. Beverly Hills, CA: Sage.

_____. 1979. "Dirty hands and deviant subjects." In Carl B. Klockars and Finnbarr W. O'Connor, eds., *Deviance and Decency,* pp. 261–82. Beverly Hills, CA: Sage.

Liebow, Elliott. 1967. *Tally's Corner.* Boston: Little, Brown.

McCall, Michal. 1980. "Who and where are the artists?" In William B. Shaffir, Robert A. Stebbins, and Allan Turowetz, eds., *Fieldwork Experience,* pp. 145–58. New York: St. Martin's.

Polsky, Ned. 1969. *Hustlers, Beats, and Others.* New York: Doubleday.

Rainwater, Lee R., and David J. Pittman. 1967. "Ethical problems in studying a politically sensitive and deviant community." *Social Problems* 14: 357–66.

Riemer, Jeffrey W. 1977. "Varieties of opportunistic research." *Urban Life* 5: 467–77.

Rochford, E. Burke, Jr. 1985. *Hare Krishna in America.* New Brunswick, NJ: Rutgers University Press.

Rosenbaum, Marsha. 1981. *Women on Heroin.* New Brunswick, NJ: Rutgers University Press.

Warren, Carol A. B., and Paul K. Rasmussen. 1977. "Sex and gender in field research." *Urban Life* 6: 349–69.

Wax, Rosalie. 1952. "Reciprocity as a field technique." *Human Organization* 11: 34–37.

_____. 1957. "Twelve years later: An analysis of a field experience." *American Journal of Sociology* 63: 133–42.

West, W. Gordon. 1980. "Access to adolescent deviants and deviance." In William B. Shaffir, Robert A. Stebbins, and Allan Turowetz, eds., *Fieldwork Experience,* pp. 31–44. New York: St. Martin's.

Whyte, William F. 1955. *Street Corner Society.* Chicago: University of Chicago Press.

PART II

✦

Sociological Blueprints

A sociological blueprint is like an architectural map of the structure of society. In Part II, we begin by looking at the layout and nature of society and flesh this in with the meat that makes up our culture. These social forms shape people into members of society through the processes of socialization discussed next. Socialization teaches us some of the ways and forms of interaction, shown in the third section. Finally, the boundaries of society and what falls at its fringes are discussed in the section on deviance.

In examining the blueprint of society, we look at both the structure that makes it up and the trends that fall within it. Social structure is comprised of core *social institutions,* abstract concepts describing social features that cannot be seen or touched. Some of the major social institutions in our society are politics, economics, religion, education, and the family. These social institutions have organizations, groups, and concrete aspects operating within them that carry out the business of each sphere. For example, politics consists partly of governing bodies, political parties, elections, and campaigning. Within the institution of economics we find the stock market, banks, businesses, and the way people manage their budgets. Religion encompasses churches, spirituality, dogma, sacred objects, and beliefs. In education we find schools, classrooms, books, tests, ideas, and the passing on of knowledge from one generation to the next. The family has extended kinship networks, new family forms, inheritance, dominance, geographic location, and the strength of blood relations.

At the same time, each social institution and its organizations and component parts have *statuses* and *roles* that people play within them. Politics has politicians, speech-writers, press secretaries, fund-raisers, elected officials, and managers. Economics has market traders, brokers, bankers, tellers, investors, salespeople, and cashiers. Religion has ministers, the faithful, disbelievers, heretics, congregation presidents, and youth group leaders. Education has teachers, students, principals, administrators, writers, publishers, and secretaries. The family has mothers, fathers, children, grandparents, aunts, uncles, and pseudo-kin. All of these roles and statuses are filled by people whose position in relation to each other is defined by the structure of society and its organizations. We learn the social scripts and we enact the behaviors associated with them, occupying our position in society.

A sociological blueprint also lays out the **Culture** of society. Culture consists of the way people share a given space, their language, their relations to each other, the way they feel about each other, and their self-identities as part of the group. Its basic building blocks are norms and values, those behavioral "recipes" that tell us how to act (norms), and our shared beliefs about what is good (values). Values can also be thought of as the ends toward which we strive, and norms as the standards or means that we use to work our way there. Each group has its own culture that makes it distinct, and makes it similar to or different from others. In our contemporary society we all belong to more than one group, each having its own slightly different culture. Most of us are, most fundamentally, Americans, and have some sense of the broad, overarching American culture into which we are socialized. We learn to cherish life, liberty, and the pursuit of happiness because this is how the early settlers established the credo of our country. We value individualism, freedom, materialism, and creativity because they are important to our culture. We know what it means to be an American, although we often take it for granted because we are surrounded by it daily. The true nature of American culture often becomes most obvious to us when we leave the United States and meet other Americans abroad, discovering our basic elements of shared culture. In the popular movie, *Pulp Fiction,* John Travolta summed this up nicely, when he said that "it's the little things that count," such as whether people eat their french fries with ketchup or mayonnaise, whether they use pounds and inches or the metric system, if they expect ice in their beverages, if they sleep on the floor or on a bed, or if they see their food alive before they eat it. These differences all mark the distinctiveness of one culture and lifestyle as compared to others.

Yet even within our society, there are marked cultural differences. We live in a large, highly pluralistic nation, populated by many diverse *subcultures.* Each subgroup has its own subculture with its particular variety of norms and values. You may notice that you belong to several different subcultures that place their emphases on certain ways of thinking and acting. In addition to being an American, you may also be a member of your ethnic group, your religious group, your social class, your gender, and your age group. Each of these may give you some distinct ideas and a characteristic outlook. Beyond this, you may belong to subcultures relating to Greek life (a fraternity or sorority), musical tastes (Phish followers), theatrical performances (thespians), clean-living punks (straight-edgers), athletic endeavors (jocks), or drug use (ravers). Your membership in any of these groups shapes your beliefs, the way you present yourself, and the way you assess others. Subcultures have their own distinct language and jargon, norms and values, ideologies or beliefs, famous characters and stories about them, support systems, key information that is passed around, and system of status stratification, through which people can assess the relative position of members in relation to each other. People's status among their peers is a key subcultural element, and their adherence to critical group standards is basic to determining this.

In the Culture section we see some of the norms and values characterizing different groups in American society. We begin with a selection by Horace Miner that describes a particular tribe and its preoccupation with health, fitness, and their bodies. Miner suggests that tribal members engage in elaborate rituals to overcome their naturally occurring physicality, and worship both abstract and physical gods in search of bodily perfection, although they often attain little satisfaction from their efforts. Yet these practices he observed appear widespread and unquestioned. We next present a selection on subculture, Elijah Anderson's stunning depiction of the "code of the streets" evident in many black, underclass youth groups living in urban centers in the United States. This study shows how smaller groups have their own codes, roles, statuses, terms, and behaviors through which they accord recognition and respect. Finally, we look at Kevin Young and Laura Craig's portrayal of skinheads, a "counterculture" that deliberately sets some of its norms and values against those of the dominant culture, and exists in a form of opposition to it.

People are born into society with the basic features that nature has given them. Sociology contends that they are shaped and formed into individuals capable of functioning within our world by their socialization. **Socialization** involves individuals in a life-long process of learning the norms, values, roles, boundaries, and beliefs of a culture so that they become members of the group and develop a sense of self, or *identity* within it. Critical *agents of socialization* that shape and mold young people include family members, friends, teachers, and the media. The early part of our lives is characterized by an especially heavy dose of socialization, as we learn what it means to be a member of all our social categories and groups. This process, though, does not end after childhood, as we continue to develop and change throughout our entire lives, taking on new roles and adding them to our repertoire. Thus, we may start out by learning what is involved in being a child, being a boy or girl, and following the rules of public behavior. Over time, we may augment this with socialization to the role of student, boyfriend or girlfriend, employee, and parent or grandparent. We never actually discard our old roles as we age out of them, but tuck them away somewhere and draw on them when it is appropriate. We even learn roles that we never enact, as this helps us to understand and interact with people filling those roles. Thus, we know how to respond to teachers, doctors, ministers, and salespeople, even if we've never been one, because we know what their roles entail and what they need to do to fill them. This helps us to act competently in society, because we learn how to anticipate other people's needs and behavior in order to coordinate our actions with theirs.

George Herbert Mead said that the apex of socialization is learning the role and perspective of a social abstraction, "the generalized other" (people in general), to understand how numerous others, or society, will act or react to what we do. We become socialized to society's perspective so that we can anticipate how others will react to our behavior and thereby stay within social norms and values, be in synch with society's expectations and rhythms, and act as a successful person within the larger group. This ability becomes especially important as we increasingly interact, via electronic means, with individuals who we never even meet face-to-face.

Learning how to interact competently in society shapes our sense of self by enabling us to recognize our place in the world. Our identity consists of a set of perceptions about who and what kind of person we are. This is partly formed by our social position or status, and partly by the way we enact that status. It involves a dynamic tension between our developing awareness about a particular role (say, college athlete), the role-set members who interact with us in that role (coaches, teammates, other students, teachers, boosters, the media), and the way each of us enacts that role individually (hot-dog, team player, star, regular guy, spirit-raiser, follower, role model, clown, leader, or rebel). There may be many people who learn a particular role, but each may enact it differently.

The selections in this section take us through a small portion of the way people become socialized and form their identities. We begin by looking at Spencer Cahill's portrayal of a pivotal aspect of early childhood socialization: gender learning. Socialization at these younger ages is particularly fascinating because it is so normative. Children struggle furiously to learn the dimensions of the roles they must enact. They then apply them stringently onto others, chastising other children who attempt to deviate. When our daughter was four, all she would ever wear was dresses. She was engaged in learning what it meant to be a girl. She rejected all the blue jeans and overalls we offered her, preferring, instead, to wear lace, costume jewelry, nail polish, patent leather shoes, and bows in her hair. She noticed gender divisions everywhere and remarked on them. Watching the Thanksgiving Day festivities on television, she commented on the divisions between football players and cheerleaders. She seriously pronounced to us one day that "men are doctors, women are nurses." This caused us some consternation, as we had been working to achieve an egalitarian marriage with role-sharing. We decided right away that the next time we took her to the pediatrician, it would be to a female doctor. When we made the appointment and took her in, we carefully watched how she interacted with the sweet doctor who called her "sugar bear" and tickled her. After we left we casually asked her what she thought of her visit to the doctor. "I like that doctor-nurse!" she exclaimed joyfully. Carefully entrenched gender socialization did not dislodge easily.

It took awhile before she and her peers felt comfortable enough in their roles as young girls to forge individual adaptations to their roles. Children move through elementary school engaged in a series of conformity exercises. Moving too close to the outer boundaries invokes censure, which the peer culture is quick to dispense. While young girls may shed their dresses earlier, it is not until high school that many adolescents first begin to gain the courage to seek out their own individuality, to find their identity. The way they differentiate themselves is subtle, and often patterned by the collective. You may find yourself relating to David Karp, Lynda Lytle Holmstrom, and Paul S. Gray's selection about moving out of your parents' house and on to college. This transition offers a great opportunity to differentiate yourself from your group of high school friends and to develop your own identity. You may have brought a set of "identity props" along with you and left others behind, intentionally setting the stage to move some aspects of your self into the past and further develop others. College is a time when you may experiment with your social roles and identities, when you may try out membership in various groups, lifestyles, and behaviors without necessarily making a commitment to any of them. You may question or challenge your parents' key values and try out others. Everyone seems to emerge from high school into older life with some strong beliefs about ways that they want to do things differently from the way their parents did and things that they want to retain. This is all part of our lifelong process of socialization and identity formation. Our last selection, Melissa Milkie's article on the way black and white girls' self-images are differently shaped by media influences, highlights the effect of peer influence on socialization. She shows how people's identities are mediated by the way their subculture intervenes between them as individuals and the larger societal media messages they encounter. White and black girls react in dissimilar ways to the potentially destructive images displayed of women's ideal bodies in teen magazines, largely because their racial subcultures accept or reject these images as relevant. Some groups develop a greater ability to resist the effects of social labeling and its identity consequences than others.

Social Interaction, or the way people act individually and in groups in everyday life, forms the foundation of our social blueprint. Socialization heavily influences interaction, as it teaches people how to behave and how to anticipate others' reactions to various behaviors. Part of learning how to interact in society is learning to read the response of others to your actions. Charles Cooley's concept of the "looking glass self" is particularly germane to helping us understand how we act. It is amazing, when we think about it, that we can simultaneously do something, think about how others are reacting to it, and think about how we might adjust our ongoing actions in light of our perception of their interpretations. That we can think separately and independently from our actions while we are engaged in them represents very sophisticated and skilled social competence.

Symbolic interactionists differ in the role they attribute to emotion in behavior. Some see our actions as involving a simple cost-benefit calculation, where we select the path that will net us the most gain. Others attribute a greater role to feelings, suggesting that we often choose paths that fall outside of these logical-rational economic models for emotional reasons. We may take a less advantageous path to further our self-feelings as good people, because of strong, driving feelings of jealousy, fear, or hatred, or just because we feel like it at the moment. Our behavior represents a fusion of these kinds of motivations.

Erving Goffman wrote extensively about the micro-interactional norms that govern behavior in public places. While it may be clear why we want to look good in the eyes of significant others, Goffman pointed out that we also seem to care very much about the way we appear to complete strangers. He suggested that we are very conscious of our actions and the anticipated responses of all kinds of people, and that our lives represent our acting out our private selves on the public stage of life. In making these self-presentations, we are often assisted by others in one of two ways: they may be insiders who help us to pull off our performances in front of outsiders, or they may be members of the target audience who help us to maintain a positive sense of self by not pointing out our public gaffes (our fly is open, we have a speck of food in our teeth, we're calling someone by the wrong name). Goffman's collection of the micro-social rules of public interaction are a fascinating topic of study.

Applying some of Goffman's key concepts, we see Cahill and his students peering into a place we have all visited: public bathrooms. Their analyses of these "backstage" regions and the norms of demeanor within them will be sure to ring home with your experience. The dramaturgical analogy is further carried out in our next selection by Vered Vinitzky-Seroussi and Robert Zussman, who write about the way people attempt to manage their self-presentations at high school reunions. No matter how long it is since you got out of high school, a reunion may pull you right back to the same set of feelings and identity that you held back then. It is amusing to see the strategies people employ to foster the best possible impressions of themselves. Finally, our last selection looks at the dynamics of social interaction without focusing especially at the self-presentational aspect. Social group membership is one of the most critical features of young people's lives, and their struggles to gain and retain inclusion in the popular clique are revealed in our study of preadolescent clique dynamics. While it may appear to portray children's behavior as excessively vicious, most people who read it find that it offers key insight into the micro-politics of everyday life throughout adulthood.

Last, we look at the role of **Deviance and Social Boundaries** in shaping the contours and character of the sociological blueprint. Although we often think of deviance as coming from some element intrinsic to the offense or offender, sociological research has shown that definitions of deviance are highly relative. No act, no matter how heinous, cannot be found as legitimate under certain circumstances. Even murder is deemed acceptable when it is committed by the State, in self-defense, in war, or, in "make my day" states, when someone threatens your property. Further, what one society, organization, or group considers deviant may well be considered offensive by another, and something that is appropriate at one time and place may not be so at another. So, getting drunk and brawling may be commonplace in some bars, but not in church, and smoking pot may be normative among your friends, but not among your extended family members. Even cigarette smoking, a legal act, has become quite stigmatized in some communities so that those who engage in this activity are ostracized and segregated from others.

When we examine what lies at the root of social definitions of deviance, we discover the importance of social power. Definitions of deviance are forged through "morality campaigns," where some groups are successful in raising awareness, accumulating the testimonials of experts and the endorsements of influential opinion leaders, spreading this information, and recruiting various organizations to support their drives. They may vie against other politicized groups to determine whose opinions become legislated into morality and law, such as we see with the continuing struggle over abortion legislation and its funding. In the final analysis, groups that have the power to mobilize these critical resources are able to enforce their opinions onto others, thus legitimating their behavior and criminalizing that of weaker groups. The legal status of the drugs preferred by the white middle classes (alcohol, tobacco, marijuana, Prozac) compared to those preferred by inner-city, lower class, minority populations (crack, heroin) is testament to this situation, as the latter drugs are not necessarily inherently more harmful than the former, yet they are significantly more heavily criminalized and enforced.

Social norms and values exist in a state of relative flux, as they evolve and change over time. Some things that used to be considered deviant have been normalized, such as divorce, tattoos, and Ritalin, while others that were acceptable have now come to be redefined in a deviant light, such as sexual harassment, date rape, and smoking cigarettes. Deviance serves many functions in society, some of which include laying the groundwork for social change and fostering full employment (for lawyers, judges, prison officials, treatment administrators, police, private investigators, and criminologists). Deviance fosters boundary maintenance by reinforcing normative behavior and sanctioning behavior outside the acceptable limits.

People are led to deviance by many factors, some lodged within the social structure, some culturally derived, and some learned through interaction with knowledgeable others. In our first selection, Dean Dabney and Richard Hollinger's examination of drug-abusing pharmacists shows us the effect of structural access on the likelihood of deviance, furthering the belief that drug availability fuels use. Ayres Boswell and Joan Spade focus on the cultural dimension of causality, looking at how certain fraternity norms and values foster the acceptance and commission of rape. Finally, Adina Nack takes us through an interactionist examination of sexually transmitted diseases, showing how women who contract these try to manage people's knowledge about their conditions to mitigate against the effects of social stigma and developing a spoiled sexual identity.

8

Body Ritual among the Nacirema

HORACE MINER

From an anthropological perspective, Miner offers a stranger's view on the members of a North American tribe. He describes some rituals common to tribal members and their invidious relationship to the natural appearances and processes of their bodies. Disdaining their ordinary functions, Naciremans collect and revere objects through which they transform and disguise their ordinary sight, smell, and feel, going to great length to conform to mainstream cultural norms. They also dogmatically share common cultural beliefs about the existence of spiritual bodies, healing powers, and important sacred practices, particularly those concerned with their physical essence. Miner's description and analysis, although not a complete portrayal of the group, offer fascinating insight into the way outsiders perceive and interpret distinct native cultures.

The anthropologist has become so familiar with the diversity of ways in which different peoples behave in similar situations that he is not apt to be surprised by even the most exotic customs. In fact, if all of the logically possible combinations of behavior have not been found somewhere in the world, he is apt to suspect that they must be present in some yet undescribed tribe. This point has, in fact, been expressed with respect to clan organization by Murdock (1949: 71). In this light, the magical beliefs and practices of the Nacirema present such unusual aspects that it seems desirable to describe them as an example of the extremes to which human behavior can go.

Professor Linton first brought the ritual of the Nacirema to the attention of anthropologists twenty years ago (1936: 326), but the culture of this people is still very poorly understood. They are a North American group living in the territory between the Canadian Cree, the Yaqui and Tarahumare of Mexico, and the Carib and Arawak of the Antilles. Little is known of their origin, although tradition states that they came from the east. According to Nacirema mythology, their nation was originated by a culture hero, Notgnihsaw, who is otherwise known for two great feats of strength—the throwing of a piece of wampum across the river Pa-To-Mac and the chopping down of a cherry tree in which the Spirit of Truth resided.

Nacirema culture is characterized by a highly developed market economy which has evolved in a rich natural habitat. While much of the people's time is devoted to economic pursuits, a large part of the fruits of these labors and a considerable portion of the day are spent in ritual activity. The focus of this activity is the human body, the appearance and health of which loom as a dominant concern in the ethos of the people. While such concern is certainly not unusual, its ceremonial aspects and associated philosophy are unique.

The fundamental belief underlying the whole system appears to be that the human body is ugly and that its natural tendency is to debility and disease. Incarcerated in such a body, man's only hope is to avert these characteristics through the use of the powerful influences of ritual and ceremony. Every household has one or more shrines devoted to this purpose. The more powerful individuals in this society have several shrines in their houses, and, in fact, the opulence of a house is often referred to in terms of the number of such ritual centers it possesses. Most houses are of wattle and daub construction, but the shrine rooms of the more wealthy are walled with stone. Poorer families imitate the rich by applying pottery plaques to their shrine walls.

While each family has at least one such shrine, the rituals associated with it are not family ceremonies but are private and secret. The rites are normally only discussed with children, and then only during the period when they are being initiated into these mysteries. I was able, however, to establish sufficient rapport with the natives to examine these shrines and to have the rituals described to me.

The focal point of the shrine is a box or chest which is built into the wall. In this chest are kept the many charms and magical potions without which no native believes he could live. These preparations are secured from a variety of specialized practitioners. The most powerful of these are the medicine men, whose assistance must be rewarded with substantial gifts. However, the medicine men do not provide the curative potions for their clients, but decide what the ingredients should be and then write them down in an ancient and secret language. This writing is understood only by the medicine men and by the herbalists who, for another gift, provide the required charm.

The charm is not disposed of after it has served its purpose, but is placed in the charm-box of the household shrine. As these magical materials are specific for certain ills, and the real or imagined maladies of the people are many, the charm-box is usually full to overflowing. The magical packets are so numerous that people forget what their purposes were and fear to use them again. While the natives are very vague on this point, we can only assume that the idea in retaining all the old magical materials is that their presence in the charm-box, before which the body rituals are conducted, will in some way protect the worshipper.

Beneath the charm-box is a small font. Each day every member of the family, in succession, enters the shrine room, bows his head before the charm-box, mingles different sorts of holy water in the font, and proceeds with a brief rite of ablution. The holy waters are secured from the Water Temple of the community, where the priests conduct elaborate ceremonies to make the liquid ritually pure.

In the hierarchy of magical practitioners, and below the medicine men in prestige, are specialists whose designation is best translated "holy-mouth-men." The Nacirema have an almost pathological horror of and fascination with the mouth, the condition of which is believed to have a supernatural influence on all social relationships. Were it not for the rituals of the mouth, they believe that their teeth would fall out, their gums bleed, their jaws shrink, their friends desert them, and their lovers reject them. They also believe that a strong relationship exists between oral and moral characteristics. For example, there is a ritual ablution of the mouth for children which is supposed to improve their moral fiber.

The daily body ritual performed by everyone includes a mouth-rite. Despite the fact that these people are so punctilious about care of the mouth, this rite involves a practice which strikes the uninitiated stranger as revolting. It was reported to me that the ritual consists of inserting a small bundle of hog hairs into the mouth, along with certain magical powders, and then moving the bundle in a highly formalized series of gestures.

In addition to the private mouth-rite, the people seek out a holy-mouth-man once or twice a year. These practitioners have an impressive set of paraphernalia, consisting of a variety of augers, awls, probes, and prods. The use of these objects in the exorcism of the evils of the mouth involves almost unbelievable ritual torture of the client. The holy-mouth-man opens the client's mouth and, using the above-mentioned tools, enlarges any holes which decay may have created in the teeth. Magical materials are put into these holes. If there are no naturally occurring holes in the teeth, large sections of one or more teeth are gouged out so that the supernatural substance can be applied. In the client's view, the purpose of these ministrations is to arrest decay and to draw friends. The extremely sacred and traditional character of the rite is evident in the fact that the natives return to the holy-mouth-man year after year, despite the fact that their teeth continue to decay.

It is to be hoped that, when a thorough study of the Nacirema is made, there will be careful inquiry into the personality structure of these people. One has but to watch the gleam in the eye of a holy-mouth-man, as he jabs an awl into an exposed nerve, to suspect that a certain amount of sadism is involved. If this can be established, a very interesting pattern emerges, for most of the population shows definite masochistic tendencies. It was to these that Professor Linton referred in discussing a distinctive part of the daily body ritual which is performed only by men. This part of the rite involves scraping and lacerating the surface of the face with a sharp instrument. Special women's rites are performed only four times during each lunar month, but what they lack in frequency is made up in barbarity. As part of this ceremony, women bake their heads in small ovens for about an hour. The theoretically interesting point is that what seems to be a preponderantly masochistic people have developed sadistic specialists.

The medicine men have an imposing temple, or *latipso,* in every community of any size. The more elaborate ceremonies required to treat very sick patients can only be performed at this temple. These ceremonies involve not only the thaumaturge but a permanent group of vestal maidens who move sedately about the temple chambers in distinctive costume and headdress.

The *latipso* ceremonies are so harsh that it is phenomenal that a fair proportion of the really sick natives who enter the temple ever recover. Small children whose indoctrination is still incomplete have been known to resist attempts to take them to the temple because "that is where you go to die." Despite this fact, sick adults are not only willing but eager to undergo the protracted ritual purification, if they can afford to do so. No matter how ill the supplicant or how grave the emergency, the guardians of many temples will not admit a client if he cannot give a rich gift to the custodian. Even after one has gained admission and survived the ceremonies, the guardians will not permit the neophyte to leave until he makes still another gift.

The supplicant entering the temple is first stripped of all his or her clothes. In everyday life the Nacirema avoids exposure of his body and its natural functions. Bathing and excretory acts are performed only in the secrecy of the household shrine, where they are ritualized as part of the body-rites. Psychological shock results from the fact that body secrecy is suddenly lost upon entry into the *latipso.* A man, whose own wife has never seen him in an excretory act, suddenly finds himself naked and assisted by a vestal maiden while he performs his natural functions into a sacred vessel. This sort of ceremonial treatment is necessitated by the fact that the excreta are used by a diviner to ascertain the course and nature of the client's sickness. Female clients, on the other hand, find their naked bodies are subjected to the scrutiny, manipulation, and prodding of the medicine men.

Few supplicants in the temple are well enough to do anything but lie on their hard beds. The daily ceremonies, like the rites of the holy-mouth-men, involve discomfort and torture. With ritual precision, the vestals awaken their miserable charges each dawn and roll them about on their beds of pain while performing ablutions, in the formal movements of which the maidens are highly trained. At other times they insert magic wands in the supplicant's mouth or force him to eat substances which are supposed to be healing. From time to time the medicine men come to their clients and jab magically treated needles into their flesh. The fact that these temple ceremonies may not cure, and may kill, the neophyte, in no way decreases the people's faith in the medicine men.

There remains one other kind of practitioner, known as a "listener." This witch-doctor has the power to exorcise the devils that lodge in the heads of people who have been bewitched. The Nacirema believe that parents bewitch their own children. Mothers are particularly suspected of putting a curse on children while teaching them the secret body rituals. The counter-magic of the witch-doctor is unusual in its lack of ritual. The patient simply tells the "listener" all his troubles and fears, beginning with the earliest difficulties he can remember. The memory displayed by the Nacirema in these exorcism sessions is truly remarkable. It is not uncommon for the patient to bemoan the rejection he felt upon being weaned as a babe, and a few individuals even see their troubles going back to the traumatic effects of their own birth.

In conclusion, mention must be made of certain practices which have their base in native esthetics but which depend upon the pervasive aversion to the natural body and its functions. There are ritual fasts to make fat people thin and ceremonial feasts to make thin people fat. Still other rites are used to make women's breasts larger if they are small, and smaller if they are large. General dissatisfaction with breast shape is symbolized in the fact that the ideal form is virtually outside the range of human variation. A few women afflicted with almost inhuman hypermammary development are so idolized that they make a handsome living by simply going from village to village and permitting the natives to stare at them for a fee.

Reference has already been made to the fact that excretory functions are ritualized, routinized, and relegated to secrecy. Natural reproductive functions are similarly distorted. Intercourse is taboo as a topic and scheduled as an act. Efforts are made to avoid pregnancy by the use of magical materials or by limiting intercourse to certain phases of the moon. Conception is actually very infrequent. When pregnant, women dress so as to hide their condition. Parturition takes place in secret, without friends or relatives to assist, and the majority of women do not nurse their infants.

Our review of the ritual life of the Nacirema has certainly shown them to be a magic-ridden people. It is hard to understand how they have managed to exist so long under the burdens which they have imposed upon themselves. But even such exotic customs as these take on real meaning when they are viewed with the insight provided by Malinowski when he wrote (1948: 70):

> Looking from far and above, from our high places of safety in the developed civilization, it is easy to see all the crudity and irrelevance of magic. But without its power and guidance early man could not have mastered his practical difficulties as he has done, nor could man have advanced to the higher stages of civilization.

REFERENCES

Linton, R. 1936. *The study of man.* New York: Appleton-Century.

Malinowski, B. 1948. *Magic, science and religion.* Glencoe, IL: Free Press.

Murdock, G. P. 1949. *Social structure.* New York: Macmillan.

9

The Code of the Street

ELIJAH ANDERSON

The richness of inner city life is revealed in Anderson's description of the norms of black street life in America. Anderson differentiates between the "decent," mainstream residents and those who adhere to the "street" subculture, finding risk, excitement, and respect there. A clear set of values and behavioral guidelines ensure that these latter community members follow the subcultural standards governing violence and other forms of street comportment. This "code" allows members to measure each others' status in the group and influences the overall behavior of large numbers of inner city black youth. Anderson highlights the role of specific material possessions, the willingness to fight, concerns over manhood, and the elaborate posturing people use to both assert themselves over others and to enhance their safety from being victimized by others. This emphasis on living on the edge, risking death, and showing no fear often puts members of this community at odds with the police, as they are more concerned with maintaining their position among peers than with breaking the law or avoiding arrest and incarceration. Anderson discusses how this street code and subculture contributes to the vicious cycle of young black men and women's hopelessness, alienation, and opposition to the dominant mainstream society and culture. Can you identify a subculture to which you belong? Does it have its own slang, customs, and ways of acting? Do the values of your subculture agree or disagree with the values of larger, mainstream American culture?

Of all the problems besetting the poor inner-city black community, none is more pressing than that of interpersonal violence and aggression. It wreaks havoc daily with the lives of community residents and increasingly spills over into downtown and residential middle-class areas. Muggings, burglaries, carjackings, and drug-related shootings, all of which may leave their victims or innocent bystanders dead, are now common enough to concern all urban and many suburban residents. The inclination to violence springs from the

From *Code of the Street: Decency, Violence, and the Moral Life of the Inner City* by Elijah Anderson. Used by permission of W.W. Norton & Company, Inc.

circumstances of life among the ghetto poor—the lack of jobs that pay a living wage, the stigma of race, the fallout from rampant drug use and drug trafficking, and the resulting alienation and lack of hope for the future.

Simply living in such an environment places young people at special risk of falling victim to aggressive behavior. Although there are often forces in the community which can counteract the negative influences, by far the most powerful being a strong, loving, "decent" (as inner-city residents put it) family committed to middle-class values, the despair is pervasive enough to have spawned an oppositional culture, that of "the streets," whose norms are often consciously opposed to those of mainstream society. These two orientations—decent and street—socially organize the community, and their coexistence has important consequences for residents, particularly children growing up in the inner city. Above all, this environment means that even youngsters whose home lives reflect mainstream values—and the majority of homes in the community do—must be able to handle themselves in a street-oriented environment.

This is because the street culture has evolved what may be called a code of the streets, which amounts to a set of informal rules governing interpersonal public behavior, including violence. The rules prescribe both a proper comportment and a proper way to respond if challenged. They regulate the use of violence and so allow those who are inclined to aggression to precipitate violent encounters in an approved way. The rules have been established and are enforced mainly by the street-oriented, but on the streets the distinction between street and decent is often irrelevant; everybody knows that if the rules are violated, there are penalties. Knowledge of the code is thus largely defensive; it is literally necessary for operating in public. Therefore, even though families with a decency orientation are usually opposed to the values of the code, they often reluctantly encourage their children's familiarity with it to enable them to negotiate the inner-city environment.

At the heart of the code is the issue of respect—loosely defined as being treated "right," or granted the deference one deserves. However, in the troublesome public environment of the inner city, as people increasingly feel buffeted by forces beyond their control, what one deserves in the way of respect becomes more and more problematic and uncertain. This in turn further opens the issue of respect to sometimes intense interpersonal negotiation. In the street culture, especially among young people, respect is viewed as almost an external entity that is hard-won but easily lost, and so must constantly be guarded. The rules of the code in fact provide a framework for negotiating respect. The person whose very appearance—including his clothing, demeanor, and way of moving—deters transgressions feels that he possesses, and may be considered by others to possess, a measure of respect. With the right amount of respect, for instance, he can avoid "being bothered" in public. If he is bothered, not only may he be in physical danger but he has been disgraced or "dissed" (disrespected). Many of the forms that dissing can take might seem petty to middle-class people (maintaining eye contact for too long, for example), but to those invested in the street code, these actions become serious indications of the other person's intentions. Consequently, such people become very sensitive to advances and slights, which could well serve as warnings of imminent physical confrontation.

This hard reality can be traced to the profound sense of alienation from mainstream society and its institutions felt by many poor inner-city black people, particularly the young. The code of the streets is actually a cultural adaptation to a profound lack of faith in the police and the judicial system. The police are most often seen as representing the dominant white society and not caring to protect inner-city residents. When called, they may not respond, which is one reason many residents feel they must be prepared to take extraordinary measures to defend themselves and their loved ones against those who are inclined to aggression. Lack of police accountability has in fact been incorporated into the status system: The person who is believed capable of "taking care of himself" is accorded a certain deference, which translates into a sense of physical and psychological control. Thus the street code emerges where the influence of the police ends and personal responsibility for one's safety is felt to begin. Exacerbated by the proliferation of drugs and easy access to guns, this volatile situation results in the ability of the street-oriented minority (or those who effectively "go for bad") to dominate the public spaces.

DECENT AND STREET FAMILIES

Although almost everyone in poor inner-city neighborhoods is struggling financially and therefore feels a certain distance from the rest of America, the decent and the street family in a real sense represent two poles of value orientation, two contrasting conceptual categories. The labels "decent" and "street," which the residents themselves use, amount to evaluative judgments that confer status on local residents. The labeling is often the result of a social contest among individuals and families of the neighborhood. Individuals of the two orientations often coexist in the same extended family. Decent residents judge themselves to be so while judging others to be of the street, and street individuals often present themselves as decent, drawing distinctions between themselves and other people. In addition, there is quite a bit of circumstantial behavior—that is, one person may at different times exhibit both decent and street orientations, depending on the circumstances. Although these designations result from so much social jockeying, there do exist concrete features that define each conceptual category.

Generally, so-called decent families tend to accept mainstream values more fully and attempt to instill them in their children. Whether married couples with children or single-parent (usually female) households, they are generally "working poor" and so tend to be better off financially than their street-oriented neighbors. They value hard work and self-reliance and are willing to sacrifice for their children. Because they have a certain amount of faith in mainstream society, they harbor hopes for a better future for their children, if not for themselves. Many of them go to church and take a strong interest in their children's schooling. Rather than dwelling on the real hardships and inequities facing them, many such decent people, particularly the increasing number of grandmothers raising grandchildren, see their difficult situation as a test from God and derive great support from their faith and from the church community.

Extremely aware of the problematic and often dangerous environment in which they reside, decent parents tend to be strict in their child-rearing practices, encouraging children to respect authority and walk a straight moral line. They have an almost obsessive concern about trouble of any kind and remind their children to be on the lookout for people and situations that might lead to it. At the same time, they are themselves polite and considerate of others, and teach their children to be the same way. At home, at work, and in church, they strive hard to maintain a positive mental attitude and a spirit of cooperation.

So-called street parents, in contrast, often show a lack of consideration for other people and have a rather superficial sense of family and community. Though they may love their children, many of them are unable to cope with the physical and emotional demands of parenthood and find it difficult to reconcile their needs with those of their children. These families, who are more fully invested in the code of the streets than the decent people are, may aggressively socialize their children into it in a normative way. They believe in the code and judge themselves and others according to its values.

CAMPAIGNING FOR RESPECT

[The] realities of inner-city life are largely absorbed on the streets. At an early age, often even before they start school, children from street-oriented homes gravitate to the streets, where they "hang"—socialize with their peers. Children from these generally permissive homes have a great deal of latitude and are allowed to "rip and run" up and down the street. They often come home from school, put their books down, and go right back out the door. On school nights eight- and nine-year-olds remain out until nine or ten o'clock (and teenagers typically come in whenever they want to). On the streets they play in groups that often become the source of their primary social bonds. Children from decent homes tend to be more carefully supervised and are thus likely to have curfews and to be taught how to stay out of trouble.

When decent and street kids come together, a kind of social shuffle occurs in which children have a chance to go either way. Tension builds as a child comes to realize that he must choose an orientation. The kind of home he comes from influences but does not determine the way he will ultimately turn out—although it is unlikely that a child from a thoroughly street-oriented family will easily absorb decent values on the streets. Youths who emerge from street-oriented families but develop a decency orientation almost always learn those values in another setting—in school, in a youth group, in church. Often it is the result of their involvement with a caring "old head" (adult role model).

In the street, through their play, children pour their individual life experiences into a common knowledge pool, affirming, confirming, and elaborating on what they have observed in the home and matching their skills against those of others. And they learn to fight. Even small children test one another, pushing and shoving, and are ready to hit other children over circumstances not to their liking. In turn, they are readily hit by other children, and the child who is toughest prevails.

Thus the violent resolution of disputes, the hitting and cursing, gains social reinforcement. The child in effect is initiated into a system that is really a way of campaigning for respect.

In addition, younger children witness the disputes of older children, which are often resolved through cursing and abusive talk, if not aggression or outright violence. They see that one child succumbs to the greater physical and mental abilities of the other. They are also alert and attentive witnesses to the verbal and physical fights of adults, after which they compare notes and share their interpretations of the event. In almost every case the victor is the person who physically won the altercation, and this person often enjoys the esteem and respect of onlookers. These experiences reinforce the lessons the children have learned at home: Might makes right, and toughness is a virtue, while humility is not. In effect they learn the social meaning of fighting. When it is left virtually unchallenged, this understanding becomes an ever more important part of the child's working conception of the world. Over time the code of the streets becomes refined.

Those street-oriented adults with whom children come in contact—including mothers, fathers, brothers, sisters, boyfriends, cousins, neighbors, and friends—help them along in forming this understanding by verbalizing the messages they are getting through experience: "Watch your back." "Protect yourself." "Don't punk out." "If somebody messes with you, you got to pay them back." "If someone disses you, you got to straighten them out." Many parents actually impose sanctions if a child is not sufficiently aggressive. For example, if a child loses a fight and comes home upset, the parent might respond, "Don't you come in here crying that somebody beat you up; you better get back out there and whup his ass. I didn't raise no punks! Get back out there and whup his ass. If you don't whup his ass, I'll whup your ass when you come home." Thus the child obtains reinforcement for being tough and showing nerve.

While fighting, some children cry as though they are doing something they are ambivalent about. The fight may be against their wishes, yet they may feel constrained to fight or face the consequences—not just from peers but also from caretakers or parents, who may administer another beating if they back down. Some adults recall receiving such lessons from their own parents and justify repeating them to their children as a way to toughen them up. Looking capable of taking care of oneself as a form of self-defense is a dominant theme among both street-oriented and decent adults who worry about the safety of their children. There is thus at times a convergence in their child-rearing practices, although the rationales behind them may differ.

SELF-IMAGE BASED ON "JUICE"

By the time they are teenagers, most youths have either internalized the code of the streets or at least learned the need to comport themselves in accordance with its rules, which chiefly have to do with interpersonal communication. The code revolves around the presentation of self. Its basic requirement is the display of a certain predisposition to violence. Accordingly, one's bearing must send the unmistakable if sometimes subtle message to "the next person" in public that one is capable of violence and mayhem when the situation requires it, that one can take care of oneself. The nature of this communication is largely determined by the demands of the circumstances but can include facial expressions, gait, and verbal expressions—all of which are geared mainly to deterring aggression. Physical appearance, including clothes, jewelry, and grooming, also plays an important part in how a person is viewed; to be respected, it is important to have the right look.

Even so, there are no guarantees against challenges because there are always people around looking for a fight to increase their share of respect—or "juice," as it is sometimes called on the street. Moreover, if a person is assaulted, it is important, not only in the eyes of his opponent but also in the eyes of his "running buddies," for him to avenge himself. Otherwise he risks being "tried" (challenged) or "moved on" by any number of others. To maintain his honor he must show he is not someone to be "messed with" or "dissed." In general, the person must "keep himself straight" by managing his position of respect among others; this involves in part his self-image, which is shaped by what he thinks others are thinking of him in relation to his peers.

Objects play an important and complicated role in establishing self-image. Jackets, sneakers, gold jewelry, reflect not just a person's taste, which tends to be tightly regulated among adolescents of all social classes, but also a willingness to possess things that may require defending. A boy wearing a fashionable, expensive jacket, for example, is vulnerable to attack by another who covets the jacket and either cannot afford to buy one or wants the added satisfaction of depriving someone else of his. However, if the boy forgoes the desirable jacket and wears one that isn't "hip," he runs the risk of being teased and possibly even assaulted as an unworthy person. To be allowed to hang with certain prestigious crowds, a boy must wear a different set of expensive clothes—sneakers and athletic suit—every day. Not to be able to do so might make him appear socially deficient. The youth comes to covet such items—especially when he sees easy prey wearing them.

In acquiring valued things, therefore, a person shores up his identity—but since it is an identity based on having things, it is highly precarious. This very precariousness gives a heightened sense of urgency to staying even with peers, with whom the person is actually competing. Young men and women who are able to command respect through their presentation of self—by allowing their possessions and their body language to speak for them—may not have to campaign for regard but may, rather, gain it by the force of their manner. Those who are unable to command respect in this way must actively campaign for it—and are thus particularly alive to slights.

One way of campaigning for status is by taking the possessions of others. In this context, seemingly ordinary objects can become trophies imbued with symbolic value that far exceeds their monetary worth. Possession of the trophy can symbolize the ability to violate somebody—to "get in his face," to take something of value from him, to "dis" him, and thus to enhance one's own worth by stealing someone else's. The trophy does not have to be something material. It can be another person's sense of honor, snatched away with a derogatory remark. It can be the outcome of a fight. It can be the imposition of a certain standard, such as a girl's getting herself recognized as the most beautiful. Material things, however, fit easily into the pattern. Sneakers, a pistol, even somebody else's girlfriend, can become a trophy. When a person can take something from another and then flaunt it, he gains a certain regard by being the owner, or the controller, of that thing. But this display of ownership can then provoke other people to challenge him. This game of who controls what is thus constantly being played out on inner-city streets, and the trophy—extrinsic or intrinsic, tangible or intangible—identifies the current winner.

An important aspect of this often violent give-and-take is its zero-sum quality. That is, the extent to which one person can raise himself up depends on his ability to put another person down. This underscores the alienation that permeates the inner-city ghetto community. There is a generalized sense that very little respect is to be had, and therefore everyone competes to get what affirmation he can of the little that is available. The craving for respect that results gives people thin skins. Shows of deference by others can be highly soothing, contributing to a sense of security, comfort, self-confidence, and self-respect. Transgressions by others which go unanswered diminish these feelings and are believed to encourage further transgressions. Hence one must be ever vigilant against the transgressions of others or even *appearing* as if transgressions will be tolerated. Among young people, whose sense of self-esteem is particularly vulnerable, there is an especially heightened concern with being disrespected. Many inner-city young men in particular crave respect to such a degree that they will risk their lives to attain and maintain it.

The issue of respect is thus closely tied to whether a person has an inclination to be violent, even as a victim. In the wider society people may not feel required to retaliate physically after an attack, even though they are aware that they have been degraded or taken advantage of. They may feel a great need to defend themselves *during* an attack, or to behave in such a way as to deter aggression (middle-class people certainly can and do become victims of street-oriented youths), but they are much more likely than street-oriented people to feel that they can walk away from a possible altercation with their self-esteem intact. Some people may even have the strength of character to flee, without any thought that their self-respect or esteem will be diminished.

In impoverished inner-city black communities, however, particularly among young males and perhaps increasingly among females, such flight would be extremely difficult. To run away would likely leave one's self-esteem in tatters. Hence people often feel constrained not only to stand up and at least attempt to resist during an assault but also to "pay back"—to seek revenge—after a successful assault on their person. This may include going to get a weapon or even getting relatives involved. Their very identity and self-respect, their honor, is often intricately tied up with the way they perform on the streets during and after such encounters. This outlook reflects the circumscribed opportunities of the inner-city poor. Generally people outside the ghetto have other ways of gaining status and regard, and thus do not feel so dependent on such physical displays.

BY TRIAL OF MANHOOD

On the street, among males these concerns about things and identity have come to be expressed in the concept of "manhood." Manhood in the inner city means taking the prerogatives of men with respect to strangers, other men, and women—being distinguished as a man. It implies physicality and a certain ruthlessness. Regard and respect are associated with this concept in large part because of its practical application: If others have little or no regard for a person's manhood, his very life and those of his loved ones could be in jeopardy. But there is a chicken-and-egg aspect to this situation: One's physical safety is more likely to be jeopardized in public *because* manhood is associated with respect. In other words, an existential link has been created between the idea of manhood and one's self-esteem, so that it has become hard to say which is primary. For many inner-city youths, manhood and respect are flip sides of the same coin; physical and psychological well-being are inseparable, and both require a sense of control, of being in charge.

The operating assumption is that a man, especially a real man, knows what other men know—the code of the streets. And if one is not a real man, one is somehow diminished as a person, and there are certain valued things one simply does not deserve. There is thus believed to be a certain justice to the code, since it is considered that everyone has the opportunity to know it. Implicit in this is that everybody is held responsible for being familiar with the code. If the victim of a mugging, for example, does not know the code and so responds "wrong," the perpetrator may feel justified even in killing him and may feel no remorse. He may think, "Too bad, but it's his fault. He should have known better."

So when a person ventures outside, he must adopt the code—a kind of shield, really—to prevent others from "messing with" him. In these circumstances it is easy for people to think they are being tried or tested by others even when this is not the case. For it is sensed that something extremely valuable is at stake in every interaction, and people are encouraged to rise to the occasion, particularly with strangers. For people who are unfamiliar with the code—generally people who live outside the inner city—the concern with respect in the most ordinary interactions can be frightening and incomprehensible. But for those who are invested in the code, the clear object of their demeanor is to discourage strangers from even thinking about testing their manhood. And the sense of power that attends the ability to deter others can be alluring even to those who know the code without being heavily invested in it—the decent inner-city youths. Thus a boy who has been leading a basically decent life can, in trying circumstances, suddenly resort to deadly force.

Central to the issue of manhood is the widespread belief that one of the most effective ways of gaining respect is to manifest "nerve." Nerve is shown when one takes another person's possessions (the more valuable the better), "messes with" someone's woman, throws the first punch, "gets in someone's face," or pulls a trigger. Its proper display helps on the spot to check others who would violate one's person and also helps to build a reputation that works to prevent future challenges. But since such a show of nerve is a forceful expression of disrespect toward the person on the receiving end, the victim may be greatly offended and seek to retaliate with equal or greater force. A display of nerve, therefore, can easily provoke a life-threatening response, and the background knowledge of that possibility has often been incorporated into the concept of nerve.

True nerve exposes a lack of fear of dying. Many feel that it is acceptable to risk dying over the principle of respect. In fact, among the hard-core street-oriented, the clear risk of violent death may be preferable to being "dissed" by another. The youths who have internalized this attitude and convincingly display it in their public bearing are among the most threatening people of all, for it is commonly assumed that they fear no man. As the people of the community say, "They are the baddest dudes on the street." They often lead an existential life that may acquire meaning only when they are faced with the possibility of imminent death. Not to be afraid to die is by implication to have few compunctions about taking another's life. Not to be afraid to die is the quid pro quo of being able to take somebody else's life—for the right reasons, if the situation demands it. When others believe this is one's position, it gives one a real sense of power on the streets. Such credibility is what many inner-city youths strive to achieve, whether they are decent or street-oriented, both because of its practical defensive value and because of the positive way it makes them feel about themselves. The difference between the decent and the street-oriented youth is often that the decent youth makes a conscious decision to appear tough and manly; in another setting—with teachers, say, or at his part-time job—he can be polite and deferential. The street-oriented youth, on the other hand, has made the concept of manhood a part of his very identity; he has difficulty manipulating it—it often controls him.

GIRLS AND BOYS

Increasingly, teenage girls are mimicking the boys and trying to have their own version of "manhood." Their goal is the same—to get respect, to be recognized as capable of setting or maintaining a certain standard. They try to achieve this end in the ways that have been established by the boys, including posturing, abusive language, and the use of violence to resolve disputes, but the issues for the girls are different. Although conflicts over turf and status exist among the girls, the majority of disputes seem rooted in assessments of beauty (which girl in a group is "the cutest"), competition over boyfriends, and attempts to regulate other people's knowledge of and opinions about a girl's behavior or that of someone close to her, especially her mother.

A major cause of conflicts among girls is "he say, she say." This practice begins in the early school years and continues through high school. It occurs when "people," particularly girls, talk about others, thus putting their "business in the streets." Usually one girl will say something negative about another in the group, most often behind the person's back. The remark will then get back to the person talked about. She may retaliate or her friends may feel required to "take up for" her. In essence this is a form of group gossiping in which individuals are negatively assessed and evaluated. As with much gossip, the things said may or may not be true, but the point is that such imputations can cast aspersions on a person's good name. The accused is required to defend herself against the slander, which can result in arguments and fights, often over little of real substance. Here again is the problem of low self-esteem, which encourages youngsters to be highly sensitive to slights and to be vulnerable to feeling easily "dissed." To avenge the dissing, a fight is usually necessary.

Because boys are believed to control violence, girls tend to defer to them in situations of conflict. Often if a girl is attacked or feels slighted, she will get a brother, uncle, or cousin to do her fighting for her. Increasingly, however, girls are doing their own fighting and are even asking their male relatives to teach them how to fight. Some girls form groups that attack other girls or take things from them. A hard-core segment of inner-city girls inclined toward violence seems to be developing. As one 13-year-old girl in a detention center for youths who have committed violent acts told me, "To get people to leave you alone, you gotta fight. Talking don't always get you out of stuff." One major difference between girls and boys: Girls rarely use guns. Their fights are therefore not life-or-death struggles. Girls are not often willing to put their lives on the line for "manhood." The ultimate form of respect on the male-dominated inner-city street is thus reserved for men.

"GOING FOR BAD"

In the most fearsome youths such a cavalier attitude toward death grows out of a very limited view of life. Many are uncertain about how long they are going to live and believe they could die violently at any time. They accept this fate; they live on the edge. Their manner conveys the message that nothing intimidates them; whatever turn the encounter takes, they maintain their attack—rather like a pit bull, whose spirit many such boys admire. The demonstration of such tenacity "shows heart" and earns their respect.

This fearlessness has implications for law enforcement. Many street-oriented boys are much more concerned about the threat of "justice" at the hands of a peer than at the hands of the police. Moreover, many feel not only that they have little to lose by going to prison but that they have something to gain. The toughening-up one experiences in prison can actually enhance one's reputation on the streets. Hence the system loses influence over the hard core who are without jobs, with little perceptible stake in the system. If mainstream society has done nothing *for* them, they counter by making sure it can do nothing *to* them.

At the same time, however, a competing view maintains that the true nerve consists in backing down, walking away from a fight, and going on with one's business. One fights only in self-defense. This view emerges from the decent philosophy that life is precious, and it is an important part of the socialization process common in decent homes. It discourages violence as the primary means of resolving disputes and encourages youngsters to accept nonviolence and talk as confrontational strategies. But "if the deal goes down," self-defense is greatly encouraged. When there is enough positive support for this orientation, either in the home or among one's peers, then nonviolence has a chance to prevail. But it prevails at the cost of relinquishing a claim to being bad and tough, and therefore sets a young person up as at the very least alienated from street-oriented peers and quite possibly a target of derision or even violence.

Although the nonviolent orientation rarely overcomes the impulse to strike back in an encounter, it does introduce a certain confusion and so can prompt a measure of soul-searching, or even profound ambivalence. Did the person back down with his respect intact or did he back down only to be judged a "punk"—a person lacking manhood? Should he or she have acted? Should he or she have hit the other person in the mouth? These questions beset many young men and women during public confrontations. What is the "right" thing to do? In the quest for honor, respect, and local status—which few young people are uninterested in—common sense most often prevails, which leads many to opt for the tough approach, enacting their own particular versions of the display of nerve. The presentation of oneself as rough and tough is very often quite acceptable until one is tested. And then that presentation may help the person pass the test, because it will cause fewer questions to be asked about what he did and why. It is hard for a person to explain why he lost the fight or why he backed down. Hence many will strive to appear to "go for bad," while hoping they will never be tested. But when they are tested, the outcome of the situation may quickly be out of their hands, as they become wrapped up in the circumstances of the moment.

AN OPPOSITIONAL CULTURE

The attitudes of the wider society are deeply implicated in the code of the streets. Most people in inner-city communities are not totally invested in the code, but the significant minority of hard-core street youths who are have to maintain the code in order to establish reputations, because they have—or feel they have—few other ways to assert themselves. For these young people the standards of the street code are the only game in town. The extent to which some children—particularly those who through upbringing have become most alienated and those lacking in strong and conventional social support—experience, feel, and internalize racist rejection and contempt from mainstream society may strongly encourage them to express contempt for the more conventional society in turn. In dealing with this contempt and rejection, some youngsters will consciously invest themselves and their considerable mental resources in what amounts to an oppositional culture to preserve themselves and their self-respect. Once they do, any respect they might be able to garner in the wider system pales in comparison with the respect available in the local system; thus they often lose interest in even attempting to negotiate the mainstream system.

At the same time, many less alienated young blacks have assumed a street-oriented demeanor as a way of expressing their blackness while really embracing a much more moderate way of life; they, too want a nonviolent setting in which to live and raise a family. These decent people are trying hard to be part of the mainstream culture, but the racism, real and perceived, that they encounter helps to legitimate the oppositional culture. And so on occasion they adopt street behavior. In fact, depending on the demands of the situation, many people in the community slip back and forth between decent and street behavior.

A vicious cycle has thus been formed. The hopelessness and alienation many young inner-city black men and women feel, largely as a result of endemic joblessness and persistent racism, fuels the violence they engage in. This violence serves to confirm the negative feelings many whites and some middle-class blacks harbor toward the ghetto poor, further legitimating the oppositional culture and the code of the streets in the eyes of many poor young blacks. Unless this cycle is broken, attitudes on both sides will become increasingly entrenched, and the violence, which claims victims black and white, poor and affluent, will only escalate.

CULTURE:
Counterculture

10

Canadian Male Street Skinheads

KEVIN YOUNG
AND LAURA CRAIG

This study of a Canadian youth subculture focuses on the meanings and values associated with membership in a nonpolitical skinhead group. It examines the skinhead style, with its aggressive, working-class white supremacist image that includes hair cut to the skull, menacing boots, and multiple tattoos and body piercings. This group postures itself in intentional counterpoint to the (middle class) hippie love and acceptance ideology, featuring racist, sexist, homophobic, anti-Semitic ideologies grounded in a white, male, christian, working class ethos. We meet members of "the crew," and see some of their attitudes and beliefs. In a "free" society, is it best to let these groups express their ideas or should we do something to curb them? Should all democratic societies foster the perpetuation of countercultural groups or, if dominant members of the society deem them dangerous, should these groups be controlled?

Unlike other flamboyant post-war youth subcultures—teddy boys, mods and rockers, new wave, rap or grunge—for a variety of people and generations throughout the world the label "skinhead" conjures an immediate, if often stereotypical, sense of what the group represents. Much of what is known of the skinhead movement both in Canada and abroad stems from the aggressive behavior of the first generation of British skinheads in the late 1960s and the 1970s, or from the more organized and xenophobic practices of the present generation of ultra-right skinhead gangs that are active across Europe. The current moral panic associated with skinheads largely derives from the fact that countries such as the United Kingdom, Germany, the Netherlands, Belgium and France are known to have experienced an increase in white-supremacist

From "Beyond White Pride: Identity, Meaning, and Contradiction in the Canadian Male Skinhead Subculture," CRSA Vol. 34, No. 2, pp. 175–206, by Kevin Young and Laura Craig. Reprinted by permission.

skinhead attacks on refugees, immigrants and gays (*Maclean's,* Nov. 23, 1992; Bjorgo and Wilte, 1993).

Of course, the deviant reputation of skinheads stems not only from their behavior but also from the very menacing image they project or, in other words, their *style.* According to Brake, style is made up of three main ingredients: image, which includes costume and accessories; demeanor, which includes gait, posture and practice; and argot, or the use of a distinct vocabulary (1985: 12). With such identifying characteristics or style in mind, Brake described the early British skinheads as:

> Aggressive working-class puritans in big industrial boots, jeans rolled up high to reveal them, hair cut to the skull, braces, and a violence and racism [that] earned for them the title "bovver boys," "boot-boys" on the look out for "aggro" (aggravation). Stylistically they have roots in the hard mods, forming local gangs called after a local leader or an area. Ardent football fans, they were involved in violence on the terraces against rival supporters. They espoused traditional conservative values, hard work, patriotism, defence of local territory, which led to attacks on hippies, gays and minorities. They became a metaphor for racism. . . . "Puritans in boots," they opposed hippy liberalism, subjectivity and disdain for work, attempting to "magically recover the traditional working-class community." (75–76)

Contrary to its portrayal in the popular media, then, the Canadian skinhead movement is both complex and multi-dimensional, and accommodates, albeit in often intersecting and contradictory ways, a range of behavioral and ideological opportunities for the members of its various branches. . . .

IDENTITY, MEANING AND CONTRADICTION IN THE CANADIAN SKINHEAD SUBCULTURE

Introduction to the Crew

Without exception, the skinheads in this study had been members of other youth subcultures including punks, skaters and White Power skinheads before becoming "non-political" Oi! skins. Although the specific focus of their interests may have been different, the general reasons for joining these groups were unsurprising. These were articulated as a sense of belonging and security, winning "instant" friends, and being able to share similar values and experiences. One feature of their backgrounds seemed to have consistently predisposed them to join a non-political skinhead group—their involvement in, or desire to distance themselves from, White Power skinhead groups. As we discuss below, all subjects actively refuted any ongoing connection with the organization or ideology of far-right skinheads.

Becoming Oi! skinheads provided our respondents with a status more accurately described as deviant than criminal. Membership allowed them to enjoy, as they described it, the rebelliousness and symbolic intimidation of the skinhead uniform, while at the same time maintaining regular work and feeling positive about their role as contributing members of the community. Following Brake's description above, this is, of course, consistent with traditional skinhead norms.

We learned early on in the fieldwork that membership really meant being a weekend deviant. As K——explained: "[Being a non-political skinhead] is about going out, working all day, and then being able to . . . go out to the pubs on the weekend, play darts and pool, and hang out as a group together." In other words, membership entailed a certain amount of mobility between the subculture and the mainstream, or what one subject called "normal life." This was clearly demonstrated in the case of the much-discussed work ethic. Since it is subculturally meaningful to be employed in conventional jobs, temporary modification of many of the traditional accoutrements of the subculture while pursuing work (wearing shoes other than the intimidating Doc Marten boots, concealing tattooed body parts, allowing the requisite cropped hair to grow out) is condoned behavior.

The crew met collectively once or twice a month. Most meetings occurred at a downtown drinking establishment in a pedestrian mall. The frequency of meetings seemed to be dictated by the current disposable income of the members. When most of the members were working, they socialized more regularly because money to be spent on alcohol could be pooled. The general attitude was that there was no point in getting together unless copious amounts of beer could be consumed. For cultural and economic reasons, beer, perceived as an inexpensive "working man's drink," was by far the alcoholic beverage of choice.

The crew had been meeting at one particular pub for over a year for three reasons: it attracted a range of similarly disreputable "outsiders" (street people, dipsomaniacs, bikers); it provided leisure activities viewed as appropriate by members (pool tables and darts); and its proprietors rewarded skinheads' frequent attendance by providing drink discounts and by occasionally playing skinhead music. In fact, the crew's options for alternative drinking establishments had been severely limited by its prior involvement in violent encounters in other pubs and bars in the downtown area.

Besides meeting once or twice a month as a whole, groups of two to six members of the crew would meet for coffee or beer at least once a week. Common meeting places included a particular area of a food court in a downtown mall, coffee shops, a crew member's house (usually their parents') or apartment, and local nightclubs. Driven by a disdain for what was viewed as middle-class pretension, the skinheads preferred locations perceived as disreputable, such as coffee shops and restaurants with panelled walls, mirrors, "mini"-jukeboxes at the tables, vinyl booths and large smoking sections.

Judging from the occupational and educational backgrounds of their parents, the subjects represented two quite distinct class backgrounds. Using Veltmeyer's (1986) class typology, one-third of the skinheads and ex-skinheads came from middle-class homes, and two-thirds came from working-class backgrounds. Despite the presence of middle-class participants, all subjects expressed pride in being "working-class."

Comparison of the respondents across educational and employment levels also revealed disparate backgrounds. Three of the practicing skinheads had some university or college education, and three had completed Grade 12, while five had not completed high school. The working skinheads were employed in a variety of service and manual labor jobs. None of the subjects aspired to professional or managerial positions, or work otherwise defined as white-collar. Occupational ambitions were vague; some expressed interest in going back to school to learn a trade. All stated that they desired a comfortable and "traditional" style of life, including a house and a family. . . .

The style adopted by the participants in the current study included cropped hair,[1] suspenders, black or cherry red boots that were usually steel-toed and mid-calf in length, and jeans rolled up to reveal the intimidating boots. Tops included pressed white or pin-striped dress shirts, white t-shirts, and t-shirts with Canadian insignia. Several members customized t-shirts by taking skinhead album covers to clothing stores where images are transferred onto clothing. Participants accorded special status to such unique t-shirts as well as to hard-to-acquire brands and items such as "Ben Sherman" and "Fred Perry" shirts—both part of the original skinhead attire. Often, the crew would spend many hours discussing stores that were rumored to keep such items in stock, or the possibility of saving money to mail-order them from England.

"I Hate the Commie Scum": Skinheads, Social Class and the Protestant Work Ethic[2] It was imperative for all members to be seen as being independent. For example, those who lived with their parents were quick to emphasize that they paid room and board. Many defined themselves as working-class: they claimed not to have professional or managerial aspirations and were satisfied with a modest, rather than extravagant, style of life. As T——,a veteran member, reported:

> To me, "working class" means you basically work for a living. You're not rich, you're not on the edge of poverty. You're just in between. . . . You work. You have to work your whole life just to get by.

To all respondents, in fact, being working-class implied employment, making enough money to subsist, and being independent of financial assistance from the state or one's family. However, not all of the respondents were gainfully employed; almost half of them were not working at the time of the study. Despite being unemployed, these subjects claimed to be looking for full-time work. The skinhead version of involuntary joblessness stood in contrast to their perceptions of unemployed punks, for example, for whom unemployment and welfare were viewed as a voluntary lifestyle choice.

By claiming to be working-class, skinheads were able to differentiate themselves from members of other youth subcultures. One subject went so far as to argue that in this respect the skinhead subculture was "classier than most subcultures." When asked for clarification, he explained by comparing skinheads to punks:

> I mean, I'm clean-cut, I work hard, and I've gotta pay for this asshole [a punk] out of my taxes that fucking spends all his money on drugs and everything else, you know. I mean at least he could work for them [the drugs], but he doesn't even do that. . . .

Consistent with Brake's characterization of skinhead values cited at the outset, subjects also advocated the importance of "hard work" and being independent of social assistance. In other words, they felt a sense of pride in what one subject called "a day's pay for a job well done." Further, subjects claimed not to use or sell drugs because, as they explained, drugs inevitably led to the destruction of one's own life and that of others. Drug use was viewed as being a left-wing "hippie kind of activity." Paradoxically, skinheads, while harboring few reservations about using violence to articulate subcultural commitment, condemned passive violence to oneself through drug use. Being unkempt, on welfare, and "doing drugs" were perceived as characteristics of punks and hippies—both of which were broadly viewed as parasites living off the state.

In sum, subjects' views on social class were consistent with traditional skinhead values, but were replete with contradictions. Despite their advocation of a so-called working-class ideology, one-third of the subjects were clearly from quite comfortable middle-class backgrounds. However, they believed current economic conditions only offered them service jobs of the low-end, minimum-wage variety. This seemed to provide them with some justification for viewing themselves as socially immobile as well as "working-class."

"This Land Is Ours": Ambiguous Positions on Race[3] While some members were not openly racist, and appeared guarded in—even uncertain of—their views on race, others discussed their beliefs about such things as immigration and the meaning of being Canadian in quite cavalier ways. A typical case was M—— who, having professed commitment to the present "non-political" skinhead group, went on to confess:

> I'm a racist. I don't make my way of life about it or anything. Like, I can say "nigger" and not feel guilty about it, you know what I mean? But I don't say "well there's a nigger, let's go beat him up, I hate his guts, he deserves to die," you know. I mean, it's not his fault.

Rather simplistically, M——argued that the difference between neo-Nazi skinheads and so-called "non-political" skinheads was that neo-Nazis were openly hostile toward non-whites, whereas those who were non-political kept any controversial views they might harbor strictly between members.

Further evidence of skinheads' contradictory position on race could be seen in their musical tastes. Generally, subjects did not view their "non-political" stance as being incongruous with listening to traditional skinhead music, much of which has clear White Power tones and incentives. In fact, more than half of the crew members identified this as the music they listened to most often. White Power, a genre of music favored by neo-Nazi skinheads, is also a term that refers to the possibility of world domination by the "white race." J——, who had been a skinhead for roughly a year and a half, explained that White Power music was his favorite because "[i]t gets my blood going. It makes me feel mean. . . . I don't go and beat somebody up after I listen to it; it's just a rush for me. Some people do drugs and I listen to my music." All of the respondents reported they were attracted to the beat of White Power music, that it made them feel "aggro" (aggressive), and many said that they achieved an "adrenaline rush" by listening to it. Most denied that the (often explicitly racist) lyrical content was an important dimension of the music. . . .

In fact, all respondents were deprecatory toward Canadian multi-culturalism policies and traditions. While many professed not to harbor anti-Semitic attitudes or other forms of bigotry, several expressed concerns with government policies regarding recent waves of Asian immigration. . . . In addition to a perceived challenge posed to traditional Canadian institutions by a multiracial society, concern was also expressed regarding an alleged negative impact of immigration policy on employment opportunities. P——, a veteran member, claimed that he had Black and Jewish friends, but that this did not prevent him from having what he called "white separatist beliefs":

> You know . . . they should stop the immigration. There's not enough work here for the Canadians, and they're letting I don't know how many people in every year, 200,000 immigrants, so I think that's wrong. Foreign aid I think is wrong. We should take care of our country first. . . .

"Kings of the Jungle": Skinheads and the Gender Order[4] The conventional rules of skinhead behavior in many ways represent an attempt by members to conform to traditional versions of working-class masculinity or "masculinism." In his work on youth subcultures, Brake (1985: 178) refers to the celebration of masculinism in predominantly male groups as the lionizing of "appropriate" manly performance. This involves displays of force and aggression and implies derogatory attitudes toward women and gays, as well as displays of crudity and excess of the kind alluded to in the by-now-considerable literature on gendered sports identities (Dunning, 1986; Messner, 1992; Young, 1993) and men's studies more broadly (Connell, 1987; Kimmel and Messner, 1992; Morgan, 1992).

The skinhead uniform itself and the symbols and rituals adopted by group members combine to achieve obvious hyper-aggressive and hyper-masculine effects. However, as Marshall observed in his study of the skinhead movement, one does not have to be physically large to experience the empowering potential of skinhead membership:

> Anyone who has ever had a crop and pulled on a pair of boots can tell you story after story about why being a skinhead made them feel ten feet tall when they were five foot nothing. . . . (1991: 3)

Unsurprisingly, the kinds of roles females were expected to play in the Oi! skinhead subculture tended to be subservient and demeaning. Looking at female roles in European soccer hooliganism, Marshall notes that when police searches at the turnstiles became standard practice, Chelseas could smuggle weapons into stadiums with relative ease (1991: 29). In performing such a task, Chelseas provided an instrumental service for their male hooligan partners and friends. . . .

All subjects identified key differences in the meanings and implications of being male or female in the skinhead subculture. For example, R——, who had been living with K——, a Chelsea, for three years, noted:

> Mentally, I think there are big differences. As much as skingirls are hanging out with the skinheads and things like that, they aren't as prone to be as violent. Women just aren't violent like men are, even when they're drinking. . . . Sometimes it's a good thing, sometimes it's a drawback. I haven't really fought much since I've been with K——, but any fights that I have really been in have been because of her. I mean, some person picking on her.

R——explained that while Chelseas sometimes dissipated potentially violent incidents between skinheads and non-members, their presence also tended to instigate fights. If a female member was being verbally or physically harassed by a male non-member, R——patronizingly argued that a male member would be obliged to come to her defense, and that this would invariably lead to a fistfight.

There were also female participants who, while not labelled as Chelseas, occupied distinct roles within the subculture. These females included both long-term girlfriends of skinheads who had not adopted the skinhead look, as well as transitory females—"crew sluts"—who associated with the crew. Male skinheads' image of women, both inside and outside the group, tended to be congruous with patriarchal structures observed in Canadian working-class culture by Dunk (1991). For example, crew members with non-skinhead girlfriends contended that they treated "their" women well because they "brought them out" once in a while and because they avoided discussing graphic details about fights, sexual conquests and other practices in the presence of girlfriends.

By excluding them from such unpleasantness, male crew members believed they were showing their girlfriends respect. In fact, girlfriends and Chelseas seemed to condone and justify the aggressive and flagrantly chauvinistic behavior of their boyfriends.

Dunk has identified a paradox that, he argues, exists between women whom some men choose to sexually objectify as opposed to the women they choose as partners: the "madonna/whore" dichotomy (1991: 99). Again, consistent with this view, the role of the transitory female in the skinhead subculture, according to veteran T——, was to "buy food, buy beer and have sex." Journeyman B—— emphasized that "crew sluts" would never be granted more than peripheral status in the group. Unlike most male and female members, "crew sluts" were usually middle-class, and lived with their parents in the suburbs. Most of them had met crew members while they were in high school.

Because of their often very obvious middle-class upbringing, and because they did not generally demonstrate loyalty to any particular member, male members treated "crew sluts" with very little respect—as their stigmatizing label would suggest. It was not uncommon for these young women to be cut off and insulted ("Shut up, you little slut!") by a male skinhead when attempting to contribute to a conversation. There were clear distinctions between women who were eligible to become long-term girlfriends or Chelseas and those who would remain "crew sluts." For example, once a female had been sexually intimate with more than one crew member, she was destined to remain on the margins of the subculture as a "crew slut."

McRobbie and Garber (1976) have argued that female membership in a masculine subculture often depends on the nature of the girl's/woman's attitude toward males in the group. Despite the widespread sexism of the group and the skinhead culture in general, the females in the current study seemed to derive security and pleasure from their association with skinheads. For example, one "crew slut" remarked that she considered it beneficial to know skinheads because:

> I used to worry about being downtown. This week someone pulled a knife on me downtown and asked me for my wallet, and about five minutes later three of them [crew members] came walking along and I didn't have to worry anymore.

Whether this "near-knifing" actually occurred is unimportant: the fact that this young woman wanted to believe that her safety was enhanced by knowing male skinheads demonstrates that she perceived her involvement with the crew to be both rewarding and practical.

In brief, both male and female participants in the subculture embraced traditional gender roles and tended to replicate wider structures of gender inequality. Males constructed "performative" (Rutherford, 1992: 186) identities based on displays of crudity, excess and physical dominance; there was little or no resistance by females within the group to the subordinate role they were expected to play.

"In Pride We Stride": Subcultural Identity and the Image of the Victimized Skinhead[5] Although none of the respondents claimed to be far-right at the time of the interview, all stated that one of the drawbacks to being a skinhead was being mistaken for a neo-Nazi skinhead. As one informant explained: "You can't go anywhere without people thinking 'wow man, there's one of them fascist-neo-Nazi-skinheads.'" Another respondent lamented the risks to reputation linked to being a skinhead: "You know, when you're a skinhead, everybody out there wants to kill you because everybody wants to be a hero. Everybody wants to say 'I beat up a skinhead.'"

Clearly, being continually mistaken for neo-Nazi skins frustrated Oi! members. As Coplon has observed with respect to Ska skinheads: "They inhabit the worst of all worlds; they're attacked by the Nazis for refusing to enter the fold, and they're attacked by the police and the public, who presume them to be Nazis" (1988: 94). Because their position resulted in misunderstanding, even harassment, many Oi! skinheads found metaphoric significance in the image and plight of a "crucified skinhead." The representation of a skinhead hanging on a cross was found to be a popular tattoo among the subjects. It seems that, on the one hand, Oi! skinheads enjoy the shock value that many "deviants" are known to pursue (Becker, 1963), but at the same time lament being confused by members of the public for white-supremacist skins. . . .

Like members of the heavy metal subculture, then, skinheads find themselves in a contradictory social position. Having people regard them with trepidation is enabling. As one rather diminutive skinhead explained, being able to intimidate people had a certain appeal: "A lot of people back down from you . . . when they see you're a skinhead. It's just a mystique." While special interest groups monitoring the behavior of skinheads acknowledge that not all skinheads are neo-Nazis, they also tend to assume that membership in the skinhead subculture serves the function of shocking mainstream society. As Relin puts it: "Many skinheads are not racist. Plenty of young men and women become skinheads simply because they enjoy getting a rise out of mainstream America with the shock value of the skinhead scene" (1989: 1).

While Oi! skinheads enjoy the attention they often elicit from mainstream society, shock value is certainly not the only incentive to join up. In addition to winning power through intimidation, just being a member of a group with obscure traditions, rules and meanings provided the subjects with a strong sense of identity not shared by the general population. And despite concerns with public perception, all respondents described their "inside" knowledge of the skinhead way of life as empowering.

The so-called "prestige" associated with being a skinhead was clearly illustrated in the case of J——, who was discussing how he had been "given the finger," spat on, and had objects thrown at him because he was a skinhead. When asked why he thought people reacted to him in these ways, he replied:

> Because they're uneducated, they don't know. They're doing just what they think we do. By generalizing us that way, they are doing exactly what Nazis do. So those kind of people are just totally uneducated. They should wake up. . . .

In sum, public anxiety stemming from skinheads' often daunting physical appearance and their notoriety was a cornerstone element in the construction of subcultural identity, and was actively pursued. On the other hand, the confirmation phase of the identity process often brought disadvantages, specifically in the form of public mislabelling of Oi! members as far-right or neo-Nazi skins.

NOTES

1. Skinheads usually qualify hair length using the number of the setting on an electric razor used to cut hair. A "number one" gives the shortest crop, a "number five" the longest. A "number one" cut is extreme, with the scalp being highly visible. Most of the participants in this study had hair lengths ranging from "number two" to "number four."

2. "I Hate the Commie Scum" is the title of a song performed by the skinhead band Fortress.

3. "This Land Is Ours" is the title of a song performed by the skinhead band No Remorse.

4. "Kings of the Jungle" is the title of a song performed by the skinhead band The Last Resort.

5. "In Pride We Stride" is the title of a song performed by the skinhead band Bound for Glory.

REFERENCES

Becker, H. 1963. *Outsiders.* NY: Free Press.

Bjorgo, T. and R. Wilte (eds). 1993. *Racist Violence in Europe.* New York: St. Martin's.

Brake, M. 1985. *Comparative Youth Culture: The Sociology of Youth Culture and Youth Subcultures in America, Britain, and Canada.* London: Routledge and Kegan Paul.

Connell, R. 1987. *Gender and Power.* Stanford: Stanford University Press.

Coplon, J. 1988. "Skinhead Nation." *Rolling Stone,* December 1, pp. 59–65, 94.

Dunk, T.W. 1991. *It's a Working Man's Town: Male Working Class Culture in Northwest Ontario.* Montreal and Kingston: McGill-Queen's University Press.

Dunning, E. 1986. "Sport as a Male Preserve: Notes on the Social Sources of Masculine Identity and Its Transformation." In *Quest for Excitement,* N. Elias and E. Dunning (eds.). New York: Blackwell, pp. 267–84.

Kimmel, M. and M. Messner. 1992. *Men's Lives.* Toronto: Maxwell McMillan.

Marshall, G. 1991. *The Spirit of '69: A Skinhead Bible.* Doonan, Scotland: S.T. Publishing.

McRobbie, A. and J. Garber. 1976. "Girls and Subcultures." In *Resistance through Rituals.* S. Hall and T. Jefferson (eds.). London: Hutchinson, pp. 209–22.

Messner, M. 1992. *Power at Play.* Boston: Beacon Press.

Morgan, D. 1992. *Discovering Men.* New York: Routledge.

Relin, D. 1989. "Harvesting Young People's Hate." *Scholastic Update,* April.

Rutherford, J. 1992. *Men's Silences: Predicaments in Masculinity.* New York: Routledge.

Veltmeyer. H. 1986. *Canadian Class Structure.* Toronto: Garamond.

Young, K. 1993. "Violence, Risk, and Liability in Male Sports Culture." *Sociology of Sport Journal* 10:373–96.

11

Fashioning Gender Identity

SPENCER E. CAHILL

Based on observations in public places, Cahill notes the importance of socialization to the establishment of gender identity in children. People are taught, from their earliest years, how to recognize, interact with, and behave in ways that are appropriate to their gender. Many sociologists have argued that individuals' gender, as Cahill illustrates, is not socially lodged in their genitalia but in the way they act, look, and dress. Learning gender is a process that occupies a significant portion of parent-child interaction and is clearly visible among young age peers as well. You may have encountered this debate as nature vs. nurture, or the extent to which we are determined by biological predisposition or socially learned environmental conditions. While the answer may lie as a combination of the two, a sociological perspective generally looks more at nurture as a determining factor. Can you relate this to your own experiences as a child? Can you recall ways in which your parents may have dressed you, decorated your room, or treated you that helped to paint a gender image for you? Can you imagine a world that is truly androgenous, one that pays no attention to sex differences in people? How might that change society?

The transsexual . . . Agnes changed her identity nearly three years before undergoing sex reassignment surgery. After five years of covertly consuming synthetic estrogens (Garfinkel 1967, pp. 285–288), two months of dieting, and much rehearsal, Agnes transformed herself into a female on a late August day in 1956.

> Taking a room in a downtown hotel, she changed into female clothes and went to a local beauty shop where her hair, which was short, was cropped and rearranged in the Italian cut Sophia Loren had made popular. (Garfinkel 1967, p. 145)

On her return home by bus that evening, Agnes was the proud recipient of several soldiers' attentions. Although still haunted by her past life as a male and the secret of her masculine genitalia, Agnes was a female for many if not most intents and purposes from that day forward.

Like the rest of Agnes's story, the manner of her identity transformation is of more than passing interest to students of social life. Agnes's masculine genitalia did not prevent her from being seen as a female nor did her pharmaceutically produced feminine form automatically make her a female in other's eyes. Rather, she secured her claim to a female identity by changing her clothing and hairstyle. The more general sociological lesson is obvious. In everyday social life, the identification of people as male or female has less to do with anatomical characteristics than with what Goffman (1963, p. 25) termed "personal front"—"the complex of clothing, make-up, hairdo and other surface decorations" the individual carries on his or her person. That is the complex of materials out of which male and female identities are commonly fashioned in our society. By implication, they may also be among the materials out of which self-identified males and females are biographically fashioned.

This article concerns the biographical fashioning of self-identified males and females in early childhood. During my 18 months as a volunteer staff member of both a university-affiliated and a parent cooperative preschool in Southern California, I observed the children who attended those schools under a variety of circumstances and interviewed a number of them and their parents. In addition, I subsequently recorded informal discussions with the parents of other children in fieldnotes. The following empirical exploration of the contributions of appearance management to young children's gender socialization is based upon these fieldnotes and others' observations and findings. . . .

ESTABLISHING GENDER IDENTITIES

In our society and probably most others, an infant's external genitalia are visually inspected moments after birth, and, in most cases, he or she is immediately identified as a boy or a girl. . . . [Although] parents are subsequently reassured of their infant's ascribed sex-class identity every time they change the infant's diapers or bathe him or her . . . others who have contact with the infant seldom have the benefit of that anatomical reassurance. Moreover, as the often reported "Baby X" studies suggest, we are not very adept at ascertaining the ascribed sex-class identity of a clothed infant in the absence of other identifying information. The adult participants in those two studies played with one of three infants who were dressed in either "a yellow jumpsuit" (Seavey, Katz and Zalk 1975, p. 105) or an undershirt and diapers (Sidorowicz and Lunney 1980, p. 70), and some were given no hint as to the sex-class identity of the infant with whom they played. When subsequently asked, the overwhelming majority of these uninformed participants misidentified the infant's ascribed sex-class placement. Such presumed indications of masculinity and femininity as body shape, physical strength (Seavey, Katz and Zalk 1975, p. 107) and frequency of smiling (Sidorowicz and Lunney 1980, p. 71) proved unreliable.

Yet . . . we implicitly consider an infant's ascribed sex-class identity as prescribing how we should view and treat him or her. When, for example, college students were shown a videotape of an infant crying in response to the opening and closing of a jack-in-the-box, those who had been told that the infant was a boy attributed the crying to anger while those who had been told the infant was a girl reported that "she" was frightened (Condry and Condry 1976). It would seem that we respond not so much to infants but to sex-class identified infants. That is apparently why, despite the obvious risk of clever retorts at our expense, we sometimes ask an infant's accompanying caregiver the literally ambiguous question: "What is it?" We are not simply asking whether the infant is a boy or a girl but thereby also requesting guidance in how we should respond to and talk about him or her. It is seldom necessary to request such guidance, however.

For the most part, parents and other caregivers in our society silently announce their infants' sex-class identities "to whom it may concern" by draping and decorating infants in what might best be termed "sartorial symbols" of sex-class identities. They often color code infants in terms of the traditional masculine blue and feminine pink, commonly dress them in miniaturized versions of adults' sex-class associated costumes, and sometimes even tape bows to female infants' hairless heads. Such conventional, sartorial symbols of sex-class identities enable anyone who comes into contact with an infant to immediately identify the infant as a boy or a girl. In Gregory Stone's (1962, p. 106) phrase, they "invest" the infant with a sex-class identity.

Moreover, the sartorial investiture of infants with sex-class identities also serves indirectly to invest them with presumed male or female human natures. To borrow from Stone (1962, p. 106), "the responses of the world toward the child are differently mobilized" depending on whether there is a bow taped to or a baseball cap resting upon the child's head. For example, as the findings of [a] previously mentioned stud[y] suggest, we tend to view an infant who has a bow taped on her head as . . . frightened rather than angry when she cries. In contrast, we tend to view an infant who is wearing a baseball cap as . . . angry rather than frightened when he cries. Because of these divergent views of differently dressed infants, we tend to treat them differently and thereby encourage them to behaviorally express their presumed male or female human natures. The psychiatrist Robert Stoller (1968, pp. 62–63) once observed that "one can see evidence" of children's "unquestioned femininity or masculinity" by the time they begin to walk. That may be so, but much effort goes into producing this evidence not the least of which is the effort devoted to the sex-class management of infant's personal fronts. It is because of such efforts that we look for evidence of infants' masculinity or femininity and act so as to insure that they will provide behavioral evidence of such presumed male or female human natures.

RECOGNIZING GENDERED IDENTITIES

Although young children may be behaviorally expressing their presumed masculinity or femininity by the time they begin to walk, they are undoubtedly unaware that they are doing so until somewhat later in their biographies. Before they can appreciate the gender expressive significance of their behavior to others, they must first learn that their social environment is populated by two distinct categories of persons. It is not until they begin to acquire their native language that they start to learn this fundamentally important lesson about the world into which they were born.

Young children's exposure to the everyday usage of such identifying verbal labels as "mommy" and "daddy," "girl" and "boy," and "lady" and "man" encourages them to sort people into sex-class related categories (Cahill 1986, pp. 299–302). Although it may be some time before they understand that such two-term collections of identifying verbal labels all point to a single, underlying system of dichotomous classification, children as young as two years of age identify clothed individuals in photographs as "mommies," "daddies," "boys," and "girls" with a high degree of accuracy. . . . (Thompson 1975). However, it is doubtful whether young children would do so unless there were obvious perceptible similarities among and differences between these categories of persons. As a number of students of language acquisition have concluded . . . perceptible similarities are the most important determinates of young children's categorical applications of their rudimentary vocabularies.

It seems that as children begin to acquire their native language they develop tentative hypotheses about the criteria on which others' application of sex-class related identifying terms is based. They then empirically test those hypotheses, heed others' responses to their own applications of sex-class related identifying terms, and thereby acquire a practical understanding of common associations between various aspects of personal appearance and sex-class identification. For example, the following occurred on a preschool playground. I (C) was sitting on the side of a sandbox, and a 35-month-old boy (S) was standing in between my legs. He reached up and tugged my beard.

S: That daddy! That daddy!

C: My beard?

S: Yeah. That daddy.

I later learned that this boy's father had been clean-shaven since before the boy's birth. Thus, rather than indicating that his "daddy" and I shared this perceptible characteristic, the boy was apparently testing his hypothesis that a beard was a sign of "daddiness," of membership in that class of persons called "daddy."

As might be expected, some individuals' personal fronts are confusing to young children in this regard. For example, I (C) was holding a 39-month-old girl (K) when a male preschool teacher (J) with a full beard and mid-back length hair which was gathered together into a "pony tail" approached. The young girl looked at me and then at the teacher.

K: You a boy. He a girl 'cause got a pony tail.

C: Oh yeah?

K: (looking at J) You got a pony tail.

J: Yes.

The young girl then looked back at me and grinned. While this girl seemed to recognize that the teacher was not "really" a girl, she obviously considered a "pony tail" a sign of "girlness." Indeed, she may well have been attempting to elicit a response from either the teacher or me which would clarify the confusing sex-class identifying implications of the teacher's personal front.

For the most part, however, our sex-class related management of both our own and our children's personal fronts does heighten the perceptual similarity of males and of females and the perceptual dissimilarity between males and females. It is primarily these perceptual similarities and differences that direct young children's application of sex-class related identifying terms in our society. . . . Although adults sometimes do instruct young children about the defining anatomical characteristics of males and females, those instructions are often more confusing than enlightening in a society in which bodies are typically clothed. For example, I was once approached by a 37-month-old girl on a preschool playground who informed me "you a girl." When I asked why I was a girl, she replied: "Cause no got penis." This girl was apparently applying a recently learned but misleading lesson when no one has a visible penis as is commonly the case in our society.

Like this girl, children apparently do take adults' instructions regarding the sex-class identifying implications of anatomical characteristics to heart, but those instructions are simply of little practical utility or significance to them. When, for example, four- to six-year-olds were asked what was the most important consideration in deciding whether someone was a boy or a girl, many referred to the genitals, yet hair length had the greatest influence on their sex-class related identifications of "anatomically correct" dolls with different body shapes and wigs (Thompson and Bentler 1971). We adults may implicitly assume that anatomical characteristics are the most obvious grounds for sex-class identification, but that is not obvious to young children as the following anecdote dramatically illustrates.

> A colleague's four-year-old son and his father enrolled in a father-son swimming class at the local YMCA. The participants swam in the nude, but some of the fathers wore bathing caps. On the way home from the first session, the boy asked his father why so many "women" had attended the class. When his father inquired "what women," the boy replied: "You know, in the hats."

Regardless of what adults may tell young children, they know that it is bathing caps, hairstyle, clothing and other surface decorations which make someone either a mommy or a daddy, a woman or a man, or a boy or a girl in everyday social life. That is the lesson they learn through observation and practical experimentation with the identity transforming power of appearance management.

EXPLORING GENDERED IDENTITIES

Soon after children acquire their native language, as George Herbert Mead first suggested, they start behaviorally to explore the social identities or "roles" which are implicitly encoded in the everyday usage of that language. In Mead's (1934, pp. 150–151, emphasis added) words, the play of young children . . . is play *at* something. A child plays at being a mother, at being a policeman: that is, it is taking different roles.

Moreover, as Stone (1962, p. 109) noted some years later, this role playing commonly involves "dressing out" of the roles or social identities which others consider the child's own and "dressing into" those which he or she is temporarily assuming. For example, the younger children in the preschools at which I observed often assumed the identities of so-called "superheroes" such as "Superman," "Batman," and "Wonderwoman." Appearance management was an integral part of this role playing. The children would fashion a "superhero" cape out of paper and tape or by tying the sleeves of their jackets around their necks. When the materials necessary to fashion such a cape were not available, the children would protest that they could not play "superhero" despite reassurances from teachers and other adults to the contrary. Moreover, children typically would not answer to their given names when wearing one of these makeshift capes but only to one of the identifying terms associated with a "superhero" cape. To adult eyes, these children may have only been playing, but it seems in their own eyes they were magically transforming themselves into different kinds of persons by altering their personal fronts.

It is particularly notable in the context of this analysis that the younger children at these preschools paid little attention to inconsistencies between the gendered identities which they sartorially assumed and their ascribed sex-class identities. For example, it was not uncommon for young boys to assume the identity of "Wonderwoman" nor for girls to assume the identity of "Superman." It was also not uncommon for these children to engage in what adults call "cross-dressing." Most did so occasionally, and some did so routinely as illustrated by the following excerpt from an interview of a mother (M) of a 35-month-old girl.

M: You know my daughter S——has this short hair cut and people are always saying what a nice boy she is.

C: Does she get upset?

M: No. In fact, some days she comes down and says that she wants to be a boy today. It's amazing how she already knows about clothes and all. When she wants to be a boy she puts on jeans and finds dirty socks. Not dirty, but older white socks that are . . .

C: Dingy?

M: Yeah, dingy.

Like Agnes, this young girl was already a sophisticated, practical sociologist. She knew that by altering her personal front she could transform her sex-class identity in others' eyes, and as Cooley (1922) reminds us, our identities are little more than reflections in others' eyes.

However, those with whom a child has regular contact are typically informed of his or her ascribed sex-class identity. Although they may temporarily indulge the child's sartorial assumption of gendered identities which are inconsistent with his or her ascribed sex-class identity, in most cases they will eventually discourage him or her from doing so. For example, the following occurred in a preschool classroom. Two 40-month-old boys (S,T) were playing doctor when a 38-month-old boy (E) who was wearing a "dress-up" dress and high heeled shoes approached.

E: Fix me (pointing to the unfastened zipper in the dress).

S: You're not a girl.

T: You're a boy.

S: Those are girl things.

E hurriedly slips out of the dress and kicks off the shoes.

Through experiences such as this, most children quickly learn that the alchemy of appearance management is limited. Ultimately, it does not enable them to escape the sex-class identity with which others have and continue to invest them.

In addition to discouraging "cross-dressing," others also encourage young children to "dress into" their ascribed sex-class identity. For example, a preschool aide (A) encountered a 39-month-old girl (S) who was dressed in a bright yellow sunsuit bordered with lace and matching sun bonnet. The girl snapped the straps of the sunsuit with her thumbs and looked at the aide.

S: I got lace.

A: You're all girl aren't you S?——You're so sweet.

At other times this encouragement of children's sartorial expression of ascribed sex-class identities takes the form of invidious comparisons as the following illustrates.

> A 43-month-old and 37-month-old girl who are both wearing summer dresses are sitting on a preschool playground. Another 37-month-old girl who is dressed in jeans and a smock is standing nearby. A preschool aide walks by and addresses the two girls in summer dresses. "There's a couple of pretty girls." The other girl pulls her smock away from her body, looks at the aide, and remarks: "My dress." In response, the aide asks: "K——, why doesn't your mom ever put you in a real dress?"

In a variety of ways, therefore, both adults and older peers implicitly instruct young children that they are obliged to manage their personal front so that it clearly announces their ascribed sex-class identity. Others thereby implicitly inform young children that they have little choice but to embrace that identity as their own. . . .

EMBRACING GENDERED IDENTITIES

Some years ago, Nelson Foote (1951, p. 7) noted that self-identification involves both appropriation of and commitment to an identity. He then observed that

> . . . the compulsive effect of identification upon behavior must arise from absence of alternatives, from unquestioned acceptance of the identities cast

> upon one by circumstances beyond his control (or thought to be). (Foote 1951, p. 19)

In most cases, others' responses to a child's experimentation with the identity transforming power of appearance management prevents the child from escaping his or her ascribed sex-class identity. From the child's perspective, that identity is cast upon him or her by circumstances that are beyond his or her sartorial control. Thus . . . ascribed sex-class identities do begin to have a "compulsive" effect on most children's behavior by the end of the preschool-age period of their biographies. Having been fashioned into self-identified males and females by others, they begin to fashion themselves into gendered persons.

One of the most obvious indications of older preschool-age children's commitment to their ascribed sex-class identity is their unswerving dedication to its sartorial expression. For example, a 60-month-old boy who had unusually long hair which was often gathered together into a pony tail attended one of the preschools at which I observed. One morning the boy's mother visited his preschool teacher to protest the school's dress code. When the boy's mother started to gather his hair into a pony tail that morning, he told her that his teacher had said that he could not wear a pony tail at school anymore. The teacher informed the mother that she had never said anything of the kind. The boy later admitted that he simply did not want "girl's hair" anymore. On another occasion at the same preschool, I observed a 55-month-old boy refuse a woman's offer to help him put on a necklace because, in his words, it was "for girls." When the woman told him that he could wear the necklace and pretend that he was a king, he again refused. He emphatically reminded her that he was not a "king" but a "boy." Perhaps kings could wear necklaces as the woman suggested, but this boy was well aware that doing so was no way to confirm his identity as a boy.

Many mothers have also told me of their frustration with their preschool age daughters' sudden refusal to wear slacks and insistence upon wearing dresses regardless of the weather or impracticality of engaging in certain activities when doing so. However, girls of this age do not seem as concerned as boys about avoiding sartorial symbols of the other sex-class identity. For example, I often saw the older girls at the preschools at which I observed wearing one of the boy's caps or jackets but never saw an older boy wearing a girl's hat or jacket. . . .

For whatever reason, it seems that we consider a greater diversity of personal fronts compatible with a female identity than with a male identity in this society. A girl with short hair who is wearing jeans and a flannel shirt will commonly be recognized by others as a girl if she also wears earrings or a necklace. That is exactly why boys must not wear earrings or a necklace if they hope socially to confirm their identity as a boy. Men may wear an earring or a necklace or a pony tail and still socially confirm their male identity, but boys do not have that luxury. Excluding the genitalia which are typically concealed, young boys and young girls are commonly indistinguishable from one another except for some small, sartorial badge of female identity. Thus, young boys must vigilantly avoid any and all sartorial symbols of female identity in order socially to confirm their identities as boys. . . .

ALIGNING APPEARANCE WITH GENDERED IDENTITIES

Although most children are clearly committed to their ascribed sex-class identities by the end of the preschool-age period of their biographies, they do not simply conform to conventional standards of sex-class related appearance management as a result. Rather, they continue to experiment with the management of their personal fronts while simultaneously attempting to "align" (Stokes and Hewitt 1976) the resulting sartorial expressions with their presumed masculinity or femininity. For example, I observed a 55-month-old boy (S) slip into a red "dress-up" dress in a preschool classroom and then walk over to a 51-month-old girl (M).

S: TA-DA-DA (in an affected high pitched voice).

M: Silly.

S: TA-DA-DA.

M: You're silly.

S: DA-TA-TA (in an affected low pitched voice while slipping out of the dress) SUUUPerman!

As M's comments indicate, S's behavior while wearing the dress was accountably "silly" or playful and, consequently, the sex-class identifying implications of the dress were not taken seriously. As if that were not enough to establish the expressive unseriousness of the dress, S then emerged from that feminine cocoon in an unmistakably masculine form. He thereby maintained his social claim to the identity of boy despite having worn the dress.

Children's experimentation with their personal fronts also takes unexpected expressive turns requiring improvisation in order to confirm their presumed masculinity or femininity. For example, I overheard the following playground conversation between two five-year-old girls (F, R) and a five-year-old boy (N), all of whom were painting their faces with watercolors.

R: I have to put on make-up 'cause I'm on a date.

F: I'm wearing make-up 'cause I'm going to the doctor.

N: I got mine on 'cause it's Halloween.

Once the girls defined the watercolors as "make-up," the boy apparently needed an explanation or account to neutralize the sex-class identifying implication of its use. His solution was ingenious. He declared that it was the one day of the year on which males can wear make-up with impunity. It is apparently through experiences such as these that children gain an increasingly sophisticated, practical understanding of the elasticity of and points at which conventional standards of sex-class related appearance management snap back with an identity undermining force. . . .

CONCLUSION

Although primarily suggestive, the preceding examination of young children's gender socialization . . . indicate[s] that appearance management is a principal mechanism of [gender identity acquisition]. Sex-class related appearance management socially invests infants with sex-class identities and, thereby, with male and female human natures. It also promotes young children's sex-class identification of both others and themselves. In addition, others' responses to children's experimentation with the identity transforming power of appearance management encourages them to embrace behaviorally their ascribed sex-class identities. They consequently begin to align their sartorial expression with . . . conventional . . . standards of sex-class related appearance management and to manage their personal fronts so as to announce clearly their ascribed sex-class identities to others. [They thereby become gendered persons to themselves as well as to others.]

REFERENCES

Cahill, S. 1986. "Language Practices and Self-Definition: The Case of Gender Identity Acquisition." *The Sociological Quarterly* 27:295–311.

Condry, J. and S. Condry, 1976, "Sex Differences: A Study of the Eye of the Beholder." *Child Development* 47: 812–819.

Cooley, C. H. 1922. *Human Nature and Social Order.* New York: Scribner's.

Foote, N. 1951. "Identification as the Basis for a Theory of Motivation." *American Sociological Review* 16: 14–21.

Garfinkel, H. 1967. *Studies in Ethnomethodology.* Englewood Cliffs, N.J.: Prentice-Hall.

Goffman, E. 1963. *Behavior in Public Places.* New York: Basic Books.

Mead, G. H. (1934) 1962. *Mind, Self, and Society.* Edited by C. Morris. Chicago, IL: University of Chicago Press.

Seavey, C., P. Katz, and S. R. Zalk, 1975. "Baby X: The Effect of Gender Labels on Adult Responses to Infants." *Sex Roles* 1: 103–109.

Sidorowicz, L. and G. S. Lunney. 1980. "Baby X Revisited." *Sex Roles* 6: 67–73.

Stokes, R. and J. Hewitt, 1976. "Aligning Actions." *American Sociological Review* 41: 838–849.

Stoller, R. 1968. *Sex and Gender.* New York: Science House.

Stone, G. 1962. "Appearance and the Self" Pp. 86–118, in *Human Behavior and Social Processes,* edited by A. Rose. Boston: Houghton Mifflin.

Thompson, S. 1975. "Gender Labels and Early Sex Role Development." *Child Development* 46: 339–347.

——— and P. M. Bentler, 1971. "The Priority of Cues in Sex Discrimination by Children and Adults." *Developmental Psychology* 5: 181–185.

12

Leaving Home for College: Expectations for Selective Reconstruction of Self

DAVID KARP,
LYNDA LYTLE HOLMSTROM,
PAUL S. GRAY

The move from high school to college is a highly significant one, involving people leaving the security of their long-term residence and going out on their own for, usually, the first extended period of time. This offers people the chance to leave behind all the conceptions people have of them and establish a new version of themselves with people that they will soon meet. David Karp, Lynda Lytle Holmstrom, and Paul S. Gray call the period of transition a liminal one, where people have left their old-town identities behind but not yet forged complete new ones. This movement creates some nervousness, as people worry about how things will change, what they will miss and how they will be missed, how their new surroundings resonate with the identity they perceive as theirs, and how they want to re-create themselves. All of these considerations command significant reflection, and the identity transformation arises out of these thoughts and the interactions that subsequently unfold. In contrast to more sudden transformations that arise as the result of unexpected events, the move to college independence is one that permits significant anticipation, planning, and enacting. If you are living away from home to go to college, does this article resonate with you? In terms of your identity, what did you leave behind, what did you bring with you, and how have you tried to change your identity in your new surroundings? If you are a first year college student living away from home, you probably are going through these transformations right now.

In their important and much discussed critique of American culture, *Habits of the Heart* (1984), Robert Bellah and his colleagues remark that American parents are of two minds about the prospect of their children leaving home. The thought that their children will leave is difficult, but perhaps more troublesome is the thought that they might not. In contrast to many cultures, American parents place great emphasis on their children establishing independence at a relatively early age. Still, as Bellah's wry comment suggests, they are deeply ambivalent about their children leaving home. The data presented in this paper, part of a larger project on family dynamics during the year that a child applies for admission to college, show that such ambivalence is shared by the children. Our goal here is to document some of the social psychological complexities of achieving independence in America by analyzing the perspectives of 23 primarily upper-middle-class high school seniors as they moved through the college application process and contemplated leaving home.[1]

Of course, a great deal has been written about the internal conflict that surrounds any significant personal change (most obviously, Erik Erikson 1963, 1968, 1974, 1980; see also Manester 1977; O'Mally 1995). Although researchers have attended to the phenomenon of "incompletely launched young adults" (Heer, Hodge, and Felson 1985; Grigsby and McGowan 1986; Schnaiberg and Goldenberg 1989), little has been written about how relatively sheltered, middle- to upper-middle-class children think about "leaving the nest." Leaving home for college is perhaps among the greatest changes that the economically comfortable students we interviewed have thus far encountered in their lives. For them, going to college carries great significance as a coming-of-age moment, in part because it has been long anticipated and not to do so would be unacceptable from a normative stand point. Literature on students who "beat the odds" by going to college suggests that this is also an important transition for them, but one carrying fundamentally different meanings. Unlike the middle- or upper-middle-class students we interviewed, who are trying, at the least, to retain their class position, students arriving at college from less privileged backgrounds must confront wholly new cultural worlds (Rodriguez 1982; Smith 1993; Hooks 1993).

The 23 students with whom we were able to complete interviews simply assumed, as did their parents, that they would go to college.[2] Among the 30 sets of parents, all but four individuals had attended college (and two received some different training beyond high school). All of the adults, however, felt strongly about the necessity of college attendance for their children. One of the four who did not go to college, a self-made and extraordinarily successful entrepreneur, did offer some reservation about the utility of an education in the rough and tumble "real world." Even so, both he and his wife were highly invested in getting their son into a prestigious college. While all of the children knew their parents' expectations and fully expected to meet them, we did speak with two students who had some misgivings about whether they really wanted to go to college. Like their counterparts in our sample, these students knew they would go, but still entertained private doubts about their interest in and motivation for college work.[3]

What does it mean to become independent of one's parents, family, and high school friendship groups? As Anna Freud noted, "few situations in life are more difficult to cope with than the attempts of adolescent children to liberate themselves" (Bassoff 1988, p. xi). Young people are ambivalent regarding independence;

it is hard to break away. Their ambivalence embodies both symbolic and pragmatic dimensions. Symbolically, independence is the desired outcome of a necessary process of differentiation (Blos 1962). The task for adolescents is "to find their own way in the world and develop confidence that they are strong enough to survive outside the protective family circle" (Bassoff 1988, p. 3). To establish their own identity and sense of purpose, ". . . they need to wrench themselves away from those who threaten their developing selfhood" (Bassoff 1988, p. 3; see also Campbell, Adams, and Dobson 1984; Katchadourian and Boll 1994). However, independence also has a pragmatic side. In college, young people can "start over"; they can make new friends, establish intimate relationships, and develop the skills and knowledge to help them become self-supporting adults. "But the truth is that they are not sure they can take care of themselves or that they want to be left alone" (Bassoff 1988, p. 3). . . .

IDENTITY AFFIRMATION, IDENTITY RECONSTRUCTION, AND IDENTITY DISCOVERY

While the students in this study anticipated college as a time during which they would maintain, refine, build upon, and elaborate certain of their identities, they also anticipated negotiating some fundamental identity changes. The students saw college as the time for discovering who they *really* were. They anticipated finding wholly new and permanent life identities during the college years. In addition, they believed that going to college provides a unique opportunity to consciously establish some new identities. Repeatedly, students described the importance of going away to college in terms of an opportunity to discard disliked identities while making a variety of "fresh starts." Their words suggest that college-bound students look forward to re-creating themselves in a context far removed (often geographically, but always symbolically) from their family, high school, and community. The immediately following sections attend, in turn, to how upper-middle-class high school seniors (1) anticipate change, (2) strategize about solidifying certain identities, (3) evaluate identities they wish to escape, and (4) imagine the kinds of identities they might discover during college. . . .

Anticipating Change

Along with such turning points as marriage, having children, and making an occupational commitment, it is plain that leaving for college is self-consciously understood as a dramatic moment of personal transformation. The students with whom we spoke all saw leaving home as a critical juncture in their lives. One measure of consensus in the way our 23 respondents interpreted the meaning of leaving home is the similarity of their words. Students used nearly identical phrases in describing the transition to college as the time to "move on," "discover who I really am," to "start over," to "become an adult," to "become independent," to "begin a new life." The students, moreover, explicitly saw going to college as the "next stage" of their lives. That they were fully prepared for a major life change is evident in comments of the following sort:

> I'm ready [to leave home]. I've been ready for a year. I want to go [laughs]. I think it's time and I think I'd probably go stir crazy if I had to live here another year, just because I'm ready for more independence than I'm getting here. . . . You get to a point where you really don't need people following your every move and making sure that you don't mess up. You want to make your own decisions and I think it's time for that. (African-American female attending a public school)

While all the students interviewed recognized the need for change and were looking forward to it, their certainty about the appropriateness of moving on did not prevent them from feeling anxiety and ambivalence about the transition to college. Theirs is an anticipation composed of optimism, excitement, anxiety, and sometimes fear.

> [I'm] starting the rest of my life. I mean, deciding what I'm going to do and figuring out my future. I mean, that's one thing I'm looking forward to, but it's also one thing I'm not looking forward to. I have mixed feelings about that. It's exciting to figure out your future. In another sense it's scary to have all of the responsibility. (White male attending a public school)

These comments suggest that the prospect of leaving home generates an anticipatory socialization process characterized by multiple and sometimes contradictory feelings and emotions. Students long for independence, anticipate the excitement that accompanies all fresh starts, but worry about their ability to fully meet the challenge. . . .

Students are aware that going to college means changes in their friendship patterns, and, surely, part of the excitement of college relates to the anticipation of making new friends. For most students, however, the idea of making new friends is also filled with uncertainty. They are excited about new friends but worry about leaving their old friends. They know they need to make a social life for themselves in the new campus environment but worry that perhaps they will not. Like so many areas of their lives at this stage, there is a tension between desiring both change and stability. While they may worry about leaving their old friends, ultimately, they know they must move on.

> We're all kind of concerned of what it will do to our friendships and relationships, but still, I mean, everybody's excited about it. . . . Everybody I know is, you know, moving on. (Bi-racial male attending a public school). . . .

Affirming Who I Really Am

The one concrete and critical choice that college-bound students must make is which school, in fact, to attend. This decision is often an agonizing one for both students and their parents and involves very high levels of "emotion work" (Hochschild 1983). The significance of making the college choice and the anxiety that it occasions go well beyond questions of money, course curricula, or the physical amenities of the institutions themselves. What makes the decision so difficult is that the students know they are choosing the context in which their new identities will be established, . . . The fateful issue in the minds of the students is whether people with their identity characteristics and aspirations will be able to flourish. Consequently, it is not surprising that the most consistent and universal pattern in our data is the effort expended by students to find a school where "a person like me" will feel comfortable. . . .

In the most global way, prospective students were searching for a place where the students seemed friendly. On several occasions, students remarked that they were turned on or off to a school because their "tour guide" was either really nice or not friendly enough. One student was smitten with a particular school because of the unusual friendliness of the students:

> At some other colleges like you get a little nod or a hi or something, but usually they steer clear. At Boston College, one guy actually came up to me and he was like talking to me, and . . . I was like shocked. . . . We were just standing there and he like came up and he was talking to me because he recognized my [high school] jacket. . . . And we started up this whole conversation. . . . I was just shocked at how nice all these people were. . . . In like the little cafeteria I was just standing there and they're like, "Oh, are you lost?" I mean the people there have been the friendliest I've seen anywhere. (White female attending a private high school). . . .

In contrast to the students-like-me theme, an interesting sub-set of seniors expressed a strong interest in diversity. These students not only wanted to meet new people, but different kinds of new people. Students who wanted diversity were excited at the prospect of meeting people different from themselves as a critical learning experience. It is important to note that it was primarily the minority students we interviewed who looked for diversity as they contemplated colleges. An Asian student put it this way:

> The more mixed the better. I think interaction with other ethnic and racial groups is very healthy. If possible, I would not mind having, you know, like an Afro-American roommate. I'd love to. (Asian male attending a public school). . . .

While the statements immediately above illustrate that students make careful assessments about the goodness of fit between certain aspects of themselves and the character of different colleges, a dominant theme in the interviews concerned change. Students repeatedly commented that, during their college years, they expected their identities to shift in two fundamental ways. First, they anticipated discovering "who I am" in the broadest sense. Second, they saw college as providing a fresh start because they could discard some of their disliked, sticky identities, often acquired as early as grade school.

Creating the Person I Want to Be

. . . Seen in terms of Erving Goffman's (1959) dramaturgical model of interaction, going to college provides a new stage and audience, together allowing for new identity performances. Goffman notes (1959, p. 6) that "When an individual appears before others his actions will influence the definition of the situation which they come to have. Sometimes the individual will act in a thoroughly calculating manner, expressing himself in a given way solely in order to give the kind of impression to others that is likely to evoke from them a specific response he is concerned to obtain." To the extent that such impression-management is most centrally dependent upon information control, leaving home provides an unparalleled opportunity to abandon labels that have most contributed to disliked and unshakable identities. When students speak of college as providing a fresh start, they have in mind the possibility of fashioning new roles and identities. Going to college promises the chance to edit, to revise, to re-write certain parts of their biographies.

> It's sort of like starting a new life. I'll have connections to the past, but I'm obviously starting with a clean slate. . . . Because no one cares how you did in your high school after you're in college. So everyone's equal now. (White male attending a public high school). . . .

As students described their hopes about college, the theme of "fresh starts" was almost universally voiced. Although, as described earlier, leaving friends behind is difficult, cutting such ties also provides the possibility to re-create oneself. Since adolescence is a time during which a "good" personal image and appearance are deemed critical, it is sensible that a chance to start over is extremely appealing. Leaving home, friends, and community offers students the possibility to jettison identities which are the product of others' consistent definitions of them over many years. Going to college provides a unique opportunity to display new identities consistent with the person they wish to become.

The data presented thus far are meant to convey the symbolic weightiness of the transition from high school to college. Every student with whom we spoke saw leaving home as a critical biographical moment. They see it as a definitive life stage when their capacity for independence will be fully tested for the first time. Some have had a taste of independence at summer camps and the like, but the transition to college is viewed as the "real thing." Their words, we have been suggesting, indicate that they see strong connections among leaving home, gaining independence, achieving adult status, and transforming their identities. Students carefully attempt to pick a college where they will fit in, thus indicating the importance of retaining and consolidating certain parts of their identities (see Shreier 1991). In addition, they believe that they will discover, in a holistic sense, who they "really" are during the college years.

DISCOVERING THE PERSON I WAS MEANT TO BE

. . . By the time they have reached their senior year in high school, most college-bound students must begin to confront, even if in a wholly preliminary way, questions concerning identity. The choice of a college is nearly always the most significant decision these students have had to make. It is a choice, moreover, that presumes at least some consideration of the questions "Who am I and what do I want to do with my life?" Not incidentally, application essays often ask some variant of these questions. This is important to note because such essays represent the institutional perspective of colleges themselves and thus contribute heavily to the expectation that the college years ought to be a time of substantial personal change. Application essays often simultaneously acknowledge students' anticipation of identity change and feed into that anticipation.

A few of the students in our sample had already decided their occupational destinies and were choosing colleges and programs on that basis. However, most of these high school students share an abiding belief that college will be the time for learning who they "really" are and ought to become. Students repeatedly told us that part of the excitement of their college years would be the discovery of who they really are. That students have high expectations about "finding themselves" in college is clear in comments like these:

> I was a little put off because of the bizarre nature of the [essay] question [for one college]. One of them was like "Who are you?" You know, that's part of the reason why I'm going to college 'cause I don't know who I am. . . . The "Who am I [question]" really turned me off, especially at this age in my life. I have no idea who I am. Well, I have an idea, but I'm going to college to answer that question. (White male attending a public high school)

WILL THEY MISS ME?

The students in our sample are emerging from high school brimming with enthusiasm. The imperative to strike out on their own is plain in their comments. However, for high school seniors contemplating the move to college, the idea of becoming more independent also poses immediate and practical problems. Arriving on campus, they will inevitably be challenged to function, to survive, in the absence of parental oversight, advice, and support. The tension between leaving home and maintaining ties with those at home is expressed in the way students think about geography, about how far from home they will go. Many want to be "away," but not too far away. Several mentioned a "two and one-half hour rule"; they would not want to be more than a 150-minute car ride from home.[4] As one senior put it, "I definitely want to stay close enough to home that I can send the laundry, but pretty far enough so that I can be independent."

The family is a social system in which roles are interconnected and interdependent. When a child goes off to college, the system is disturbed and the family will try to adapt to the new circumstances. College-bound seniors worry about this process of adaptation. They speculate that their remaining siblings will miss them, or will be left to face the unremitting attentiveness and concern of parents. They also wonder about prospective changes in their parents' marital relationship. In particular, they are concerned for their mothers, whom they identify as being more invested than their fathers in keeping the family system *status quo ante.* Finally, and most significantly, these late adolescents manifest insecurity about their place in the family, especially now that they are leaving. Several of them remarked ruefully, "I should hope they feel some grief [laughter]!" "I think they'll be lonelier. I hope they will." "They'll miss me, I hope. . . . I hope they feel my presence being gone. . . . They don't have to be, like, mourning my departure, but just a little bit would be nice." It's not that they actually want their parents and siblings to suffer, but missing them would be proof positive that their membership in the family was valued, and that their future place in the family system is assured, in spite of their changing addresses. This insecurity extends to speculation regarding parents' visits to campus and students' future visits home while they are in college. . . .

In many of our conversations, it appeared that the worst thing about going away to college was that the young people would no longer be able to participate in many aspects of family life. However, perhaps no issue symbolizes the worry associated with leaving home as powerfully as pending decisions over space in the household. How quickly one's bedroom is claimed by other members of the family is, for many of these students, a commentary on the fragility of their position. Although Silver (1996) points out that both the home room and college dorm room are used to symbolically affirm family relations, our conversations with students were more focused on the meanings they attached to their bedrooms at home. One senior said, "They always joke around and they say, 'Oh, we're going to make your room into a den.'" Another young woman said, "When I come back home . . . it will still be my bed, but it won't really be my bed." Common sentiments about the recognition of household space are powerfully summarized by the student who made the following statement:

> [My room] it's an awful big issue . . . a confrontation. They want to turn it into a guest room, and it's the only room I have, and I want it to be my room when I come home from college, so that I can have something to come home to, you know? I mean, it's the room I spent most of my life in . . . I don't even want people to stay in it. Don't want anyone to sleep in my bed. . . . When it comes to my room, I don't want a blessed soul inside of it unless it's okayed by me . . . I've created a room that I feel comfortable in, and a room that symbolizes home for me, and if that was changed it would be very difficult, because then it wouldn't feel like I was coming home as much. (White male attending a public high school)

Some of the seniors are beginning to understand that the nature of relations with their parents will be altered forever. They will have much more discretion concerning what to reveal about themselves, and therefore much more control over the impression they choose to give their parents. As one young woman put it, "I will experience a lot of things without them there, so that they won't know that they've happened . . . [unless] I tell them or if they can see a difference in me." Others expressed shared anxieties about personal transformations and the consequent stability of their place in the family constellation. One young woman empathized with a friend's concerns:

> When I come home from college, what if I am different and my parents don't like me [giggle], you know, any more? . . . Or, what if . . . after living away from them, I've changed my personality and my views . . . so they don't appreciate me? But, I don't think that would really happen . . . might not be the same person, but I think my parents at least will be willing to accept that [giggle], and the changes they would probably think are for better anyway, 'cause . . . I think you should change a little after high school. (White female attending a private high school)

What are we to make of these worries, speculations, and musings? College-bound young adults genuinely want to remain attached to their families, even as they are yearning for true independence. Getting into college is understood as a point of departure which has the potential to alter fundamentally their relationship with their family. However, in spite of their worries, most students see the transition to college as a good thing—a positive transformation with life-long consequences. They cannot predict precisely how their relations with parents and siblings will change, but they know for sure that they have initiated a process that will alter the character of these primary relationships. Such knowledge is plainly implicated in the calculus of ambivalence they feel about leaving home:

> It's like, if you want to be treated like an adult, you have to act like an adult. If you want to be treated like a child, act like a child. If you want to be treated like an adult the rest of your life, you've got to start sometime. (White male attending a public high school)

"You've got to start sometime." That, of course, is exactly what they are doing as they embark on their great adventure of self-discovery, into college first and hopefully, thereby, toward full adulthood.

NOTES

1. We used father's occupation as a proxy for social class. We characterized our sample as predominantly upper-middle class. A sampling of the types of fathers' occupations that warrant this description includes: physician, lawyer, professor, administrator, and architect. A few occupations were either higher or lower in status.

2. Either because we could not reach them or because they declined to be interviewed, we did not speak to eight of the 31 students originally included in our sample. The number is 31 because one of the 30 families had twins.

3. One student, who declined to be interviewed, did not complete the college application process during his senior year in high school. He was the only student in our sample who did not anticipate attending college in the year following high school graduation.

4. Distance from home was an important factor for many students. Therefore, one might ask if how students negotiate their

selective identity reconstruction hinges on how far away from home they choose to go. All 23 students were headed to four-year residential colleges. Of the 22 for whom we know the outcome, 27% went to college close to home (under 1 1/4 hour drive), 36% went a relatively easy visiting distance (between 2 and 4-hour drive), and 36% went a considerable distance (requiring a 4-hour drive or perhaps suggesting the possibility of an airplane trip to visit). The expected selective reorganization of the self was remarkably the same for all three groups in most ways. Almost all the students anticipated change, wanted to maintain specific parts of their self-identity, and imagined a future self, no matter what geographic distance they were going. The groups did differ, however, on whether students hoped to jettison certain unwanted identities. Students who wanted to discard some disliked aspect of themselves were most apt to select colleges between two and four hours away and none enrolled in a college within a 1 1/4 hour drive from home. Escaping an unwanted identity presumably would be more difficult for these students if they remained in close proximity to their existing social networks.

REFERENCES

Bassoff, Evelyn. 1988. *Mothers and Daughters: Loving and Letting Go.* New York: Penguin Books.

Bellah, Robert, Richard Madsen, William Sullivan, Ann Swidler, and Steven Tipton. 1985. *Habits of the Heart: Individualism and Commitment in American Life.* Berkeley: University of California Press.

Blos, Peter. 1962. *On Adolescence: A Psychoanalytic Interpretation.* New York: Free Press.

Campbell, Eugene, Gerald Adams, and William Dobson. 1984. "Familial Correlates of Identity Formation in Late Adolescence: A Study of the Predictive Utility of Connectedness and Individuality in Family Relations." *Journal of Youth and Adolescence* 13: 509–525.

Erikson, Erik. 1963. *Childhood and Society.* 2nd ed. New York: W. W. Norton.

———. 1968. *Identity: Youth and Crisis.* New York: W. W. Norton.

———. 1974. *Dimensions of a New Identity.* New York: W. W. Norton.

———. 1980. *Identity and the Life Cycle.* New York: W. W. Norton.

Goffman, Erving. 1959. *The Presentation of Self in Everyday Life.* Garden City, NY: Doubleday Anchor.

Grigsby, Jill and Jill McGowan. 1986. "Still in the Nest: Adult Children Living with Their Parents." *Sociology and Social Research* 70: 146–148.

Heer, David, Robert Hodge, and Marcus Felson. 1985. "The Cluttered Nest: Evidence That Young Adults Are More Likely to Live at Home Now Than in the Recent Past." *Sociology and Social Research* (69): 436–441.

Hochschild, Arlie. 1983. *The Managed Heart: Commercialization of Human Feeling.* Berkeley: University of California Press.

Hooks, Bell. 1993. "Keeping Close to Home: Class and Education." Pp. 99–111 in *Working-Class Women in the Academy,* edited by Michelle Tokarczyk and Elizabeth Fay. Amherst, MA: The University of Massachusetts Press.

Katchadourian, Herant and John Boli. 1994. *Cream of the Crop: The Impact of Elite Education in the Decade After College.* New York: Basic Books.

Manaster, Guy. 1977. *Adolescent Development and the Life Tasks.* Boston: Allyn and Bacon.

O'Mally, Dawn. 1995. *Adolescent Development: Striking a Balance Between Attachment and Autonomy.* Ph.D. dissertation, Department of Psychology, Harvard University, Cambridge, MA.

Rodriguez, Richard. 1982. *Hunger of Memory: The Education of Richard Rodriguez.* Boston: David R. Godine.

Schnaiberg, Allan and Sheldon Goldenberg. 1998. "From Empty Nest to Crowded Nest: The Dynamics of Incompletely-Launched Young Adults." *Social Problems* 36: 251–269.

Schreier, Barbara. 1991. *Fitting In: Four Generations of College Life.* Chicago: Chicago Historical Society.

Silver, Ira. 1996. "Role Transitions, Objects, and Identity." *Symbolic Interaction* 19: 1–20.

Smith, Patricia. 1993. "Grandma Went to Smith, All Right, But She Went from Nine to Five: A Memoir," Pp. 126–139 in *Working-Class Women in the Academy,* edited by Michelle Tokarczyk and Elizabeth Fay. Amherst, MA: The University of Massachusetts Press.

13

The Impact of Pervasive Beauty Images on Black and White Girls' Self-Concepts

MELISSA MILKIE

Not only does our mainstream culture posit health, appearance, and body norms and values as discussed by Miner earlier, but a stream of beauty images is carried through the media that convey models of how young women are expected to look. We largely assume that these media images influence people significantly, but may not understand how this occurs. Milkie uses a combination of qualitative interviews and quantitative measures from a larger study to show that media influences on identity are mediated through others, rather than direct. She offers us a study design comparing the self-concepts of white and black ninth and tenth grade girls to examine this process. This selection shows us how the white girls are negatively affected by the media-transmitted beauty images, a fundamental precursor to obsessive exercise and dieting behaviors as well as eating disorders, while the black girls manage to deflect the internalization and self-application of these images and thereby avoid the consequent negative self-conceptions. Why do you think there are differences by racial group? What does this say about the cultural expectations of these groups? If you are a young woman, how have media images affected your sense of self? If you are a young man, do you see similar issues occurring as for the girls in this study?

Questions of whether and how media influence self-concept—both self-identities and self-evaluations—as well as their impact on beliefs, values and behaviors underlie much media research. First, researchers have shown how the social context of media use is crucial, particularly in that significant others are relevant to the way people interpret and are affected by media.

Second, they have focused on people's power to select and be critical of media content, and thus to discount media messages. Although recent qualitative work has increased our understanding of media processes by revealing the complexities of people's understandings of images, it has done so to the detriment of assessing how media *influence* people. Interpretive researchers view people as powerful in relation to media content, but do not examine this assumption directly.

In this study, I provide a way to bridge perspectives which argue either that media content is powerful or that people are powerful in interpreting media. I do so by analyzing the extent to which people's power to make critical assessments of content (for example, believing that stereotyped portrayals of one's group are unrealistic or unimportant) may prevent that content from negatively affecting the self. I take the case of feminine beauty images in media, and assess the relative power of critical interpretations in countering harmful effects on the self-concept, specifically on self-esteem. To clarify how media can affect people indirectly, I draw on basic principles in social psychology which point to the key role of *others* in self processes, and discuss these in terms of some unique properties of mass media. . . .

I examine the case of pervasive beauty ideals disseminated through mass media, which many suggest are harmful to young women. These images, particularly in regard to body shape, are extremely unlike "real" American women. The gap between the image and the reality has grown in recent years, as the media images have become slimmer and Americans have grown heavier (Wiseman et al. 1992). The in-depth interviews focus on a tangible, explicit embodiment of idealized femininity—girls' magazines—which saturate their target audience (Evans 1990). The images presented therein are also pervasive in other media such as movies and television. I address these broad research questions: How do girls interpret the female image in media, how do they critique it, and how do they perceive its influence? How do girls view peers' interpretations of these images? How important are critical views of the imagery in protecting girls' self-esteem?

RESULTS

Ethnic status sharply differentiated whether girls identified with the images, supposedly intended for and about all adolescent girls. This status created an important filter for social comparisons and reflected appraisal processes, and thus influenced the effectiveness of critical interpretations for shielding harm to self-concept. First I discuss how the magazines and the images they contain were a part of white girls' culture at both the rural and the urban school, but how black girls generally rejected the images as part of their reference group even though they occasionally read popular girls' magazines. Both white and black girls interpreted the images as largely unrealistic; many wanted more normal or more "real" girls in the images.

Reference Groups, Media Interpretation and Criticism of Images

White girls: Peer culture and media images as role models for the reference group. Explicit in cultural products labeled *Seventeen* or *'Teen* is the notion that such products provide images and information relevant to particular groups of people. . . .

Girls' magazines, like other media, were part of the white girls' peer culture in both schools. They helped these girls to assess how well they fit into, or were similar to, their reference group (also see Curne 1997). The magazines gave advice on, and were perceived to help with, girls' concerns about "fitting in" and being accepted by others. In the interviews, for example, many girls stated that any hypothetical girl who does *not* read girls' magazines does not care about others' opinions or is very independent. This comment implies that the information contained in these media pertains to conforming to the "norm" of adolescent femininity.

The respondents considered reading the magazines an enjoyable leisure activity: 95 percent of the white girls surveyed read them occasionally or more often; more than half read them "always." Magazine reading as a part of peer culture, and the relative amount of interaction centered around the cultural products themselves, differed somewhat in the two locales. At the rural school, where girls made slightly higher use of the imagery and evaluated it more positively (data not shown), cliques regularly discussed content during school hours and after school over the telephone. They read the magazines in the lunchroom, the hallways, and the school library, and even during class. Subscribers often shared their magazines with friends, reading them either together or to each other, and passing on copies to those who did not subscribe. Indeed, for rural white girls, a great deal of peer interaction surrounded these magazines. This is not surprising because in rural areas, media may be an important means of understanding the larger world and the variety of people in it, with which the rural dwellers have much less contact (Johnstone 1974; Morgan and Rothchild 1983).

Urban white girls also said they discussed the magazines or particular items in the magazines with friends, but they reported this experience less often. Perhaps because more varied activities are available to the urban girls, magazine reading is less salient. Yet in a quantitative analysis examining how often white girls read the magazines alone or with friends, I found no differences between schools. Indeed, more white girls at the urban school than at the rural school subscribed to at least one girls' magazine (64 percent versus 50 percent).

Black girls: "Maybe if there were more of us in there." Black respondents less often read mainstream girls' magazines, both individually and as a collective activity: 86 percent of the black girls surveyed read them at least occasionally, but only 11 percent always read them. Even though, in recent years, black models have appeared regularly in the four magazines with the greatest circulation, the magazines are perceived as largely for white girls. Most of the black girls read *Ebony* or *Essence,* aimed at black adults and black women respectively, magazines about music directed toward black youth, and hairstyle publications. Thus, in contrast to white girls, these respondents largely regard mainstream girls' magazines as something they do not want to or should not orient themselves toward because they view the magazines as for and about white girls. They define the images as irrelevant to their reference group for this social aspect of the self. This finding is not surprising because, as Collins (1991) and others have observed, mainstream media show an especially distorted image of black females—very thin, and with "whitened" hair and features. Tanya,[1] in responding to how people would understand "girls" if they had only girls' magazines to look at, said:

> I think this is mainly toward . . . white females . . . you really wouldn't see too many black people in here—so if this is all you saw, you'd be kinda scared when you saw one like me or something. (May 9, 1994; urban black girl)

Minority girls were quite critical about the realism of the images. Part of this critique was that normative adolescent femininity was portrayed as white femininity. Although ethnicity differentiated the respondents' use of these mainstream magazines, both African-American and white girls seemed to hold common perceptions about the unreality of the images.

Interpretations and critical interpretations of media images of females. When asked to describe the magazines to a girl who had never seen them, most of the respondents interpreted them as conveying very traditional aspects of femininity, such as appearance and romance. They mentioned fashion, makeup, styles—all related to appearance—and relationships with males. Previous analyses of these magazines (Evans et al. 1991; Peirce 1990) indicate that these reports represent accurate assessments of the content of girls' magazines. Barb's explanation is similar to how most girls described the magazines:

> They're about how girls can do their hair, what's in fashion. They give advice on boys; sometimes they give you advice on your body and stuff like that—how to get in shape. They've got how to do your makeup right, hair—I think I already said that—what's the right jewelry. They talk a lot about stars and stuff like that, they also talk a lot about boys. (May 9, 1994; urban Asian girl)

Secondarily, the respondents reported that the magazines were about girls' "problems." This view is closely related to the above observation. The information presented about appearance and relationships with boys was interpreted by the girls as advice about problems of traditional femininity which they were experiencing or which were common to adolescent females. A minority of the girls described the magazines more broadly as about "teenagers' lives" or "everything." Only two of 60 respondents described the magazines in what might be considered feminist terms, as about girls' "being independent," although these two girls also discussed appearance as an important component of the magazines.

The great majority of the respondents, even those who seldom read the magazines, liked them as a whole or liked certain parts. The girls stated that they read them because they were interesting, entertaining, and informative. An important feature of the girls' enjoyment and interest was learning about themselves and assessing their lives and their problems in relation to their peers. Linda, a grade 9 student, explained this:

> The girls will write in, and you kind of realize they have the same problems as you do . . . you know they [other girls] kinda make you feel like you're not the only one. (May 19, 1994; urban white girl)

Researchers have suggested that one reason why people are critical of media is that the media distort reality and reflect groups in distorted ways. Most of the respondents were critical in that they said media images of girls were not realistic at all, and they made negative comments about the lack of "normal" girls. In general, the respondents indicated that the feminine images in the magazines presented an unrealistic appearance, both in the styles of clothing and in the perfection of their faces, hair, and bodies in comparison with the largely imperfect local girls. A few respondents said that the girls in the magazines were somewhat realistic; sometimes they referred to the pages that focused on "real" girls' problems or compared the images with the most popular or most beautiful girls in the school. The black girls were quite critical of the magazine models' physical appearance in general and tended to be critical about the lack of ethnic diversity or representation.

In discussing how the models looked, the respondents were likely to comment that they were *too* perfect, especially in body shape, weight, hair, facial features, and complexion. Sandra, a grade 10 student, discussed the message sent by the magazines:

> They mainly focus on models . . . they make them look perfect, which nobody is. Makes everyone's expectations really high of theirself, and they don't need it. I don't think they show the true girl. You know, nobody is perfect, and they all have their mistakes, and some of these people look like they never make a mistake. (April 21, 1994; rural white girl)

In fact, many viewed the images not merely as unrealistic, but as artificial. A girl who had recently lost a good deal of weight remarked that some models shown in the magazines have altered their "true" selves:

> I think some of them might be fake. Like get contacts to change their eye color, cake on their makeup, starve themselves. Like they're really not that skinny, but they just starve themselves. (May 25, 1994; urban white girl)

Generally, the respondents disliked the fact that these pervasive media images deviated so much from reality. They remarked, as noted above by Sandra, that the media created an uneasy gap between image and reality. Barb, while looking at the title of a girls' magazine article in front of her, observed that even the so-called "problem" bodies shown in the magazines are perfect:

> Oh, if I read that "Four Weeks to a Better Body," I'd probably . . . these magazines are trying to tell you "Do this and do that." Sometimes they have . . . swimsuits and stuff, and what you can do if you have a problem

> body. If you got a big butt, big chest . . . what to do. And these girls that they are showing don't have that problem. I mean you can tell they don't, and that makes me mad. . . . They say if you got a stick figure, wear a one-piece and . . . colorful and I'm looking at the girl and she doesn't have a stick figure. If you got big hips, if you got a big stomach—she doesn't have it—you can never understand that. (May 9, 1994; urban Asian girl)

In response to open-ended questions about whether they would change anything about the magazines, particularly anything that was emphasized too much or was not included, more than one-third of the respondents specified that the magazines should change the feminine image to be more realistic or "normal." Amy believes that "normal" people are missing from the images:

> One thing I guess would be just more normal people . . . not like the models, but just average. Other people that haven't really had modeling experience. . . . (May 25, 1994; urban white girl)

In sum, most of the respondents regarded media images of females, particularly those which are common in ads or fashion pages, as unrealistic. Many disliked the images for this reason, considered them harmful to themselves or to others and advocated that media producers should alter their products to include more "real," ordinary, or "normal" girls. . . .

Emulation of media images of females verus distance from them. Both the white and the African-American respondents, but especially the white girls, liked the magazines, even though they criticized the lack of realism of the girls pictured therein. The white respondents used the images and ideas in the magazines to assess themselves. They said frequently that they "felt better" or more normal when reading about the problems and experiences of other girls their age. This feeling came from the numerous articles and advice columns that dealt with problems of relationships with boyfriends and family members, peer group pressures, and health, beauty, and fashion issues. The respondents particularly liked to assess themselves in relation to their reference group by taking quizzes that evaluated them on topics such as relationships (e.g., "How Good a Friend Are You?"). These quizzes provide scores that categorize the reader as a certain type of person and explain how she tends to act in situations in comparison with others. Jackie explains why she reads the magazines:

> I guess I like to see what . . . the clothes, like what people are wearing. And like questions-answers, like what people are . . . curious about, and see if I'm the same, I guess. . . . (May 23, 1994; urban white girl)

The minority respondents, in sharp contrast, did not emulate these images nor compare themselves as negatively with the models. Even though most of the black girls occasionally read the mainstream publications, they considered the images less relevant, belonging to "white girls'" culture and not part of a reference group toward which they oriented themselves. Strikingly, 10 of the 11 minority girls (nine black and one Asian-American) said unequivocally that they did not want to be like these girls; one mixed-race respondent (African-American and white) said that she "sometimes" did. The black girls indicated that they did not relate to the images and did not wish to emulate the rigid white beauty ideal.[2] Tamika described why she generally did not read mainstream girls' magazines:

> Well, I don't see a lot of black girls . . . don't see a lot of us . . . maybe if they had more, maybe I could relate to that. I don't know. 'Cause obviously we can't wear the same makeup or get our hair the same way. . . . things like that. So maybe if they had more. (May 11, 1994; urban black girl)

In sum, for the great majority of white girls in both locales, national media images and information about the reference group served as an additional social comparison introduced into the local context. The white girls evaluated their own behavior, problems, emotions, and importantly, physical appearance in comparison with these media others. Even though they knew that the images were unrealistic, the white girls saw themselves as part of the reference group being portrayed, and compared their "problems" with adolescent females' problems. They reported that they often (reluctantly) made social comparisons with the perfect physical appearance of media images because they knew that these images were what "everybody" wants. The minority respondents and a very few of the white respondents did not emulate these feminine images in media, did not bring them meaningfully into peer groups, and seemingly did not make social comparisons unfavorable to themselves.

Criticism, social comparisons and self-evaluations. Although the white girls liked the magazines a good deal and enjoyed finding out that they were "normal" on the basis of other girls' behaviors and problems, many said that they personally or that "girls" in general felt abnormal and inferior in relation to the idealized feminine image. A key influence of the magazines, then, is that the great majority of white respondents said they wanted to look like the girls pictured therein, *even though most saw the images as unrealistic and unattainable.* These girls necessarily experienced relative deprivation because they could not attain the valued image promoted by the pervasive display of this unique part of the reference group. Although they generally understood that the images were unrealistic, the girls perceived that other girls in the school, and especially males, valued such an appearance. Thus it was difficult for critical appraisal of media images to become meaningful in local interaction. . . .

The white respondents made negative social comparisons even while they recognized the media distortion. They indicated that the comparisons were difficult to opt out of and made them or "girls" feel worse about themselves because the girls inevitably looked worse than the glamorized, exceptional females in the media. In quantitative data from the larger group of girls surveyed (N = 210) the white girls felt significantly worse about themselves compared with the images than the minority girls. . . .

The black girls' criticisms of media imagery, in contrast to the white girls', may be effective in reducing the impact of media in this case, because the black subculture as a whole is more critical of mainstream beauty ideals. The black girls in this study, although as concerned about appearance as the white respondents, perceived themselves as better-looking and were more satisfied with their appearance than were the white girls, and their self-esteem was higher. Though the black girls objectively are farther from the mainstream ideals of beauty in skin color, hair style, and weight (see Dawson 1988), they compared themselves more favorably with mainstream media images than did the white girls. Evidence from interviews also indicates that black girls perceived the white ideal as narrow or as less applicable to them. Eliza discussed how minority girls may strive less often to be like the images of girls shown in these magazines:

> This is kind of a stereotype, but more of my white friends than my black friends are into [trying to be like feminine images in magazines]. I mean a lot of them are going on a diet or "I want that body so bad"—I don't know how anybody can be like that. . . . (May 27, 1994; urban black girl)

Thus, because media images were a part of the white girls' peer culture, and because these girls perceived that significant others—other girls and especially boys in their local networks—evaluated them on the basis of media ideals that were nearly unattainable, they were influenced regardless of how strongly they criticized the imagery. Especially important were body shape "norms" in the media, which tended to warp average-weight and thin girls' perceptions of their weight and attractiveness, or at least made them overconcerned about weight at objectively normal, healthy, weights. Even girls who articulated the distorted nature of peers' views of attractiveness seemed to feel compelled to abide by the shifted "norm" of body shape.

The wider range of physical appearances and body shapes that the black girls seemed to accept as good-looking in themselves and others was related to a more inclusive beauty ideal promoted in the "black" media. In addition, the black girls were more tentative about suggesting that males evaluated them on the basis of mainstream (white) media images. Most girls indicated that some males might do so but that others would not. This belief that males (often specified as black) rejected the "whitened" image was important in reducing black girls' negative self-evaluations especially related to body size.

REFERENCES

Collins, Patricia Hill. 1991. *Black Feminist Thought: Knowledge, Consciousness, and the Politics of Empowerment.* New York: Routledge.

Currie, Dawn H. 1997. "Decoding Femininity: Advertisements and Their Teenage Readers." *Gender & Society* 11:453–77.

Dawson, Deborah A. 1988. "Ethnic Differences in Female Overweight: Data from the 1985 National Health Interview Survey." *American Journal of Public Health* 78:1326–29.

Evans, Ellis D. 1990. "Adolescent Females' Utilization and Perception of Contemporary Teen Magazines." Presented at the biennial meetings of the Society for Research on Adolescence. Atlanta.

Evans, Ellis D., Judith Rutberg, Carmela Sather, and Chari Turner. 1991. "Content Analysis on Contemporary Teen Magazines for Adolescent Females." *Youth & Society* 23:99–120.

Johnstone, John W. C. 1974. "Social Integration and Mass Media Use among Adolescents: A Case Study." Pp. 35–48 in *The Uses of Mass Communications,* edited by J. G. Blumler and E. Katz. Beverly Hills, CA: Sage.

Morgan, Michael and Nancy Rothchild. 1983. "Impact of the New Television Technology: Cable TV, Peers, and Sex-Role Cultivation in the Electronic

Environment." *Youth & Society* 15:33–50.

Parker, Sheila, Mimi Nichter, Mark Nichter, Nancy Vickovic, Colette Sims, and Cheryl Ritenbaugh. 1995. "Body Image and Weight Concerns among African-American and White Adolescent Females: Differences and Making a Difference." *Human Organization* 54:103–14.

Peirce, Kate. 1990. "A Feminist Theoretical Perspective on the Socialization of Teenage Girls through *Seventeen* Magazine." *Sex Roles* 23:491–500.

Wiseman, Claire V., James J. Gray, James E. Mosimann, and Anthony H. Ahrens. 1992. "Cultural Expectations of Thinness in Women: An Update." *International Journal of Eating Disorders* 11:85–89.

NOTES

1. All names are fictitious.

2. Many of the black girls mentioned alternative publications focusing on hairstyles for black females. These publications did not contain articles, however—only hairstyle photos and information. (See Parker et al. 1995 for an assessment of African-American girls' culture and appearance.)

14

Meanwhile Backstage: Behavior in Public Bathrooms

SPENCER E. CAHILL ET AL.

Erving Goffman, the founder of the dramaturgical perspective in sociology, first proposed that people behave in ways that they consciously manage in order to foster the most favorable impressions of themselves. They do this by scrupulously adhering to the micro-social norms of individual and interactional behavior, using backstage regions to prepare themselves for their frontstage, public displays. To investigate these norms, Cahill and his students made systematic observations in men's and women's public bathrooms, carefully recording people's behavior patterns. In a selection that is sure to generate both recognition and amusement, Cahill et al. describe and analyze the landscape of public bathrooms, the common rituals found there, and the way people engage in backstage behavior designed to support their appearance on subsequent reemergence into the public domain. The norms upheld in this private yet public setting assert more fully how loyal members of society are to the behavioral guidelines we share and the meanings that people attribute to them. Do you recognize some of the behaviors that these authors describe? Have you ever wondered why you do these things? How is the social order maintained by acting in these ways?

YEARS AGO the anthropologist Horace Miner (1955) suggested, with tongue planted firmly in cheek, that many of the rituals that behaviorally express and sustain the central values of our culture occur in bathrooms. Whether Miner realized it or not, and one suspects that he did, there was more to this thesis than his humorous interpretation of bathroom rituals suggests. As Erving Goffman (1959: 112–113) once observed, the vital secrets of our public shows are often visible in those settings that serve as backstage regions relative to our public performances:

From Urban Life and Culture, reprinted by permission of Sage Publications, Inc.

> it is here that illusions and impressions are openly constructed. . . . Here the performer can relax; he can drop his front, forgo speaking his lines, and step out of character.

Clearly, bathrooms or, as they are often revealingly called, *rest*rooms, are such backstage regions. By implication, therefore, systematic study of bathroom behavior may yield valuable insights into the character and requirements of our routine public performances. . . .

THE PERFORMANCE REGIONS OF PUBLIC BATHROOMS

Needless to say, one of the behaviors for which bathrooms are explicitly designed is defecation. In our society, as Goffman (1959: 121) observed, "defecation involves an individual in activity which is defined as inconsistent with the cleanliness and purity standards" that govern our public performances.

> Such activity also causes the individual to disarrange his clothing and to "go out of play," that is, to drop from his face the expressive mask that he employs in face-to-face interaction. At the same time it becomes difficult for him to reassemble his personal front should the need to enter into interaction suddenly occur. [Goffman, 1959: 121]

When engaged in the act of defecation, therefore, individuals seek to insulate themselves from potential audiences in order to avoid discrediting the expressive masks that they publicly employ. Indeed, over 60 percent of the 1000 respondents to a survey conducted in the early 1960s reported that they "interrupted or postponed" defecation if they did not have sufficient privacy (Kira, 1966: 58).

In an apparent attempt to provide such privacy, toilets in many public bathrooms are surrounded by partially walled cubicles with doors that can be secured against potential intrusions. In fact, public bathrooms that do not provide individuals this protection from potential audiences are seldom used for the purpose of defecation. In the course of our research, for example, we never observed an individual using an unenclosed toilet for this purpose. If a bathroom contained both enclosed and unenclosed toilets, moreover, individuals ignored the unenclosed toilets even when queues had formed outside of the enclosed toilets. In a sense, therefore, the cubicles that typically surround toilets in public bathrooms, commonly called stalls, physically divide such bathrooms into two distinct performance regions.

Indeed, Goffman (1971: 32) has used the term "stall" to refer to any "well-bounded space to which individuals lay temporary claim, possession being on an all-or-nothing basis." Clearly, a toilet stall is a member of this sociological family of ecological arrangements. Sociologically speaking, however, it is not physical boundaries, per se, that define a space as a stall but the behavioral regard given such boundaries. For example, individuals who open or attempt to open the door of an occupied toilet stall typically provide a remedy for this act, in most cases a brief apology such as "Whoops" or "Sorry." By offering such a remedy, the offending individual implicitly defines the attempted intrusion as a delict and, thereby, affirms his or her belief in a rule that prohibits such intrusions (Goffman, 1971: 113). In this sense, toilet stalls provide occupying individuals not only physical protection against potential audiences but normative protection as well.

In order to receive this protection, however, occupying individuals must clearly inform others of their claim to such a stall. Although individuals sometimes lean down and look under the doors of toilet stalls for feet, they typically expect occupying individuals to mark their claim to a toilet stall by securely closing the door.[1] On one occasion, for example, a middle-aged woman began to push open the unlocked door of a toilet stall. Upon discovering that the stall was occupied, she immediately said, "I'm sorry," and closed the door. When a young woman emerged from the stall a couple minutes later, the older woman apologized once again but pointed out that "the door was open." The young woman responded, "it's okay," thereby minimizing the offense and perhaps acknowledging a degree of culpability on her part.

As is the case with many physical barriers to perception (Goffman, 1963: 152), the walls and doors of toilet stalls are also treated as if they cut off more communication than they actually do. Under most circumstances, for example, the walls and doors of toilet stalls are treated as if they were barriers to conversation. Although acquainted individuals may sometimes carry on a conversation through the walls of a toilet stall if they believe the bathroom is not otherwise occupied, they seldom do so if they are aware that others are present. Moreover, individuals often attempt to ignore offensive sounds and smells that emanate from occupied toilet stalls, even though the exercise of such "tactful blindness" (Goffman, 1955: 219) is sometimes a demanding task. In any case, the walls and doors of toilet stalls provide public actors with both physical and normative shields behind which they can perform potentially discrediting acts.

Toilet stalls in public bathrooms are, therefore, publicly accessible yet private backstage regions. Although same-sexed clients of a public establishment may lay claim to any unoccupied stall in the bathroom designated for use by persons of their sex, once such a claim is laid, once the door to the stall is closed, it is transformed into the occupying individual's private, albeit temporary, retreat from the demands of public life. While occupying the stall, that individual can engage in a variety of potentially discrediting acts with impunity.

When not concealed behind the protective cover of a toilet stall, however, occupants of public bathrooms may be observed by others. For the most part, as previously noted, same-sexed clients of a public establishment can enter and exit at will the bathroom designated for their use, and it may be simultaneously occupied by as many individuals as its physical dimensions allow. By implication, therefore, occupants of public bathrooms must either perform or be ready to perform for an audience. As a result, the behavior that routinely occurs in the "open region" of a public bathroom, that area that is not enclosed by toilet stalls, resembles, in many important respects, the behavior that routinely occurs in other public settings. . . .

THE RITUALS OF PUBLIC BATHROOMS

As Goffman (1971) convincingly argued, much of this behavior can best be described as "interpersonal rituals." Emile Durkheim (1965), in his famous analysis of religion, defined a ritual as a perfunctory, conventionalized act which expresses respect and regard for some object of "ultimate value." In a different context, moreover, he observed that in modern, Western societies,

> the human personality is a sacred thing; one dare not violate it nor infringe its bounds, while at the same time the greatest good is in communion with others. . . . [Durkheim, 1974: 37]

According to Durkheim, negative rituals express respect and regard for objects of ultimate value by protecting them from profanation. By implication, according to Goffman (1971: 62), negative interpersonal rituals involve the behavioral honoring of the sacred individual's right to private "preserves" and "to be let alone." For example, individuals typically refrain from physically, conversationally, or visually intruding on an occupied toilet stall. In doing so, they implicitly honor the occupying individual's right to be let alone and in this respect perform a negative interpersonal ritual.

Similarly, the queues that typically form in public bathrooms when the demand for sinks, urinals, and toilet stalls exceeds the available supply are also products of individuals' mutual performance of negative interpersonal rituals. Individuals typically honor one another's right to the turn claimed by taking up a position in such a queue, even when "creature releases" (Goffman, 1963: 69) threaten to break through their self-control. Young children provide an occasional exception, sometimes ignoring the turn-order of such queues. Yet even then the child's caretaker typically requests, on the child's behalf, the permission of those waiting in the queue. Between performances at a music festival, for example, a preschool-age girl and her mother were observed rapidly walking toward the entrance to a women's bathroom out of which a queue extended for several yards down a nearby sidewalk. As they walked past those waiting in the queue, the mother repeatedly asked: "Do you mind? She really has to go."

The interpersonal rituals that routinely occur in the open region of public bathrooms are not limited, however, to negative ones. If individuals possess a small patrimony of sacredness, then, as Durkheim (1974: 37) noted, "the greatest good is in communion" with such sacred objects. When previously acquainted individuals come into contact with one another, therefore, they typically perform conventionalized acts, positive interpersonal rituals, that express respect and regard for their previous communion with one another. In a sense, moreover, negative and positive interpersonal rituals are two sides of the same expressive coin. Whereas negative interpersonal rituals symbolically protect individuals from profanation by others, positive interpersonal rituals symbolically cleanse communion between individuals of its potentially defiling implications.[2] Although a positive interpersonal ritual may consist of no more than a brief exchange of greetings, failure to at least acknowledge one's previous communion with another is, in effect, to express disregard for the relationship and, by implication, the other individual's small patrimony of sacredness (Goffman, 1971: 62–94).

Even when previously acquainted individuals come into contact with one another in a public bathroom, therefore, they typically acknowledge their prior relationship. In fact, the performance of such positive interpersonal rituals sometimes interfered with the conduct of our research. On one occasion, for example, a member of the research team was in the open region of an otherwise unoccupied men's bathroom. While he was writing some notes about an incident that had just occurred, an acquaintance entered.

A: Hey——! (walks to a urinal and unzips his pants) Nothing like pissin.

O: Yup.

A: Wh'da hell ya doin? (walks over to a sink and washes hands)

O: Writing.

A: Heh, heh, yea. About people pissin . . . That's for you.

O: Yup.

A: Take care.

O: Mmm Huh.

As this incident illustrates, individuals must be prepared to perform positive interpersonal rituals when in the open region of public bathrooms, especially those in public establishments with a relatively stable clientele. Whereas some of these may consist of no more than a brief exchange of smiles, others may involve lengthy conversations that reaffirm the participants' shared biography.

In contrast, when unacquainted individuals come into contact with one another in the open regions of public bathrooms, they typically perform a brief, negative interpersonal ritual that Goffman (1963: 84) termed "civil inattention":

> one gives to another enough visual notice to demonstrate that one appreciates that the other is present . . . while at the next moment withdrawing one's attention from him so as to express that he does not constitute a target of special curiosity or design.

Through this brief pattern of visual interaction, individuals both acknowledge one another's presence and, immediately thereafter, one another's right to be let alone.

A variation on civil inattention is also commonly performed in the open region of public bathrooms, most often by men using adjacent urinals. Although masculine clothing permits males to urinate without noticeably disturbing their clothed appearance, they must still partially expose their external genitalia in order to do so. Clearly, the standards of modesty that govern public behavior prohibit even such limited exposure of the external genitalia. Although the sides of some urinals and the urinating individual's back provide partial barriers to perception, they do not provide protection against the glances of someone occupying an adjacent urinal. In our society, however, "when bodies are naked, glances are clothed" (Goffman, 1971: 46). What men typically give one another when using adjacent urinals is not, therefore, civil inattention but "nonperson treatment" (Goffman, 1963: 83–84); that is, they treat one another as if they were part of the setting's physical equipment, as "objects not worthy of a glance." When circumstances allow, of course, unacquainted males typically avoid occupying adjacent urinals and, thereby, this ritually delicate situation.

It is not uncommon, however, for previously acquainted males to engage in conversation while using adjacent urinals. For example, the following interaction was observed in the bathroom of a restaurant.

> A middle-aged man is standing at one of two urinals. Another middle-aged man enters the bathroom and, as he approaches the available urinal, greets the first man by name. The first man quickly casts a side-long glance at the second and returns the greeting. He then asks the second man about his "new granddaughter," and they continue to talk about grandchildren until one of them zips up his pants and walks over to a sink. Throughout the conversation, neither man turned his head so as to look at the other.

As this example illustrates, urinal conversations are often characterized by a lack of visual interaction between the participants. Instead of looking at one another while listening, as is typical among white, middle-class Americans (see LaFrance and Mayo, 1976), participants in such conversations typically fix their gaze on the wall immediately in front of them, an intriguing combination of the constituent elements of positive and negative interpersonal rituals. Although ritually celebrating their prior communion with one another, they also visually honor one another's right to privacy.

Due to the particular profanations and threats of profanations that characterize public bathrooms, moreover, a number of variations on these general patterns also commonly occur. In our society, as Goffman (1971: 41) observed, bodily excreta are considered "agencies of defilement." Although supported by germ theory, this view involves somewhat more than a concern for hygiene. Once such substances as urine, fecal matter, menstrual discharge, and flatus leave individuals' bodies, they acquire the power to profane even though they may not have the power to infect. In any case, many of the activities in which individuals engage when in bathrooms are considered both self-profaning and potentially profaning to others.[3] As a result, a variety of ritually delicate situations often arise in public bathrooms.

After using urinals and toilets, for example, individuals' hands are considered contaminated and, consequently, a source of contamination to others. In order to demonstrate both self-respect and respect for those with whom they might come into contact, individuals are expected to and often do wash their hands after using urinals and toilets. Sinks for this purpose are typically located in the open region of public bathrooms, allowing others to witness the performance of this restorative ritual.[4] Sometimes, however, public bathrooms are not adequately equipped for this purpose. Most commonly, towel dispensers are empty or broken. Although individuals sometimes do not discover this situation until after they have already washed their hands, they often glance at towel dispensers as they walk from urinals and toilet stalls to sinks. If they discover that the towel dispensers are empty or broken, there is typically a moment of indecision. Although they sometimes proceed to wash their hands and then dry them on their clothes, many times they hesitate, facially display disgust, and audibly sigh. By performing these gestures-in-the-round, they express a desire to wash their hands; their hands may remain contaminated, but their regard for their own and others' sacredness is established.

Because the profaning power of odor operates over a distance and in all directions, moreover, individuals who defecate in public bathrooms not only temporarily profane themselves but also risk profaning the entire setting. If an individual is clearly responsible for the odor of feces or flatus that fills a bathroom, therefore, he or she must rely on others to identify sympathetically with his or her plight and, consequently, exercise tactful blindness. However, this is seldom left to chance. When other occupants of the bathroom are acquaintances, the offending individual may offer subtle, self-derogatory display as a defensive, face-saving measure (Goffman, 1955). Upon emerging from toilet stalls, for example, such persons sometimes look at acquaintances and facially display disgust. Self-effacing humor is also occasionally used in this way. On one occasion, for example, an acquaintance of a member of the research team emerged from a toilet stall after having filled the bathroom with a strong fecal odor. He walked over to a sink, smiled at the observer, and remarked: "Something died in there." Through such subtle self-derogation, offending individuals metaphorically split themselves into two parts: a sacred self that assigns blame and a blameworthy animal self. Because the offending individual assigns blame, moreover, there is no need for others to do so (Goffman, 1971: 113).

If other occupants of the bathroom are unfamiliar to the offending individual, however, a somewhat different defensive strategy is commonly employed. Upon emerging from a toilet stall, individuals who are clearly responsible for an offensive odor seldom engage in visual interaction with unacquainted others. In so doing, they avoid visually acknowledging not only the presence of others but others' acknowledgement of their own presence as well. In a sense, therefore, the offending individual temporarily suspends his or her claim to the status of sacred object, an object worthy of such visual regard. The assumption seems to be that by suspending one's claim to this status, others need not challenge it and are, consequently, more likely to exercise tactful blindness in regard to the offense.

Despite Miner's humorous misidentification and interpretation of bathroom rituals, therefore, there is something to recommend the view that many of the rituals that behaviorally express and sustain the central values of our culture occur in bathrooms. Although these "central values do but itch a little," as Goffman (1971: 185) noted, "everyone scratches." And, it must be added, they often scratch in public bathrooms. However, routine bathroom behavior consists of more than the interpersonal rituals that are found in other public settings or variations on their general theme. . . .

MANAGING PERSONAL FRONTS

When in a public setting, as Goffman (1963: 24) pointed out, individuals are expected to have their "faculties in readiness for any face-to-face interaction that might come" their way. One of the most evident means by which individuals express such readiness is "through the disciplined management of personal appearance or 'personal front,' that is, the complex of clothing, make-up, hairdo, and other surface decorations" that they carry about on their person (Goffman, 1963: 25). Of course, keeping one's personal front in a state of good repair requires care and effort (Gross and Stone, 1964: 10). However, individuals who are inspecting or repairing their personal fronts in public encounter difficulties in maintaining the degree of interactional readiness often expected of them; their attention tends to be diverted from the social situations that surround them (Goffman, 1963: 66). For the most part, therefore, close scrutinization and major adjustments of personal fronts are confined to backstage regions such as public bathrooms.

Most public bathrooms are equipped for this purpose. Many offer coin-operated dispensers of a variety of "personal care products" (e.g., combs and sanitary napkins), and almost all have at least one mirror. The most obvious reason for the presence of mirrors in public bathrooms is that the act of defecation and, for females, urination, requires individuals to literally "drop" their personal fronts. In order to ensure that they have adequately reconstructed their personal front after engaging in such an act, individuals must and typically do perform what Lofland (1972) has termed a "readiness check." For example, the following was observed in the men's bathroom of a neighborhood bar:

> A young man emerges from a toilet stall and, as he passes the mirror, hesitates. He glances side-long at his reflection, gives a nod of approval and then walks out the door.

When such a readiness check reveals flaws in the individual's personal front, he or she typically makes the appropriate repairs: Shirts are often retucked into pants and skirts, skirts are rotated around the waist, and pants are tugged up and down.

Because bodily movement and exposure to the elements can also disturb a disciplined personal front, the post-defecation or urination readiness check sometimes reveals flaws in individuals' personal fronts that are the result of normal wear and tear. Upon emerging from toilet stalls and leaving urinals, therefore, individuals sometimes repair aspects of their personal fronts that are not normally disturbed in the course of defecating or urinating. For example, the following was observed in the women's bathroom of a student center on a college campus.

> A young woman emerges from a toilet stall, approaches a mirror, and inspects her reflection. She then removes a barrette from her hair, places the barrette in her mouth, takes a comb out of her coat pocket, and combs her hair while

> smoothing it down with her other hand. With the barrette still in her mouth, she stops combing her hair, gazes intently at the mirror and emits an audible "ick." She then places the barrette back in her hair, pinches her cheeks, takes a last look at her reflection and exits.

Interestingly, as both this example and the immediately preceding one illustrate, individuals sometimes offer visible or audible evaluations of their reflections when inspecting and repairing their personal front, a finding that should delight proponents of Meadian sociological psychology. Public bathrooms may protect individuals from the critical reviews of external audiences, but they do not protect them from those of their internal audience.

In any case, public bathrooms are as much "self-service" repair shops for personal fronts as they are socially approved shelters for physiological acts that are inconsistent with the cleanliness and purity standards that govern our public performances. In fact, individuals often enter public bathrooms with no apparent purpose other than the management of their personal front. For example, it is not uncommon for males to enter public bathrooms, walk directly to the nearest available mirror, comb their hair, rearrange their clothing, and then immediately exit. In our society, of course, females are often expected to present publicly a more extensively managed personal front than are males. Consequently, females often undertake extensive repairs in public bathrooms. For example, the following was observed in the women's bathroom of a student center on a college campus:

> Two young women enter, one goes to a toilet stall and the other immediately approaches a mirror. The second woman takes a brush out of her bookbag, throws her hair forward, brushes it, throws her hair back, and brushes it into place. She returns the brush to her bookbag, smooths down her eyebrows, and wipes underneath her eyes with her fingers. She then removes a tube of lipstick from her bookbag, applies it to her lips, and uses her finger to remove the lipstick that extends beyond the natural outline of her lips. As her friend emerges from the toilet stall, she puts the lipstick tube back into her bookbag, straightens her collar so that it stands up under her sweater and then exits with her friend.

Even though individuals routinely inspect and repair their personal fronts in the open regions of public bathrooms, they often do so furtively. When others enter the bathroom, individuals sometimes suspend inspecting or repairing their personal fronts until the new arrivals enter toilet stalls or approach urinals. In other cases, they hurriedly complete these activities before they can be witnessed. For example, the following was observed from inside a toilet stall in a women's bathroom:

> A young woman walks to the end of the sinks where there is a full-length mirror. She turns sideways, inspects her reflection and reaches up to adjust her clothing. The outer door of the bathroom begins to open, and the young woman quickly walks over to the sink on which her purse is laying, picks it up and heads for the door.

Despite the furtiveness that sometimes characterizes individuals' inspection and repair of their personal fronts, however, the open region of a public bathroom is often the only available setting in which they can engage in these activities without clearly undermining their frontstage performances. As Lofland (1972: 101) observed in a somewhat different context, "it is apparently preferable to be witnessed by a few . . . In a brief episode of backstage behavior than to be caught . . . with one's presentation down" on the frontstage. . . .

In short, the systematic study of routine bathroom behavior reveals just how loyal members of this society are to the central values and behavioral standards that hold our collective lives together. Whatever else they may do, users of public bathrooms continue to bear the "cross of personal character" (Goffman, 1971: 185), and, as long as they continue to carry this burden, remain self-regulating participants in the "interaction order" (Goffman, 1983).

NOTES

1. Yet a closed door is not always a reliable indicator that a toilet stall is occupied, as anyone who has cared for children is aware. The young sometimes exit toilet stalls by crawling under locked doors.

2. Along these lines, casual sex may make an individual "feel so cheap" in part because this intimate communion with another has not been adequately cleansed of its defiling implications by the performance of positive interpersonal rituals.

3. One further expression of these defiling implications is the fact that cleaning bathrooms is an almost universally despised activity in our society. Apparently, such close contact with objects that are used for the elimination and disposal of bodily excrete profanes individuals. Indeed, those who routinely clean bathrooms, janitors in particular, are often treated as if they had abdicated their claim to a small patrimony of sacredness.

4. Although we did not record the frequency of hand washing, it was our impression that individuals are more likely to wash their hands after using toilets or urinals if they think others would notice their failure to do so. When a bathroom was not otherwise occupied and we were observing from within a toilet stall, for example, it was not uncommon for individuals to neglect this practice.

REFERENCES

Durkheim, E. (1974). *Sociology and Philosophy.* Trans. by D. F. Pocock. New York: Free Press. (originally published in 1924)

———. (1965). *The Elementary Forms of the Religious Life.* (J. W. Swain, trans.) New York: Free Press. (originally published in 1915)

Goffman, E. (1983). "The interaction order." *Amer. Soc. Rev.* 48 (February): 1–17.

———. (1971). *Relations in Public: Microstudies in Public Order.* New York: Basic.

———. (1963). *Behavior in Public Places: Notes on the Social Organization of Gatherings.* New York: Free Press.

———. (1959). *The Presentation of Self in Everyday Life.* Garden City, NY: Doubleday.

———. (1955). "On face-work: An analysis of ritual elements in social interaction." *Psychiatry* 18 (August): 213–231.

Gross, E. and G. Stone. (1964). "Embarrassment and the analysis of role requirements." *Amer. J. of Sociology* 70 (July): 1–15.

Kira, A. (1966). *The Bathroom: Criteria for Design*. Ithaca, NY: Cornell University Center for Housing and Environmental Studies.

LaFrance, M. and C. Mayo. (1976). "Racial differences in gaze behavior during conversation: Two systematic observational studies." *J. of Personality and Soc. Psychology* 33 (May): 547–552.

Lofland, L. (1972). "Self-management in public settings: Part I." *Urban Life* 1 (April): 93–108.

Miner, H. (1955). "Body ritual among the Nacirema." *Amer. Anthropologist* 58 (June): 503–507.

15

High School Reunions and the Management of Identity

VERED VINITZKY-SEROUSSI
AND ROBERT ZUSSMAN

In contrast to bathrooms, high school reunions represent social settings of the most public kind, and Vinitzky-Seroussi and Zussman's fascinating study of people's actions both in and around these occasions casts insight into some of the ways that people think about and plan for their appearances on these spotlighted stages. They found that not only do people omit to tell others about significant negative aspects of their lives, but they conspire (often with others) to construct various levels of fraudulent stories and props designed to create positive self-impressions. People unable to build and sustain such favorable identities often absent themselves from these occasions, but ironically, depending on their former visibility in the school, they may not be able to escape making a self-presentation so easily, as they are the topic of conversation by others, and may even be contacted by attenders. High school reunions represent significant identity markers because they show how people have capitalized (or not) on the potential they were once thought to have, and offer them a chance to display the achievements they have earned (or not) since they left the daily contact with their high school classmates. This article stands in sharp contrast to the previous one by Karp et al. that discusses the transition to college. How do the two articles differ? What might you emphasize when you go back to your high school reunion in a few years? How will you present yourself in this highly public, ritualized setting?

High school reunions, although episodic and unevenly attended, are a critical vantage point from which to make sense of issues of identity in contemporary America. Although reunions take place in the present, they are organized around an aspect of the past. As a result, reunions are not only an invitation to account for one's life, but virtually a mandate to do so. Moreover, reunions are organized around an aspect of the past in which one's participation is unusually diffuse. High schools, at least in the United States, are not simply about academic performance, but about coming of age more generally. And, as a further result, the accounts men and women provide of themselves at reunions tend to be accounts not only of what they have done, but also of who they are. They become, in short, accounts of identity.

Identity involves, in William James' (1910) phrase, the perception that we are, at any given moment, "in some peculiarly subtle sense the same" as before. But reunions threaten this perception, confronting reunion goers with discontinuities over time: The boy and girl, unnoticed and unpopular in high school, grow prosperous and prominent; the boy voted "best looking" and the girl voted "best athlete" return overweight and out of shape. Nonetheless, we argue, many reunion goers do find (to borrow again a phrase from James) "a substantial principle of unity." Although, as Strauss (1959, p. 146) once pointed out, this "subjective feeling of continuity" may have many sources, it is not typically based on appearances. Rather, in contemporary American society, in particular, it is typically based on a belief in an inner, unchanging self (Turner 1976; Carbaugh 1989; Veroff, Douvan, and Kulka 1981). . . .

Put somewhat differently, reunions raise fundamental issues about the possibilities of the alignment and misalignment of different aspects of identity (Stokes and Hewitt 1976). It is often the case that who we believe ourselves to be depends heavily on the appearances we foster. But it is not always so. Identity refers, in part, to understandings of the self that are public in the sense that they are socially enacted and shared with others. But identity also refers, in part, to those deeply felt, inner understandings of the self that transcend both time and situation and are independent, at least in the short run, of the responses of others. These different aspects of identity are not only distinct. They are also often in conflict. If we are fortunate, if the person we believe ourselves to be corresponds with the person we seem to be, then there is no conflict. But we are rarely so fortunate, and, more often than not, there is at least some tension between who we seem to be and who we believe ourselves to be. . . .

"HIGHER TRUTHS"

Individual preparations for reunions are remarkably extensive. Not only do those who plan to attend often have their hair done, or buy new clothes, or go on extensive diets, but also occasionally quit old jobs, look for new ones, resolve relationships, and (in at least one instance known to us independently of the formal research plan) finish dissertations. All of this is testimony not only to the emotional power of reunions, but, more importantly, to the extent of self-conscious impression management. Still, the character of this impression management should not be misconstrued. From the perspective of the participants themselves, it is not deception. Rather, it constitutes an effort to select among different appearances and, often, to find that appearance which most closely approximates what the reunion goer believes aligns most closely with what he or she "really" is.

Most often, we will argue, reunion goers' concern to align their appearance with their own beliefs about who they are precludes self-consciously deceptive impression management. On some occasions, however, reunion goers may engage in deception precisely in order to align appearances with what they believe to be a kind of higher truth. Isabel, a 38-year-old housewife from a lower-middle-class neighborhood, provides a particularly dramatic example of this process.

Some time before her reunion, Isabel's diamond ring had been stolen. As she explained:

> My husband said that in time he'd get me [another] one . . . but not in time for the reunion. I said, "I don't *even* have a diamond ring." . . . So my friend Linda said "Let's go to the flea market, they sell zircons there." So we both went out . . . and we bought these humongous "diamond" rings, almost two carats, that cost $9.99.

Over the course of her reunion, Isabel worried that her fake diamond would be discovered. And she was afraid that her friend would expose her, even though that friend had not only convinced her to buy the ring, but also bought one for herself.

> [At the reunion] we went down to the bathroom and . . . [one of my former classmates] said, "Well, everybody must be doing really well because the diamonds on the fingers are killing me." It was embarrassing. Did they really know that my ring was a fake? . . . On the way home I told Linda, "While I was [in the bathroom] I thought that if you were to open your mouth and say it was a zircon I would have killed you."

Isabel feared that she would be embarrassed. But, more importantly, she feared that discovery of her fakery would involve a misrepresentation more serious than that embodied in the ring itself. Describing the incident in the ladies room, Isabel insisted that, when her former classmate concluded that "everybody must be doing well," she did not say anything. Nonetheless, she insisted, she "wouldn't lie." For Isabel, the zirconium ring was not a deception because, until she lost it, she had had a "real one . . . a perfect blue-white diamond." Thus, the zirconium ring, although a fake diamond, represented a real possession, one she had been deprived of not because of any putatively inner characteristic, but because of what she conceived of as an accident of timing in its theft. In this sense, the fake represented the higher truth of the actual diamond. In wearing the zirconium ring, Isabel was not, at least by her own construction, lying, but representing herself as she believed herself to be.

THE MANAGEMENT OF IMPRESSIONS

In popular culture, reunions are a stock element of comedy. In, for example, episodes of the television shows *Taxi* and *Laverne and Shirley,* plots revolve around the protagonists' convoluted (and unsuccessful) efforts to hide their failures in adult life from former classmates. In the curious reality of actual reunions; something of the same sort often happens, although only rarely in the same tone of comedic hysteria found on television. In particular, those who attend reunions practice various forms of what Goffman (1963), with characteristic understatement, called "information control." Information control does not, of course, alter an underlying "reality" but is critical to the management of those impressions out of which some aspects of identity are constructed. Yet, information control as a technique for the management of identity ultimately fails—not simply, as is typically the case in representations of reunions on television, because of the danger of discovery, but also because of a more basic conflict with exigencies of managing an inner sense of selfhood.

For certain purposes, it may be critical to distinguish sharply between lies and half-truths, between exaggerations and misdirections, between avoidance and omission. For our purposes, however, it is sufficient to note that these practices differ only with degree in regard to two characteristics. First, all involve limiting information that is at least potentially discoverable and that, if discovered, discredits those who have limited information. Second, all involve representations of the self the practitioner himself or herself believes to be untrue.

Perhaps the most obvious means of limiting information at a reunion is to avoid the event altogether. This strategy is perhaps most tempting to those who have recently experienced a major, potentially discrediting life event. One respondent, for example, had enjoyed earlier reunions but did not want to attend her 25th. She had just been separated from her husband but, more than that, "it would be too emotional to discuss the death of my oldest son who had been almost 23" (executive, 25th reunion, Central High School). Another did not attend either her 10th or her 20th reunion because, in both cases, she was in the middle of a divorce: "I was embarrassed," she said, "by my marital status and life status at those times" (secretary, 25th reunion, Central High School).

Information control, however, is by no means restricted to those who avoid the reunion altogether. It is also a conversational strategy. Randy, for example, had owned a restaurant and an inn, "three or four million dollars worth of property." But, as Randy explained, "it got involved in all kinds of legal problems," to a point where he "had to back off, let it close down, let it be taken back actually." Thus, before his 20th reunion, Randy (to whom we will return later) found himself with no property at all, taking temporary jobs to provide for his wife and newborn baby. One of the ways he chose to cope with the loss was by not giving people the opportunity to ask him crucial questions regarding his work and financial situation:

> I didn't want to be asked, "What are you doing?" or "How's it going?" I avoided that type of questioning and just asked, "How are you doing?" "Good to see you." (Randy, bartender, 20th reunion, Garden High School)

Similarly, another respondent, a football player and one of the most popular boys in his class, had by the time of his 25th reunion developed an alcohol problem which had caused him a great deal of trouble, both personally and professionally. Fighting the temptation to skip his reunion, he attended but told no one of his problem:

> I didn't tell them that I live in a pig sty. I said, "I live upstairs at my father's, trying to get my act together. I moved back from Texas." I think I used "moved back from Texas" as an excuse that I live upstairs at my father's. Nobody wants to be a failure. (Mario, garbage collector, 25th reunion, Central High School)

Other respondents reported avoiding former classmates they thought especially likely to probe, as well as steering conversations away from sensitive topics and toward topics they found comfortable.

At reunions, the control of information is, at its core, an effort to safeguard a conception of the self by safeguarding the past. For the former cheerleader who now finds herself divorced and overweight, or for the man once voted most likely to succeed who now finds his career foundering—both of whom appeared among the respondents—there is an obvious problem. This problem, however, is not simply a problem of the present. The former cheerleader is not only overweight and divorced now, and the boy once voted most likely to succeed is not only foundering now. Their current problems also suggest that they were not what they seemed then. As one respondent, who had attended both her 10th and her 20th reunion accompanied by a husband "who looks like Tom Cruise" but now found herself divorced and penniless, reported of her decision not to attend her 25th reunion: "I didn't want people to know that my marriage had failed. I wanted people to think that I was still at that [20th reunion] point" (graduate student and part-time administrator, 25th reunion). The present, of course, does not change the past; it does, however, change its meaning. In this sense, at reunions, the present threatens constantly to undo the past and the identities constructed out of it. By limiting information about the present, those who attend reunions attempt to maintain identities based on past accomplishments.

The ability to limit information is itself undermined, however, by the dense network of relations that typically characterize former classmates. Those who are absent, especially if they were "class celebrities"—the class president, a star athlete, a popular girl or boy—are also part of the conversation. As one reunion organizer explained:

> People ask where so and so is and I knew and said, "I don't think she wanted to come because she could not afford it or she is not happy." And then we talked about another girl that we hadn't seen. Somebody else had been in touch with her and said that she was a lesbian and so she didn't want to come and be with the crowd. (Maryellen, clerical worker, 25th reunion, Central High School)

Thus, only those who have broken away from old friends altogether—or those in whom no one was much interested to begin with—are able to limit information effectively.

Even more fundamentally, the effort to limit information itself arouses suspicion. Even if some invitees do not attend their reunions because of a more or less genuine lack of interest or because of inconvenient scheduling, their former classmates are likely to suspect them of hiding something: "If you are a loser," said

Greg firmly, "you don't come to see people" (owner of a plumbing company, 20th reunion, Garden High School). And another echoed those sentiments: "If I was not happy with myself . . . why would I want to put myself in the position of meeting 150 or 200 people that I knew 25 years ago?" For this reason, attendance, accompanied by an active effort to control information, is typically a more successful technique of impression management than is avoidance of the event altogether. . . .

MAINTAINING INNER IDENTITY

No less than the construction of appearances, maintaining an inner identity involves impression management. However, because an inner identity involves something internally felt as well as socially enacted, it cannot be constructed out of material the actor himself or herself believes to be untrue. Thus, while the construction of inner identities may involve display, this display is more likely to be directed at oneself than toward others.

In regard to appearance, an admission—an open acknowledgment of flaws—does have a strategic advantage. It creates the impression, if not always the reality, of integrity. But it is not in the service of appearance that reunion-goers are most likely to admit their failures; it is, instead, in the service of maintaining inner beliefs. They are less interested in creating the impression of integrity than in sustaining their own belief in that integrity. The admission of failure is not easy. It may constitute a threat to an impression carefully constructed over many years. Yet, to some returnees, the admission of failure appears the only way to avoid the internal conflict that would accompany an account that, even aside from any fear of discovery, the storyteller could not accept as the "real me."

Consider the following example related by a reunion organizer:

> Our class president was Richard Tylor. . . . Two weeks before the reunion, his business went bust and he was on the verge of personal bankruptcy. And he didn't want to go [to the reunion]. He said to me, "It is really hard for me to tell people that I'm doing badly," and I said, "Why in the world are you saying that, are you nuts? . . . you still own your business, at least formally. Yes, it's bankrupt, but who the hell knows that and who the hell is going to ask? Just tell them what you do. You still do it . . . it is not a lie . . . nobody is asking for a loan and interested to see your financials. They just want to know what you do . . . and I think you want to present yourself in an upbeat way. Maybe you are putting on a rosier face, but there is nothing wrong with that, because even if you go bankrupt tomorrow, that doesn't mean that two weeks from tomorrow you are not going to bounce back again, so what does one week mean? You should just . . . go as Richard Tylor, the class president." [So what did he do at the reunion?] He just moped around, telling everybody about his personal problems. (Michael, physician, 20th reunion)

Michael, the reunion organizer, advised that Richard manage impressions by ignoring the previous two weeks. He did not, to be sure, suggest that Richard should exactly lie, but that he should treat the events of his life selectively, abandoning what Michael called the "truth for the moment" in the name of what might be thought of, with some generosity, as what we have called a "higher truth." But Richard's strategy was different. Although we do not have his own account of the reunion, we do know that, as a former class president, Richard had a past worth protecting. Nonetheless, he endangered the meanings of that past in the service of what appears to be his own felt sense of integrity.

In the case of Randy, whom we met earlier, we have more direct evidence. Randy had suffered a serious business setback prior to his reunion. He hesitated to tell his former classmates and in some cases chose to avoid specific conversations. But, he reported proudly, "I wound up telling the truth because I'm that kind of person" (Randy, bartender, 20th reunion, Garden High School). In both cases, the admission of failure may serve to create an impression of integrity. But, for Randy and (we speculate) Richard, the impression of integrity made on others was far less important than the impression of integrity made on themselves. . . .

Like many others who attend reunions, Jeffrey attempts to discredit his former classmates as suitable partners in the establishment of his identity: "Who are these people?" To do this, Jeffrey questions the values of his former classmates, suggesting that "they slowly get more and more messed up" and proposing an alternative mode of valuation as "an old hippie kind of guy." Similarly, Matthew, a 28-year-old teacher from a middle-class background, felt vulnerable in a room filled with doctors, lawyers, and business people because of his low income. Nonetheless, he insisted that his work was important, even if not well paid, and pointed to his successful marriage:

> . . . if a person is not married or has no serious relationship at 28 or 29 . . . I would consider it a bit weird . . . out of the norm. (Matthew, teacher, 10th reunion, Central High School). . . .

Perhaps more fundamentally, Jeffrey evokes a notion of selfhood quite different from that implicit in impression management. His identity, he argues, is "independent" of his "so-called position in life." Nor does his identity depend on what he has achieved: "that's a lot of bullshit." Something of the same notion appears in Evelyn's comments. A 38-year-old teacher, who had been voted "most likely to succeed" in high school, Evelyn realized that she had "never made it." She had neither family nor an exciting career and had gained a good deal of weight in the two decades since graduation. Indeed, many of her former classmates singled her out as one of the sadder stories of their reunion. Yet, while hardly immune to her former classmates' evaluations, Evelyn found another set of meanings at the event. "I went there," she concluded, "and I saw what people are doing. *It is good to see that people are just people*" [emphasis added]. Implicitly, both Evelyn and Jeffrey are invoking a notion of selfhood that is internal and inviolable—"people are just people." And, critically, it is a notion of selfhood that allows those who subscribe to it to detach themselves from the efforts of others to identify them at reunions. . . .

CONCLUSION

Impression management holds many risks. Frauds may be exposed; conversations may prove uncontrollable; avoidance cannot stop gossip. Those who attempt to manage impressions seldom know if their efforts are successful. More importantly, impression management fails to resolve internal conflicts. Widespread skepticism undermines the perceived truth of virtually everything seen and heard throughout the reunion. Appearances begin to seem, to the reunion goers themselves, little more than polite fictions without relevance to identity.

REFERENCES

Carbaugh, Donal. 1989. *Talking Americans: Cultural Discourses on Donahue.* Norwood, NJ: Ablex.

Goffman, Erving. 1963. *Stigma.* New York: Simon and Schuster.

James, William. 1910. *Psychology: The Briefer Course.* New York: Henry Holt.

Stokes, Randall and John P. Hewitt. 1976. "Aligning Actions." *American Sociological Review* 40:1–11.

Strauss, Anselm. 1959. *Mirrors and Masks.* New York: Free Press.

Turner, Ralph. 1976. "The Real Self: From Institution to Impulse." *American Journal of Sociology* 81:989–1016.

Veroff, Joseph, Elizabeth Douvan, and Richard Kulka. 1981. *The Inner American: A Self-Portrait from 1957–1976.* New York: Basic.

16

Inclusion and Exclusion in Preadolescent Cliques

PATRICIA A. ADLER
AND PETER ADLER

This study of third to sixth graders takes an insider's look at children's social worlds, focusing on the cliques that stand atop their social hierarchy and dominate their interest and attention. In a study we conducted in the 1980s and 1990s using our children and their friends, we probe the question of where cliques and their leaders get so much power, enabling them to coerce followers to engage in such outrageous and self-demeaning behavior while commanding high status and submission. We conclude that this power stems from the inclusionary/exclusionary dynamic found in these elite and highly restricted friendship circles. By tracing the parameters of how leaders both include and exclude others (and induce their followers to follow suit), and by showing how nearly all children's social standing is, at one time or another, systematically undercut, we illustrate how children learn that the price of supporting others is social exclusion themselves, and that, in this dog-eat-dog world, they must sacrifice their friends to save themselves. These interactional dynamics are where children learn the essence of in-group and out-group behavior that carries into later adulthood, and forms the foundation of the micro-politics of everyday life. Do these forms look familiar to you? Were you a member of such a clique or an outcast from these powerful groups? What are the implications of these groups for adult life? Do adults engage in this type of behavior or is this just indicative of the relations between young children?

One of the dominant features of children's lives during the later elementary school years (fourth through sixth grades) is the popular-clique structure that organizes their social worlds. The fabric of their relationships with others, their levels and types of activity, their participation in friendships, and their feelings about themselves are tied to their involvement in, around, or outside the cliques organizing their social landscape. Cliques are basically friendship circles, encompassing a high likelihood that members will identify each other sociometrically as mutually connected (Hallinan 1979; Hubbell 1965; Peay 1974). Yet cliques are more than that: they have a hierarchical structure, being dominated by leaders, and are exclusive, so that not all individuals who desire membership are accepted. They function as bodies of power within grades, incorporating the most popular individuals, offering the most exciting social lives, and commanding the most interest and attention from classmates (Eder and Parker 1987). As such they represent a vibrant component of the childhood experience. . . .

TECHNIQUES OF INCLUSION

Cliques maintained exclusivity through careful membership screening. Cliques are not static entities; they shifted irregularly and evolved their membership as individuals moved away or were ejected from the group and others took their place. In addition, cliques were characterized by frequent group activities designed to foster some individuals' inclusion while excluding others. They had embedded, although often unarticulated, modes of considering and accepting or rejecting potential new members. These modes were linked to the leaders' critical power in making vital group decisions. Leaders derived power from their popularity and used it to influence membership and social stratification within the group. This stratification manifested itself in tiers and subgroups within cliques, composed of people who were ranked as leaders, followers, and wannabes (see Adler and Adler 1996). Cliques embodied systems of dominance whereby individuals with more status and power exerted control over others' lives. They accomplished this by alternately applying the dynamics of inclusion and exclusion to various people, both within and outside the boundaries of the clique (see Elkin and Handel 1989).

Recruitment

Potential new members could be brought to the group by established members who had met and liked them. The leaders then decided whether these individuals would be granted a probationary period of acceptance in which they could be informally evaluated. If the newcomers were liked, they were allowed to remain in the friendship circle; if they were rejected, they were forced to leave. Alexis, a popular, dominant seventh-grade girl, reflected on the boundary maintenance that she and her best friend, Hope, two clique leaders, had exercised in sixth grade:

> Q: Who defines the boundaries of who's in or who's out?
>
> Alexis: Probably the leader. If one person might like them they might introduce them, but if one or two people didn't like them, then they'd start

> to get everyone up. Like in sixth grade, there was Dawn Bolton and she was new. And the girls in her class that were in our clique liked her, but Hope and I didn't like her, so we kicked her out. So then she went to the other clique, the Margo clique.

Timing was critical to recruitment. The beginning of the year, when classes were reconstructed and people were shuffled into new social configurations, was the major time when cliques considered new additions. Once these alliances were set, cliques tended to close their boundaries once again and to socialize primarily within the group. Kara, a fifth-grade girl, offered her view:

> In the fall, right after school starts, when everyone's lining up and checking each other out, is when people move up, but not during the school year. You can move down during the school year, if people decide they don't like you, but not up.

Most individuals felt that an invitation to membership in the popular clique was irresistible. They asserted repeatedly that the popular group could get any people they wanted to join with them. One of the strategies used by the cliques was to try to select new desirables and seek them out. This usually entailed separating those people from their established friends. Melody, an unpopular fourth-grade girl, described her efforts to hold on to her best friend, who was being targeted for recruitment by the popular clique:

> She was saying that they were really nice and stuff. I was really worried. If she joined their group she would have to leave me. She was over there and she told me that they were making fun of me, and she kind of sat there and went along with it. So I kind of got mad at her for doing that. "Why didn't you stick up for me?" She said, "Because they wouldn't like me anymore."

Melody subsequently lost her friend to the clique.

When clique members wooed someone to join them, they usually showed only the better side of their behavior. The shifts in behavior associated with leaders' dominance and status stratification activities did not begin until the new person was firmly committed to the group. Julie recalled her inclusion in the popular clique, and its aftermath:

> In fifth grade I came into a new class and I knew nobody. None of my friends from the year before were in my class. So I get to school a week late, and Amy comes up to me and she was like, "Hi Julie, how are you? Where were you? You look so pretty." And I was like, wow, she's so nice. And she was being so nice for like, two weeks, kiss-ass major. And then she started pulling her bitch moves. Maybe it was for a month that she was nice. And so then she had clawed me into her clique and her group, and so she won me over that way, but then she was a bitch to me once I was inside it, and I couldn't get out because I had no other friends. 'Cause I'd gone in there and already been accepted into the popular clique, so everyone else in the class didn't like me, so I had nowhere else to go.

Eder (1985) also has noted that popular girls are often disliked by unpopular people because of their exclusive and elitist manner (befitting their status).

Application

A second way in which individuals gain initial membership into a clique is through actively seeking entry (Blau 1964). Several factors influence the likelihood that a person will be accepted as a candidate for inclusion. . . .

According to Brian, a fifth-grade boy who was in the popular clique but not a central member, application for clique entry was accomplished more easily by individuals than by groups. He described how individuals found routes into cliques:

> It can happen any way. Just you get respected by someone, you do something nice, they start to like you, you start doing stuff with them. It's like, you just kind of follow another person who is in the clique back to the clique and he says, "Could this person play?" So you kind of go out with the clique for a while and you start doing stuff with them, and then they almost like invite you in. And then soon after, like a week or so, you're actually in. It all depends. . . . But you can't bring your whole group with you, if you have one. You have to leave them behind and just go in on your own.

Successful membership applicants often experienced a flurry of immediate popularity because their entry required clique leaders' approval, which gave them associational status.

Realignment of Friendships

Status and power in a clique were related to stratification; those who remained more closely tied to the leaders were more popular. Individuals who wanted to be included in the inner circle often had to work regularly to maintain or improve their position.

Like initial entry, this was sometimes accomplished by people striving on their own for upward mobility. Danny was brought into the clique by Tim, a longtime member who went out of his way to befriend him. Soon after joining the clique, however, Danny abandoned Tim when Jesse, the clique leader, took an interest in him. Tim discussed the feelings of hurt and abandonment caused by this experience:

> I felt really bad, because I made friends with him when nobody knew him and nobody liked him, and I put all my friends to the side for him, and I brought him into the group, and then he dumped me. He was my friend first, but then Jesse wanted him. . . . He moved up and left me behind, like I wasn't good enough anymore.

The hierarchical structure of cliques, and the shifts in position and relationships within them, caused friendship loyalties within these groups to be less reliable than they might have been in other groups. People looked toward those above them, and were more susceptible to being wooed into friendship with individuals more popular than themselves. When courted by a higher-up, they could easily drop their less popular friends.

The stratification hierarchies in cliques might motivate lower-echelon members to seek greater inclusion by propelling themselves toward the elite inner circles, but membership in these circles was dynamic; active effort was required to sustain it. More popular individuals also had to invest repeated effort in their friendship alignments to maintain their central positions relative to people just below them, who might rise and gain in group esteem. Efforts to protect themselves from potential incursions by others took several forms, including cooptation, position maintenance, realignment of followers, and challenge to membership, only some of which draw on inclusionary dynamics.

In cooptation, leaders diminished other members' threats to their position by drawing them into their orbit, thus increasing their loyalty and diminishing their independence. Clique members who were gaining in popularity thus sometimes received special attention. At the same time, leaders might try to cut out their rivals' independent base of support from other friends. Melanie, a fourth-grade girl, had occupied a second-tier leadership position with Kristy, her best friend. She explained what happened when Denise, the clique leader, came in and tore apart their long-standing friendship:

> Denise split me and Kristy up. Me and Kristy used to be best friends but she hated that, 'cause even though she was the leader, we were popular and we got all the boys. She didn't want us to be friends at all. But me and Kristy were, like, getting to be a threat to her, so Denise came in the picture and tore me and Kristy apart, so we weren't even friends. She made Kristy make totally fun of me and stuff. And they were so mean to me. . . .

In friendship realignment, clique members abandoned previous friendships or destroyed existing ones in order to assert themselves as part of relationships with those in central positions. All of these actions were geared toward improving the instigators' position and thus assuring their inclusion. The outcome, whether anticipated or not, was often the separation of people and the destruction of their relationships.

Ingratiation

In addition to being wooed into the elite strata and breaking up friendships to consolidate or use power in the group, currying favor with people in the group was another dynamic of inclusion found in clique behavior. Like the previously described inclusionary endeavors, ingratiation can be directed either upward (supplication) or downward (manipulation). Addressing the former, Dodge et al. (1983) noted that children often begin their attempts at entry into groups with low-risk tactics; rather than ingratiating themselves directly with the leader, they first attempt to become accepted by more peripheral members. They elevate their gaze and their attempts at inclusion only later. The children we observed did this as well, making friendly overtures toward clique followers and hoping to be drawn by them into the career. More often, however, group members curried favor with the leader to enhance their popularity and obtain greater respect from other group members. One way of doing this was by imitating the group leaders' style and interests. Marcus and Andy, two fifth-grade boys, described how borderline people fawned on their clique and its leader to try to gain inclusion:

> Marcus: Some people would just follow us around and say, "Oh yeah, whatever he says, yeah, whatever his favorite kind of music is, is my favorite kind of music."
>
> Andy: They're probably in a position where they want to be more *in* because if they like what we like, then they think more people will probably respect them. Because if some people in the clique think this person likes their favorite group, say it's REM or whatever, so it's, say, Bud's [the clique leader's], this person must know what we like in music and what's good and what's not, so let's tell him that he can come up and join us after school and do something.

Not only outsiders and peripherals fawned on more popular people. This was also common practice among regular clique members, even those with high standing. Melanie, the second-tier fourth-grade girl mentioned earlier, described how, in fear, she used to follow the clique leader and parrot her opinions:

> I was never mean to the people in my grade because I thought Denise might like them and then I'd be screwed. Because there were some people that I hated that she liked and I acted like I loved them, and so I would just be mean to the younger kids, and if she would even say, "Oh she's nice," I'd say, "Oh yeah, she's really nice!"

Clique members, then, had to stay abreast of the leader's shifting tastes and whims if they were to maintain status and position in the group. Part of their membership work involved a constant awareness of the leader's fads and fashions, so that they could align their actions and opinions accurately with the current trends, in a timely manner (also see Eder and Sanford 1986).

The art of ingratiating oneself with a clique was not practiced only upward, however. Besides outsiders' supplicating insiders and insiders' supplicating those of higher standing, individuals at the top had to consider the effects of their actions on their standing with those below them. Although leaders did not have to imitate their followers' style and taste, they had to act so as to hold their adulation and loyalty. To begin this process, people at the top made sure that those directly below them remained firmly placed where they could count on them. Any defection, especially by the more popular members of a clique, could threaten their standing. Leaders often employed manipulation to hold clique members' attention and loyalty. Oswald et al. (1987) noted that one way in which children assert superiority over others and obligate them with loyalty is to offer them "help," either materially or socially.

Another technique involved acting in different ways toward different people. Bill, a sixth-grade boy, recalled how the clique leader in fifth grade used this strategy to maintain his position of centrality:

> Mark would always say that Trevor is so annoying. "He is such an idiot, a stupid baby." and everyone would say, "Yeah, he is so annoying. We don't like

> him." So they would all be mean to him. And then later in the day, Mark would go over and play with Trevor and say that everyone else didn't like him, but that he did. That's how Mark maintained control over Trevor.

Mark employed similar techniques of manipulation to ensure that all the members of his clique were similarly tied to him. Like many leaders, he shifted his primary attention among the different clique members, so that everyone enjoyed the power and status associated with his favor. Then, when his followers were out of favor, they felt relatively deprived and strove to regain their privileged status. This process ensured their loyalty and compliance.

To a lesser degree, clique members curried friendship with outsiders. Although they did not accept them into the group, they sometimes included them in activities and tried to influence their opinions. While the leaders had their in-group followers, lower-status clique members could look to outsiders for respect, admiration, and imitation if they cultivated them carefully. This attitude and this behavior were not universal, however, because some popular cliques were so disdainful and so unkind to outsiders that nonmembers hated them. Diane, Jennifer, and Alyssa, three popular junior high school girls who had gone to two different elementary schools, described how the grade school cliques to which they had belonged displayed different relationships with individuals of lesser status:

> Diane: We hated it if the dorks didn't like us and want us to be with them. 'Cause then we weren't the popularest ones, 'cause we always had to have them look up to us, and when they wouldn't look up to us we would be nice to them.
>
> Jennifer: The medium people always hated us.
>
> Alyssa: They hated us royally and we hated them back whenever they started.

Thus, despite notable exceptions (as described by Eder 1985), many popular-clique members strove from time to time to ingratiate themselves with people less popular than themselves, to ensure that their dominance and adulation extended beyond their own boundaries, throughout the grade.

TECHNIQUES OF EXCLUSION

Individuals enhanced their own and others' status by maneuvering into more central and more powerful positions and/or recruiting others into such positions. These inclusionary techniques reinforced their popularity and prestige while maintaining the group's exclusivity and stratification. Yet the inclusionary dynamics failed to contribute to other, essential clique features such as cohesion and integration, the management of in-group and out-group relationships, and submission to the clique's leaders. These features are rooted, along with other sources of domination and power, in the exclusionary dynamics of cliques.

Exclusionary techniques illuminate how clique leaders enhanced their elite positions by disdaining and deriding others lower in the prestige hierarchy both inside and outside their cliques, thus supporting their power and authority on the foundation of others' subservience. These very techniques fostered clique solidarity, however, because members developed internal cohesion through their collective domination over others, and were tied to the leaders by their fear of derision and exclusion by the leader-dominated group.

Subjugation of the Out-Group

When clique members were not being nice to outsiders to try to keep them from straying too far outside their influence, they largely subjected them to exclusion and rejection. Insiders were entertained by picking on these lower-status individuals. As one clique follower remarked, "One of the main things is to keep picking on unpopular kids because it's just fun to do." Eder (1991) observed that this kind of ridicule, in which the targets are excluded and are not encouraged to join in the laughter, contrasts with teasing, in which friends make fun of each other in a more lighthearted manner but permit the targets to remain in the group by also jokingly making fun of themselves. Hilary, a fourth-grade clique leader, described how she acted toward outsiders:

> Me and my friends would be mean to the people outside of our clique. Like, Eleanor Dawson, she would always try to be friends with us, and we would be like, "Get away, ugly."

Interactionally sophisticated clique members not only treated outsiders badly, but managed to turn others in the clique against them. Parker and Gottman (1989) observed that gossip is one way of doing this. Hilary recalled how she turned all the members of her class, boys as well as girls, against an outsider:

> I was always mean to people outside my group like Crystal, and Emily Fiore; they both moved schools. . . . I had this gummy bear necklace, with pearls around it and gummy bears. She came up to me one day and pulled my necklace off . . . It was my favorite necklace, and I got all of my friends, and all the guys even in the class, to revolt against her. No one liked her. That's why she moved schools, because she tore my gummy bear necklace off and everyone hated her. They were like. "That was mean. She didn't deserve that. We hate you."

Turning people against an outsider solidified the group and asserted the power of the strong over the vulnerability of the weak. Other classmates tended to side with the dominant people over the subordinates, not only because they admired their prestige but also because they respected and feared the power of the strong.

In the ultimate manipulation in leading the group to pick on outsiders, insiders instigated the bullying and caused others to take the blame. Robert, a fourth-grade clique follower, described with some mystification and awe the skilled maneuvering of Scott, his clique leader:

> He'd start a fight and then he would get everyone in it, 'cause everyone followed him, and then he would get out of it so he wouldn't get in trouble.
>
> *Q: How'd he do that?*
>
> Robert: One time he went up to this kid Hunter Farr, who nobody liked, and said, "Come on Farr, you want to talk about it?" and started kicking him, and then everyone else started doing it. Scott stopped and started watching, and then some para-pro[fessional] came over and said "What's going on here?" And then everyone got in trouble except for him.
>
> *Q: Why did he pick on Hunter Farr?*
>
> Robert: 'Cause he [Farr] couldn't do anything about it, 'cause he was a nerd.

Being picked on instilled outsiders with fear, grinding them down to accept their inferior status and discouraging them from rallying together to challenge the power hierarchy (see Eder and Sanford 1986). In a confrontation between a clique member and an outsider, most people sided with the clique member. They knew that clique members banded together against outsiders, and that they themselves could easily become the next target of attack if they challenged them. Clique members picked on outsiders with little worry about confrontation or repercussion. They also knew that their victims would never carry the tale to teachers or administrators (as they might in dealing with other targets: see Sluckin 1981) for fear of reprisal. As Matt, a fifth-grade clique follower, observed, "They know if they tell on you, then you'll beat them up, and so they won't tell on you. They just kind of take it in, walk away."

Subjugation within the In-Group

A second form of domination occurred through picking on people within the clique. More central clique members commonly harassed and were cruel to those with lesser standing.[1] Many of the same factors that prompted the ill-treatment of outsiders motivated high-level insiders to pick on less powerful insiders. Craig, a sixth-grade clique follower, articulated the systematic organization of downward harassment:

> Basically the people who are the most popular, their life outside in the playground is picking on other people who aren't as popular, but are in the group. But the people just want to be more popular so they stay in the group. They just kind of stick with it, get made fun of, take it. . . . They come back every day, you do more ridicule, more ridicule, more ridicule, and they just keep taking it because they want to be more popular, and they actually like you but you don't like them. That goes on a lot, that's the main thing in the group. You make fun of someone, you get more popular, because insults is what they like. They like insults.

The moving finger of ridicule was capricious, and could stop at any individual but the leader. It might turn toward a person because he or she did something deserving insult; it might be directed toward someone who the clique leader felt had become an interpersonal threat; or it might fall on someone for no apparent reason (see Eder 1991). Melanie, the second-tier fourth-grade girl discussed earlier, described the ridicule she encountered and told of her feelings of mortification when the clique leader derided her hair.[2]

> Like I remember, she embarrassed me so bad one day. Oh my God, I wanted to kill her! We were in music class and we were standing there and she goes. "Ew! What's all that shit in your hair?" in front of the whole class. I was so embarrassed 'cause I guess I had dandruff or something.

Derision against insiders often followed a pattern: leaders started a trend and everyone followed it. This multiple force intensified the sting of the mockery. Jeff, a fifth-grade boy, compared the behavior of people in cliques to the links on a chain:

> Like it's a chain reaction. You get in a fight with the main person, then the person right under him will not like you, and the person under him won't like you, and etcetera, and the whole group will take turns against you. A few people will still like you because they will do their own thing, but most people will do what the person in front of them says to do, so it would be like a chain reaction. It's like a chain. One chain turns and the other chain has to turn with them, or else it will tangle.

Compliance

When leaders or other high-status clique members initiated such negative and wounding power dynamics, others followed, participating either actively or passively in the derision. Active participation occurred when instigators persuaded other clique members to become involved in picking on their friends. This often happened in telephone prank calling, when leaders conceived the idea of making trick calls and convinced their followers to do the dirty work. They might start the call and then place followers on the line to finish it, or they might pressure others to make the entire call, thus keeping one step away from becoming implicated if the victim's parents should complain.

In passive participation, followers went along when leaders were mean and manipulative, as when Ryan acquiesced in Brad's scheme to convince Larry that Rick had stolen his money. Ryan knew that Brad was hiding the money, but he watched while Brad whipped Larry into a frenzy, pressing him to deride Rick, destroy Rick's room and possessions, and threaten to expose Rick's alleged theft to others. Only when Rick's mother came home, interrupting the bedlam, was the money revealed and Larry's onslaught stopped. The following day at school, Brad and Ryan could scarcely contain their glee. Rick was demolished by the incident and was cast out by the clique. Ryan was elevated to the status of Brad's best friend by his conspiracy in the scheme.

Many clique members relished the opportunity to go along with such exclusive activities, welcoming the feelings of privilege, power, and inclusion. Others appreciated the absence of ridicule towards themselves. This situation sometimes was valued by new members, who often feel unsure about their standing in a group (Sanford and Eder 1984). Two fifth-grade clique followers expressed their different feelings about such participation:

> *Q: What was it like when someone in your group got picked on?*
>
> Gary: If it was someone I didn't like or who had picked on me before, then I liked it. It made me feel good.

Nick: I didn't really enjoy it. It made me feel better if they weren't picking on me. But you can't do too much about it, so you sort of get used to it.

Like outsiders, clique members knew that complaining to persons in authority did them no good. Quite the reverse: such tactics made their situation worse. So did showing their vulnerabilities to the aggressors. Kara, a popular fifth-grade girl, explained why such declarations had the opposite of the intended effect:

> Because we knew [it] bugged them, so we could use [it] against them. And we just did it to pester 'em, aggravate 'em, make us feel better about ourselves. Just to be shitty.

When people saw their friends in tenuous situations, they often reacted passively. Popular people who got into fights with other popular people might be able to count on some of their followers for support, but most people could not command such loyalty. Jeff, the fifth-grade boy discussed earlier, explained why people went along with hurtful behavior.

> It's a real risk if you want to try to stick up for someone because you could get rejected from the group or whatever. Some people do and nothing happens because they're so high up that other people listen to them. But most people would just find themselves in the same boat. And we've all been there before, so we know what that's like.

Clique members thus cooperated in picking on their friends, even though they knew it hurt, because they were afraid (also see Best 1983). They became accustomed to living in a social world where the power dynamics could be hurtful, and accepted it.

Stigmatization

Beyond individual incidents of derision, clique insiders often were made the focus of stigmatization for longer periods. Unlike outsiders, who commanded less enduring interest, clique members were much more deeply involved in picking on their friends, whose discomfort held their attention more readily. Jeff described how negative attention could focus on a person for longer than a single incident:

> Usually at certain times, it's just a certain person you will pick on all the time, if they do something wrong. I've been picked on for a month at a time, or a week, or a day, or just a couple of minutes, and then they will just come to respect you again.

When people became the focus of stigmatization, as happened to Rick above, they were rejected by all their friends. The entire clique rejoiced in celebrating their disempowerment. They were made to feel alone whenever possible. Their former friends might join hands and walk past them at recess, physically demonstrating their union and the discarded person's aloneness.

Worse than being ignored was being taunted. Anyone who could create a taunt was favored with attention and imitated by everyone else (also see Fine

1981). Taunting included verbal insults, put-downs, and sing-song chants. Even outsiders, who normally were not privileged to pick on a clique member, could elevate themselves by joining such taunting (also see Sanford and Eder 1984).

The ultimate degradation was physical. Although girls generally confined themselves to verbal humiliation, the culture of masculinity allowed boys to injure each other (Eder and Parker 1987; Oswald et al. 1987; Thorne 1993). Fights occasionally broke out in which boys were punched in the ribs or stomach, kicked, or given black eyes. When this happened at school, adults were quick to intervene. After hours or on the school bus, however, boys could be hurt. Physical abuse was also heaped on people's homes or possessions. People spat on each other or on others' books or toys, threw eggs at their families' cars, and smashed pumpkins in front of their houses.

Expulsion

Most people returned to a state of acceptance after a period of severe derision (see Sluckin 1981 for strategies used by children to help attain this end), but this was not always the case. Some people were excommunicated permanently from the clique. Others could be cast out directly without undergoing a transitional phase of relative exclusion. This could happen to clique members from any stratum of the group, although it was more likely among people with lower status. Jason, a sixth-grade boy, described how expulsion could occur as a natural result of the hierarchical ranking, in which a person at the bottom rung of the popularity ladder was pushed off. He described the ordinary dynamics of clique behavior.

> *Q: How do they decide who they are going to insult that day?*
>
> Jason: It's just basically everyone making fun of everyone. The small people making fun of smaller people, the big people making fun of the small people. Nobody is really making fun of people bigger than them because they can get rejected . . . then they can say, "Oh yes, he did this and that, this and that, and we shouldn't like him anymore." And everybody else says, "Yeah, yeah, yeah," 'cause all the lower people like him, but all the higher people don't. So the lower-case people just follow the higher-case people. If one person is doing something wrong, then they will say "Oh yeah, get out, good-bye."

Being cast out could result either from a severely irritating infraction or from individuals' standing up for their rights against the dominant leaders.

Sometimes expulsion occurred as a result of breakups between friends or realignments in friendship leading to membership challenges (described earlier), in which higher-status people carried the group with them and turned their former friends into outcasts. Adam was able to undercut Kevin's rising popularity and to coopt his friendship by cutting off the support of Kevin's longtime friend Nick. Adam pretended to be best friends with both Kevin and Nick, but secretly encouraged each one to say bad things about the other so that he could carry each one's nasty remarks back to the other. By escalating these hurtful remarks, Adam finally wounded Nick severely. Nick retaliated against Adam in a way that Adam used to turn both Kevin and the whole clique against Nick, and to expel him from the group.

On much rarer occasions, high-status clique members or even leaders could be cast out of the group (see Best 1983). One sixth-grade clique leader, Tiffany, was deposed by her former lieutenants for continued petulance and self-indulgent manipulations. She recounted the moment of her expulsion:

> Do you want to know why I turned dweeb? Because they kicked me out of the clique.
>
> *Q: Who kicked you out?*
>
> Tiffany: Robin and Tanya. They accepted Heidi into their clique and they got rid of me. They were friends with her. I remember it happened in one blowup in the cafeteria. I asked for pizza and I thought I wasn't getting enough attention anymore, so I was pissed and in a bitchy mood all the time and stuff, and so I asked them for some, so she said like "Wait, hold on, Heidi is taking a bite," or something, and I got so mad I said "Give the whole fuckin' thing to Heidi" and something like that, and they got so sick of me right then, and they said, like, "Fuck you."

When clique members are kicked out of the group, they leave an established circle of friends and often seek to make new ones. Some people find it relatively easy to make what Davies (1982) called "contingency friends" (temporary replacements for their more popular friends), and were described by a fifth-grade teacher as "hot items" for the unpopular crowd. James, a sixth-grade clique follower, explained why people expelled from a popular clique might be in demand with nonclique members:

> Because they want more people, who are bigger, who have more connections, because if you get kicked out of the group, usually you still have a friend who is still going to be in the group, so then they can say, "Oh yeah, we'll be more popular even though this person isn't respected anymore. At least there is one person who still respects them in the group, so he'll get a little higher up or more popular, or we just should give him a chance."

Many cast-outs, however, found new friendships harder to establish. They went through a period when they kept to themselves, feeling rejected, stigmatized, and cut off from their former social circle and status. Because of their previous behavior and their relations with other classmates, they had trouble being accepted by unpopular children. Others had developed minimum acceptability thresholds for friends when they were in the popular crowd, and had difficulty stooping to befriend unpopular people. Todd, a fifth-grade boy who was ejected from his clique, explained why he was unsuccessful in making friends with the unpopular people:

> Because there was nobody out there I liked. I just didn't like anybody. And I think they didn't like me because when I was in the popular group we'd

make fun of everyone, I guess, so they didn't want to be around me, because I had been too mean to them in the past.

Rejects from the popular clique occasionally had trouble making friends among the remainder of the class because of interference by their former friends. If clique members became angry at one of their friends and cast that person out, they might want to make sure that nobody else befriended him or her. By soliciting friendship with people outside the clique, they could influence outsiders' behavior, causing their outcast to fall beyond the middle crowd to the status of pariah or loner. Melanie, a fourth-grade popular girl, explained why and how people performed such manipulations:

> *Q: Have you ever seen anyone cast out?*
>
> Melanie: Sure, like, you just make fun of them. If they don't get accepted to the medium group, if they see you, like "Fuck, she's such a dork," and like you really don't want them to have any friends, so you go to the medium group, and you're like "Why are you hanging out with *that* loser, she's *such* a dork, we *hate* her," and then you be nice to them so they'll get rid of her so she'll be such a dork. I've done that just so she'll be such a nerd that no one will like her. You're just getting back at them. And then they will get rid of her just 'cause you said to, so then you've done your way with them. If you want something, you'll get it.

People who were cast out of their group often kept to themselves, staying in from the playground at recess and going home alone after school. They took the bus to school, went to class, and did what they had to do, but they didn't have friends. Their feelings about themselves changed; this was often reflected in the way they dressed and carried themselves. Being ejected from the clique thus represented the ultimate form of exclusion, carrying severe consequences for individuals' social lives, appearance, and identity.

NOTES

1. Eder (1991) also noted that when insiders picked on other members of their clique, this could have good-natured overtones, an indication that they liked them.

2. Eder and Sanford (1986) and Eder and Parker (1987) discussed the importance of physical appearance, particularly hair, in adhering to group norms and maintaining popularity.

REFERENCES

Adler, Patricia and Peter Adler. 1996 "Preadolescent Clique Stratification and the Hierarchy of Identity." *Sociological Inquiry* 66: 111–142.

Best, Raphaela. 1983. *We've All Got Scars.* Bloomington: Indiana University Press.

Blau, Peter M. 1964. *Exchange and Power in Social Life.* New York: Wiley.

Davies, Bronwyn. 1982. *Life in the Classroom and Playground: The Accounts of Primary School Children.* London: Routledge & Kegan Paul.

Dodge, Kenneth A., David C. Schlundt, Iris Schocken, and Judy D. Delugach. 1983. "Social Competence and Children's Sociometric Status: The Role of Peer Group Entry Strategies." *Merrill-Palmer Quarterly* 29(3): 309–36.

Eder, Donna. 1985. "The Cycle of Popularity: Interpersonal Relations Among Female Adolescents." *Sociology of Education* 58: 154–65.

———. 1991. "The Role of Teasing in Adolescent Peer Group Culture." Pp. 181–97 in *Sociological Studies of Child Development,* Volume 4, edited by Spencer Cahill. Greenwich, CT: JAI.

Eder, Donna and Stephen Parker. 1987. "The Cultural Production and Reproduction of Gender: the Effect of Extracurricular Activities on Peer-Group Culture." *Sociology of Education* 60(3): 200–13.

Eder, Donna and Stephanie Sanford. 1986. "The Development and Maintenance of Interactional Norms Among Early Adolescents." Pp. 283–300 in *Sociological Studies of Child Development,* Volume 1, edited by Patricia A. Adler and Peter Adler. Greenwich, CT: JAI.

Elkin, Fred and Gerald Handel. 1989. *The Child and Society.* New York: Random House.

Fine, Gary Alan. 1981. "Friends, Impression Management, and Preadolescent Behavior." Pp. 29–52 in *The Development of Children's Friendships,* edited by Steven Asher and John Gottman, Cambridge: Cambridge University Press.

Hallinan, Maureen. 1979. "Structural Effects on Children's Friendships and Cliques." *Social Psychology Quarterly* 42: 43–54.

Hubbell, C. H. 1965. "An Input-Output Approach to Clique Identification." *Sociometry* 28: 377–99.

Oswald, Hans, Lothar Krappmann, Irene Chowdhuri, and Maria von Salisch. 1987. "Gaps and Bridges: Interactions Between Girls and Boys in Elementary School." Pp. 205–23 in *Sociological Studies of Child Development,* Volume 2, edited by Patricia A. Adler and Peter Adler. Greenwich, CT: JAI.

Parker, Jeffrey G. and John M. Gottman. 1989. "Social and Emotional Development in a Relational Context: Friendship Interaction from Early Childhood to Adolescence." Pp. 95–132 in *Peer Relationships in Child Development,* edited by Thomas Berndt and Gary Ladd. New York: Wiley.

Peay, Edmund R. 1974. "Hierarchical Clique Structures." *Sociometry* 37: 54–65.

Sanford, Stephanie and Donna Eder. 1984. "Adolescent Humor During Peer Interaction." *Social Psychology Quarterly* 47(3): 235–43.

Sluckin, Andy. 1981. *Growing Up in the Playground.* London: Routledge & Kegan Paul.

Thorne, Barrie. 1993. *Gender Play.* New Brunswick, NJ: Rutgers University Press.

17

Pharmacists and Illicit Prescription Drug Use

DEAN A. DABNEY AND RICHARD C. HOLLINGER

Some people have greater structural opportunity to step outside the social boundaries and commit deviance than others, as we see in Dabney and Hollinger's occupational study of drug-abusing pharmacists. We see the combined influence of availability and a medical education emphasizing the effects and curative features of prescription medications on individuals working continuously in an underregulated, drug-filled environment. Like other medical personnel, pharmacists move progressively into drug abuse due to the stresses of their work, a ready supply of stimulant and pain-killing medications, and a technical training that prompts positive attitudes toward the curative effects of drugs without adequate socialization into the risks and propensities of substance abuse and addiction. This is a study in white-collar crime, an under-investigated form of criminal activity that takes place among many seemingly legitimate citizens, such as executives, lawyers, and accountants. Do the findings in this study surprise you? Have you ever considered the opportunities available to doctors, nurses, and other medical personnel to abuse drugs? More generally, are pharmacists how you envision criminals? What can this study tell us more generally about the nature of criminal activity in society?

A heightened level of social status and respect accompanies membership in an occupational profession (Greenwood, 1957). An example of this is the pharmacy profession. Ten years of public opinion polls consistently rank pharmacy as the most honest and ethical occupation, even above the clergy (McAneny & Moore, 1994). This elevated social status can be partially attributed to the specialized training and knowledge that pharmacists possess about the intricacies of prescription medicines and their pharmacological effects on the human body. Given their level of expertise, society freely yields its respect, trust, and admiration to those who are believed to be authorities on the safe and effective use of prescription drugs and medicines.

From *Work and Occupations,* Vol. 26 (1), by D. Dabney and R. Hollinger. Reprinted by permission of Sage Publications, Inc.

Pharmacists are granted the primary jurisdiction over the day-to-day dispensing of a variety of controlled pharmaceuticals to the public. This dispensing function is accompanied by numerous privileges and responsibilities. For example, pharmacists are expected to serve as the front line of defense against the inappropriate use or unauthorized acquisition of prescription medicines. Furthermore, they serve as mediators between the prescribing doctor and the patient, guarding against potentially undesirable health consequences such as adverse drug interactions, allergic reactions, or inappropriate and unnecessary prescription medication use. Pharmacists are accorded virtually unrestricted access to a cache of potent drugs and medicines far exceeding that of any other member of society. Collectively, these responsibilities and privileges act to further solidify the pharmacist's position as society's drug expert. As the above-mentioned integrity polls suggest, Americans seem to be quite satisfied with the ways in which pharmacists exercise their knowledge, training, and access to fill 1.6 billion prescriptions annually (Wivell & Wilson, 1994).

The present analysis identifies a potential negative effect that can result from the technical training, expertise, and familiarity that pharmacists have with prescription medicines. In particular, factors related to the process of being and becoming a practicing pharmacist can be linked to the most serious form of professional misconduct that faces the profession today—illicit prescription drug use.[1] Interviews with 50 pharmacists recovering from prescription drug abuse demonstrate how a pharmacist's self-confidence, knowledge, and virtually unrestricted access to prescription drugs can contribute to the eventual abuse of these substances. . . .

PHARMACISTS' DRUG USE TRENDS AND PATTERNS

Each of the 50 pharmacists who were interviewed spoke at length and in detail about their personal drug use histories. As expected, there were many unique aspects to each individual's past drug abuse; however, the thematic content analysis revealed several consistent trends and patterns in their drug use behaviors. . . .

Titrating

An interesting trend emerged from the inquiry into the development of drug tolerance. Early in the interview process, pharmacists began speaking about a drug use practice called *titrating.* This term refers to a practice whereby individuals apply their pharmaceutical knowledge to manage their personal drug use, enhancing or neutralizing specific drug effects by ingesting counteracting drugs. In effect, they would walk a chemical tightrope that allowed them to remain high, function, and disguise the obvious physical signs of drug abuse. For example, one 44-year-old male pharmacist said,

> When I was out partying I could drink. See, I didn't drink at work. That's one of the ways that I got heavier into the benzos [benzodiazepines] and the

> barbs [barbiturates]. . . . You could take something like that and get the sedative effects, but it wouldn't smell on your breath. So I would not drink, but basically I had to wake up. . . . I would put about four Percosets [narcotic analgesic] and four biphetamine 20 mg or Dexedrine [amphetamine] with the Percoset. When it was time to roll out of bed, to try to get to work, I would swallow all that and wait a little bit until it would start to kick in. And then I would just compulsively swallow whatever Percoset I could get my hands on. And swallow amphetamines as needed to titrate my energy levels to being productive and not to appear impaired from the opiate.

This pharmacist used his titrating knowledge to achieve three different goals: (a) physical euphoria, (b) avoidance of negative side-effects, and (c) avoidance of detection. All three of these themes were found in varying degrees among the 28 pharmacists who recounted titrating practices.

Regardless of the origin, there is clear evidence that titrating behavior evolves from pharmaceutical training. Most pharmacists did not hesitate to state that they had learned how to titrate by applying what they had learned in class lectures or by reading books or articles on pharmacology. This is illustrated in the following exchange between the interviewer (I) and a 56-year-old male pharmacist (P).

> **I:** It seems like you were really putting your expertise to work there?
>
> **P:** Yep, all my knowledge of pharmacology so I could get perfectly titrated. And I'd go in there [to work], and 15 minutes later I could be snowed over.

Widespread presence of titrating offers important support for our assertion that being and becoming a pharmacist affects the individuals' drug use. Clearly, these pharmacists were exploiting their access to prescription drugs and applying their educationally acquired knowledge to enhance and inform their drug use.

PARADOX OF FAMILIARITY

Without exception, every interviewed pharmacist saw his- or herself as a drug expert. They argued that pharmacists, more so than doctors, were best prepared to dispense and counsel patients about the nature and dynamics of prescription drugs. For example, a 33-year-old male pharmacist said,

> You are the guardian of their health. The doctor might not know what they're doing. It's your idea to make sure that the right meds are used. We are supposed to question the doctors in a good way. We say "Hey look, don't come off arrogant, but hey doc, why are you doing this?"

This individual emphasizes the important role that pharmacists play in the large health care delivery system. This was a common sentiment offered by many respondents.

The naive observer might expect that a strong professional identity as well as extensive knowledge regarding the effects of prescription drugs should be the perfect deterrent against abuse. However, we found that this was not always the case. Often, addicted pharmacists described how their intimate familiarity with prescription medicines actually was a contributing factor to abuse. In particular, the interview data show that the professional socialization process exposes pharmacists to a dangerous combination of both access to drugs and detailed knowledge about them. This professional pharmacy socialization process produces what we wish to call a paradox of familiarity.

The seeds of this paradox can be traced back to the pharmacy school experience. The interview data reveal that pharmacy school offers students very limited training in the dangers of addiction. For example, when asked about his addiction education, one pharmacist offered up the following types of responses:

> I had no [drug abuse] education. I was a drug expert and knew nothing about the [addictive effects of these] substances.
>
> It was just that junkies shoot heroin.
>
> Never did we ever have a class like "addiction." Never did anyone ever come in.
>
> Absolutely not. That was not even considered. That never came up.
>
> That was the sum total of pharmacy education on substance abuse. "Keep your hands off the stuff, ha, ha, ha." That's it, that's all I ever learned.

What little drug education training these pharmacists did receive was usually quite technical and rudimentary. For each drug type, they were made aware of the addiction potential ratings that are contained in the Controlled Substance Act of 1970 (Shulgin, 1992). Pharmacy students were provided general information about the signs of abuse that accompany various controlled substances. This is not to say that pharmacists in training are not told about the dangers of drugs. On the contrary, instructors did stress that people can and do get addicted to prescription medicines. However, the message was conveyed on a very general and impersonal level. Rarely were they told that pharmacists just like them could get addicted to prescription medicines. This can be seen in the comments of a 59-year-old male pharmacist:

> In school it was cold, clinical—"Yes, this can be habit forming and addictive." But as far as the addiction process is described in detail, you know, the mental and the physical part of it and how those interact, and all those self-esteem issues, that's totally lost in it. It's just totally clinical. . . . They did touch on it, actually, but from a legal and a clinical standpoint. "If you do this, you will get in trouble. There are the penalties, this is what they'll do to you. This class of drugs have this addiction potential."

The insufficient and abstract nature of the drug addiction awareness is also expressed in the following exchange between the interviewer (I) and a 46-year-old male pharmacist (P):

> **I:** Do you think that they teach you everything that you need to know about substance abuse?
>
> **P:** Substances of abuse? Absolutely not.
>
> **I:** Did they teach you anything? I guess would be the better question?

P: Let me tell you, we went on rounds to [a hospital], which was the alcoholic ward. And we saw society's [alcohol problem], we learned how to manage it. We learned about drugs that we used for DTs [delirium tremors] and that was about it. And I think that many classes even after that received very little more than that, basic stuff such as . . . "This is addiction and how to treat it." Not even that, "These are the physical effects of alcoholism, etc. And this is the medical management of those physical effects." No treatment of the disease at all, ever.

Drug use in pharmacy school is further exacerbated by the presence of other educationally related factors. For example, we found widespread evidence that pharmacy school helped foster benign attitudes toward prescription drugs. This belief system was affected by easy access to prescription medication, relaxed attitudes toward occasional drug use, heavy social drinking and drug use in pharmacy fraternities and other social get-togethers, and exposure to drug-using pharmacy mentors (i.e., internship preceptors).

A major element of the paradox of familiarity is rooted in a pharmacist's constant exposure to prescription drugs. Throughout a pharmacist's career (both during and after college), his or her every day is filled with repeated contact to a host of prescription medicines. This continued exposure to pharmaceutical company representatives and their sample medicines effectively erodes the individual's fear of drug addiction danger. For example, a 41-year-old male pharmacist commented,

I: You say that work desensitized you further. How so?

P: Because of pharmacy school, I didn't realize the dangers of the chemicals, and I see so many prescriptions for Valium [benzodiazepine] and codeine [narcotic analgesic] . . . that over a period of years it seemed okay. I wasn't smoking marijuana, I wasn't doing anti-Baptist alcohol, so I was okay.

The combined effect of constant access to prescription drugs is also seen in the remarks of a 52-year-old female pharmacist:

> Well, the accessibility helped. I mean, I used the profession because you're accessible to all these drugs. And I mean, I know. I'm a trained pharmacist. I know what can make you feel good, and what can make you feel quiet, and all the different kinds of drugs, so my drug knowledge helped me pick the best one for me.

The situation is exacerbated by the multimillion-dollar drug marketing offensive waged by the pharmaceutical companies. Pharmacists in training are constantly being told about the powerful, therapeutic effects of the prescription medicines that they dispense. This reinforcement comes from a variety of sources, such as patient consultations, coworker discussions, professional organizations, and especially personal interactions with sales representatives from the pharmaceutical industry. A 44-year-old male pharmacist stated,

P: That's the way I grew up. . . . If there's a problem—if you have a headache—you can take a pill. If there's a stomachache, you can take a pill. There was a pill for everything.

I: Like, "better living through chemistry"?

P: I think that that's right.

I: Did you buy into that?

P: I think I definitely did. If there was something wrong, regardless if it was one of my kids or whoever it was, we made something that could take away anything that you had. I mean any problem that you had. It wasn't looked at like it is now. Now it's "say no to drugs" and blah, blah, blah. . . .

Exposure to marketing propaganda about prescription drugs was clearly one-sided in nature. The positive aspects of the drugs were never tempered by any real effort to educate the pharmacists about corresponding dangers. For example, a 46-year-old male pharmacist said,

> Yeah, what we learned back in the '70s, was "Yes, this was indeed a problem for society. Isn't it fortunate that pharmacy doesn't have it." We know too much. We know what these drugs will do, so obviously it could never happen to a pharmacist.

Respondents described how this combination of open access and positive professional reinforcement led to a feeling of familiarity and closeness toward the drugs. Pharmacists eventually let their guard down and began to adopt a benign belief system toward prescription medicines. They believed that these drugs could only improve lives; and therefore, they dismissed or minimized the dangers. Self-medication became a viable and attractive form of medicating every problem. The paradox of familiarity is clearly articulated in the following exchange that occurred between the interviewer (I) and a 40-year-old female pharmacist (P):

P: But as far as respect for the medications, yes, I had that right out of school. A lot of respect for the power of good that the medications could do.

I: What do you mean by that?

P: Well, we have a lot of people who are alive today that would not be alive without some of the pharmaceuticals that we have. And that I think is [a source] of respect.

I: What about substance abuse? Did you know anything about it at that point?

P: Zero.

I: Nothing in school or at work?

P: Right. Pretty scary, huh?

The interview data clearly demonstrate that pharmacists were able to deny the dangers and maintain an attitude of invincibility when it came to the issue of drug addiction. As professional pharmacists, they thought that they were immune to drug addiction. For example, a 33-year-old male pharmacist said,

P: I felt that I could handle it better than the average layperson. Because, after all, I'm a professional. So yeah, it was a very cavalier attitude towards drugs. Very cavalier.

I: So you bought into that "I'm a professional" thing?

P: Yeah, and I know what I'm doing. I know what the edge is, and I'm not going to go over the edge, but clearly I was well over the edge. In the end, I got very paranoid, and I got very out of control. And it's hard for me to talk about it, because it's a shameful thing. Because I do consider myself a professional, and I let myself down, and I let a lot of people down. But for me it was a very cavalier—kind of, "I can handle this"—kind of attitude. I know what I'm doing. . . .

In all, 46 of the 50 pharmacists interviewed spoke directly about this paradox of familiarity. All expressed little doubt that this dangerous combination of access and knowledge contributed to the onset of their drug use. This paradoxical mindset did not simply affect pharmacists' decisions to start using prescription drugs. It also seemed to offer the pharmacists a convenient rationalization to continue and even increase their drug use. For example, it was not at all uncommon for interviewees to describe how they continuously broadened their definitions of acceptable use levels. In many cases, pharmacists even described how they used their pharmacy knowledge to fine-tune their drug use, maximizing the drugs' pharmaceutical potential. This tendency can be seen in the comments of a 47-year-old male pharmacist.

P: Yeah, because now that I knew more, I knew how to be more careful about it. To fine-tune my taste. I knew what to stay away from and what to go towards. I knew how to keep from overloading my own system. I knew when I was starting to get toxic, and I could adjust my drug use. Because I was still at the point where I hadn't quite crossed that line yet.

I: So it didn't slow down your use, it didn't cause you to think, "Oh, this is bad." It allowed you to see more exactly how to do it and do it better.

P: Exactly right. For example, access to the pharmaceuticals. If I didn't steal Percodan [narcotic analgesic], I knew that I could take a Tylenol #3 [narcotic analgesic] and something else and enhance the high. I knew how to create that synergistic effect so that much inventory wouldn't be missing. Because it wasn't always Percodan. I would sometimes do a Tylenol #3, and those were so liberally kept, sold by the thousands for a week type of thing, it was such a big mover. So I could keep myself high and keep myself from developing the toxicity by combining other drugs and that sort of thing.

Similarly, a 41-year-old male said,

> I used my training to its fullest potential. Fullest potential. . . . Do you remember John Lily? The guy did a lot of work, experimental work with dolphins. He had a book called the *Center of the Cyclone.* In his book, he made a statement that one cannot consider oneself a true researcher unless one is willing to experiment, do the experiment on oneself. I said, "Exactly." That was one of my defining moments. So I would take that statement. . . . If you didn't try it, you can't consider yourself a true researcher. That's how I viewed myself. I always believed in constant improvement.

As these pharmacists progressed into the later phases of their drug abuse period, they were forced to ponder the significance of their drug use habit. At some point, they came to the realization that they had a drug problem. However, very few individuals voluntarily sought help. Instead, they kept their problem a secret, reasoning that their knowledge would again afford them the vehicle to get themselves out of any predicament that they were in. To them, only uneducated street addicts needed professional drug treatment. . . .

Even when they got caught for stealing or using drugs, many pharmacists still maintained their shield of overconfidence. In the following quote, a 39-year-old female describes what happened when she was hospitalized for drug related health problems.

P: I was judgmental too. You know I [thought] those [drug abusers] were people who had no self-will. In the ER [emergency room] they did treat me like an addict and I'd get very angry, very angry.

I: Why?

P: Well, because I wasn't an addict. I would say things like, "Don't you know who I am? I am a pain specialist in this hospital; I know when I need narcotics," and they, you know, looking back on it, it must have just been humorous. . . .

The above excerpts from personal interviews demonstrate how pharmacists' professional expertise contributes to the detriment of their health. These druggists were all aware that their drug use was wrong. They knew that their employers and the federal law made it illegal for them to remove drugs from pharmacy stock or ingest drugs while on the job. However, they were adept at developing vocabularies of adjustment (Cressey, 1953) or techniques of neutralizations (Sykes & Matza, 1957) to offset negative normative judgments. They came up with a series of excuses or justifications that served as post hoc rationalizations and a priori justifications for their behaviors. Faced with a growing drug problem, they convinced themselves that they were capable of controlling it without any outside assistance. Without exception, these behaviors and their accompanying vocabularies of adjustment were rooted in the experiences and expertise that they had gained while becoming a member of the pharmacy profession.

NOTE

1. In the context of the present study, the term *illicit prescription drug use/user* represents a legal distinction. This concept is meant to refer only to the illegal use of prescription medications as outlined in the Controlled Substance Act of 1970 (Shulgin, 1992). It includes the use of mind-altering, prescription medications when such use is done without a legitimate prescription order that has been signed or authorized by a licensed, FDA-approved physician. The use of prescription medications without an authorizing prescription order constitutes illicit prescription drug use regardless of whether such medications were procured from pharmacy stock, from a street level drug dealer, or any other illicit market source. Our use of the term illicit prescription drug use does not refer to the use of those mind-altering controlled substances that are deemed to

have no medicinal purposes and thus are classified as Schedule I substances under the Controlled Substance Act of 1970 (e.g., marijuana, hashish, heroin, industrial inhalants, and hallucinogens such as LSD; see Shulgin, 1992). Moreover, the term illicit prescription drug use does not include the use of prescription medications when such use is done in accordance with the instructions on a physician authorized prescription order, regardless of how substantial or prolonged the use may be. The term illicit prescription drug use does not include the use or abuse of alcohol. Also, note that this concept carries no functional distinction. It is not intended to speak directly to any physical, emotional, or mental consequences or resulting states of behavior/consciousness associated with an individual's use of any prescription medicine. Issues related to an individual's drug-related behavioral or mental functionality will be referred to under the headings of drug abuse, impairment, or problematic drug use.

REFERENCES

Cressey, D. R. 1953. *Other People's Money.* Glencoe, IL: Free Press.

Greenwood, E. 1957. "Attributes of a Profession." *Social Work* 2:45–55.

McAneny, L. and D. W. Moore. 1994. "Annual Honesty and Ethics Poll." *The Gallup Poll Monthly* 349:2–4.

Shulgin, A. T. 1992. *Controlled Substances: A Chemical Guide to the Federal Drug Laws.* Berkeley, CA: Ronin.

Sykes, G. and D. Matza. 1957. "Techniques of Neutralization: A Theory of Delinquency." *American Journal of Sociology* 22(4):664–670.

Wivell, M. K. and G. L. Wilson. 1994. "Prescription for Harm: Pharmacist Liability." *Trial* 30(5):36–39.

18

Fraternities and Collegiate Rape Culture

A. AYRES BOSWELL AND JOAN Z. SPADE

Some environments create a culture that fosters the commission of deviant behavior more than others, and all-male groups have been particularly identified recently as being prone to nourishing sexist attitudes and behaviors. In particular, fraternities have been linked with the commodification and objectification of women that leads to rape. Boswell and Spade compare the climate at two different college fraternities at the same institution that stand in contrast to each other and note clear distinctions between those fostering a high versus low risk of generating sexually assaultive behavior. Focusing on social interactions featuring both men and women, they were able to compare the differences between high-risk and low-risk fraternity party and bar scenes in their gender relations, treatment of women, and general attitudes toward rape. Although individual men are the ones who rape, Boswell and Spade were able to find that the group norms at some settings promoted certain behaviors and reinforced a rape culture. Could the behavior described in this study happen on your college campus? Why or why not? What is it about male culture that fosters these attitudes? Has society done anything in the last decade to prevent this kind of situation from occurring?

Date rape and acquaintance rape on college campuses are topics of concern to both researchers and college administrators. Some estimate that 60 to 80 percent of rapes are date or acquaintance rape (Koss, Dinero, Seibel, and Cox 1988). Further, 1 out of 4 college women say they were raped or experienced an attempted rape, and 1 out of 12 college men say they forced a woman to have sexual intercourse against her will (Koss, Gidycz, and Wisniewski 1985).

Although considerable attention focuses on the incidence of rape, we know relatively little about the context or the *rape culture* surrounding date and acquaintance rape. Rape culture is a set of values and beliefs that provide an environment conducive to rape (Buchwald, Fletcher, and Roth 1993; Herman 1984). The term

From *Gender & Society*, Vol. 10(2). Reprinted by permission of Sage Publications, Inc.

applies to a generic culture surrounding and promoting rape, not the specific settings in which rape is likely to occur. We believe that the specific settings also are important in defining relationships between men and women.

Some have argued that fraternities are places where rape is likely to occur on college campuses (Martin and Hummer 1989; O'Sullivan 1993; Sanday 1990) and that the students most likely to accept rape myths and be more sexually aggressive are more likely to live in fraternities and sororities, consume higher doses of alcohol and drugs, and place a higher value on social life at college (Gwartney-Gibbs and Stockard 1989; Kalof and Cargill 1991). Others suggest that sexual aggression is learned in settings such as fraternities and is not part of predispositions or pre-existing attitudes (Boeringer, Shehan, and Akers 1991). To prevent further incidences of rape on college campuses, we need to understand what it is about fraternities in particular and college life in general that may contribute to the maintenance of a rape culture on college campuses.

Our approach is to identify the social contexts that link fraternities to campus rape and promote a rape culture. Instead of assuming that all fraternities provide an environment conducive to rape, we compare the interactions of men and women at fraternities identified on campus as being especially *dangerous* places for women, where the likelihood of rape is high, to those seen as *safer* places, where the perceived probability of rape occurring is lower.

RESULTS

The Settings

Fraternity Parties We observed several differences in the quality of the interaction of men and women at parties at high-risk fraternities compared to those at low-risk houses. A typical party at a low-risk house included an equal number of women and men. The social atmosphere was friendly, with considerable interaction between women and men. Men and women danced in groups and in couples, with many of the couples kissing and displaying affection toward each other. Brothers explained that, because many of the men in these houses had girlfriends, it was normal to see couples kissing on the dance floor. Coed groups engaged in conversations at many of these houses, with women and men engaging in friendly exchanges, giving the impression that they knew each other well. Almost no cursing and yelling was observed at parties in low-risk houses; when pushing occurred, the participants apologized. Respect for women extended to the women's bathrooms, which were clean and well supplied.

At high-risk houses, parties typically had skewed gender ratios, sometimes involving more men and other times involving more women. Gender segregation also was evident at these parties, with the men on one side of a room or in the bar drinking while women gathered in another area. Men treated women differently in the high-risk houses. The women's bathrooms in the high-risk houses were filthy, including clogged toilets and vomit in the sinks. When a brother was told of the mess in the bathroom at a high-risk house, he replied, "Good, maybe some of these beer wenches will leave so there will be more beer for us."

Men attending parties at high-risk houses treated women less respectfully, engaging in jokes, conversations, and behaviors that degraded women. Men made a display of assessing women's bodies and rated them with thumbs up or thumbs down for the other men in the sight of the women. One man attending a party at a high-risk fraternity said to another, "Did you know that this week is Women's Awareness Week? I guess that means we get to abuse them more this week." Men behaved more crudely at parties at high-risk houses. At one party, a brother dropped his pants, including his underwear, while dancing in front of several women. Another brother slid across the dance floor completely naked.

The atmosphere at parties in high-risk fraternities was less friendly overall. With the exception of greetings, men and women rarely smiled or laughed and spoke to each other less often than was the case at parties in low-risk houses. The few one-on-one conversations between women and men appeared to be strictly flirtatious (lots of eye contact, touching, and very close talking). It was rare to see a group of men and women together talking. Men were openly hostile, which made the high-risk parties seem almost threatening at times. For example, there was a lot of touching, pushing, profanity, and name calling, some done by women.

Students at parties at the high-risk houses seemed self-conscious and aware of the presence of members of the opposite sex, an awareness that was sexually charged. Dancing early in the evening was usually between women. Close to midnight, the sex ratio began to balance out with the arrival of more men or more women. Couples began to dance together but in a sexual way (close dancing with lots of pelvic thrusts). Men tried to pick up women using lines such as "Want to see my fish tank?" and "Let's go upstairs so that we can talk; I can't hear what you're saying in here."

Although many of the same people who attended high-risk parties also attended low-risk parties, their behavior changed as they moved from setting to setting. Group norms differed across contexts as well. At a party that was held jointly at a low-risk house with a high-risk fraternity, the ambience was that of a party at a high-risk fraternity with heavier drinking, less dancing, and fewer conversations between women and men. The men from both high- and low-risk fraternities were very aggressive; a fight broke out, and there was pushing and shoving on the dance floor and in general.

As others have found, fraternity brothers at high-risk houses on this campus told about routinely discussing their sexual exploits at breakfast the morning after parties and sometimes at house meetings (cf. Martin and Hummer 1989; O'Sullivan 1993; Sanday 1990). During these sessions, the brothers we interviewed said that men bragged about what they did the night before with stories of sexual conquests often told by the same men, usually sophomores. The women involved in these exploits were women they did not know or knew but did not respect, or *faceless victims.* Men usually treated girlfriends with respect and did not talk about them in these storytelling sessions. Men from low-risk houses, however, did not describe similar sessions in their houses. . . .

Gender Relations

Relations between women and men are shaped by the contexts in which they meet and interact. As is the case on other college campuses, *hooking up* has replaced dating on this campus, and fraternities are places where many students hook up. Hooking up is a loosely applied term on college campuses that had different meanings for men and women on this campus.

Most men defined hooking up similarly. One man said it was something that happens

> when you are really drunk and meet up with a woman you sort of know, or possibly don't know at all and don't care about. You go home with her with the intention of getting as much sexual, physical pleasure as she'll give you, which can range anywhere from kissing to intercourse, without any strings attached.

The exception to this rule is when men hook up with women they admire. Men said they are less likely to press for sexual activity with someone they know and like because they want the relationship to continue and be based on respect.

Women's version of hooking up differed. Women said they hook up only with men they cared about and described hooking up as kissing and petting but not sexual intercourse. Many women said that hooking up was disappointing because they wanted longer-term relationships. First-year women students realized quickly that hook-ups were usually one-night stands with no strings attached, but many continued to hook up because they had few opportunities to develop relationships with men on campus. One first-year woman said that "70 percent of hook-ups never talk again and try to avoid one another; 26 percent may actually hear from them or talk to them again, and 4 percent may actually go on a date, which can lead to a relationship." Another first-year woman said, "It was fun in the beginning. You get a lot of attention and kiss a lot of boys and think this is what college is about, but it gets tiresome fast."

Whereas first-year women get tired of the hook-up scene early on, many men do not become bored with it until their junior or senior year. As one upperclassman said, "The whole game of hooking up became really meaningless and tiresome for me during my second semester of my sophomore year, but most of my friends didn't get bored with it until the following year."

In contrast to hooking up, students also described monogamous relationships with steady partners. Some type of commitment was expected, but most people did not anticipate marriage. The term *seeing each other* was applied when people were sexually involved but free to date other people. This type of relationship involved less commitment than did one of boyfriend/girlfriend but was not considered to be a hook-up.

The general consensus of women and men interviewed on this campus was that the Greek system, called "the hill," set the scene for gender relations. The predominance of Greek membership and subsequent living arrangements segregated men and women. During the week, little interaction occurred between women and men after their first year in college because students in fraternities or sororities live and dine in separate quarters. In addition, many non-Greek upper-class students move off campus into apartments. Therefore, students see each other in classes or in the library, but there is no place where students can just hang out together.

Both men and women said that fraternities dominate campus social life, a situation that everyone felt limited opportunities for meaningful interactions. One senior Greek man said,

> This environment is horrible and so unhealthy for good male and female relationships and interactions to occur. It is so segregated and male dominated. . . . It is our party, with our rules and our beer. We are allowing these women and other men to come to our party. Men can feel superior in their domain.

Comments from a senior woman reinforced his views: "Men are dominant; they are the kings of the campus. It is their environment that they allow us to enter; therefore, we have to abide by their rules." A junior woman described fraternity parties as

> good for meeting acquaintances but almost impossible to really get to know anyone. The environment is so superficial, probably because there are so many social cliques due to the Greek system. Also, the music is too loud and the people are too drunk to attempt to have a real conversation, anyway.

Some students claim that fraternities even control the dating relationships of their members. One senior woman said, "Guys dictate how dating occurs on this campus, whether it's cool, who it's with, how much time can be spent with the girlfriend and with the brothers." Couples either left campus for an evening or hung out separately with their own same-gender friends at fraternity parties, finally getting together with each other at about 2 A.M. Couples rarely went together to fraternity parties. Some men felt that a girlfriend was just a replacement for a hook-up. According to one junior man, "Basically a girlfriend is someone you go to at 2 A.M. after you've hung out with the guys. She is the sexual outlet that the guys can't provide you with."

Some fraternity brothers pressure each other to limit their time with and commitment to their girlfriends. One senior man said, "The hill [fraternities] and girlfriends don't mix." A brother described a constant battle between girlfriends and brothers over who the guy is going out with for the night, with the brothers usually winning. Brothers teased men with girlfriends with remarks such as "whipped" or "where's the ball and chain?" A brother from a high-risk house said that few brothers at his house had girlfriends; some did, but it was uncommon. One man said that from the minute he was a pledge he knew he would probably never have a girlfriend on this campus because "it was just not the norm in my house. No one has girlfriends; the guys have too much fun with [each other]."

The pressure on men to limit their commitment to girlfriends, however, was not true of all fraternities or of all men on campus. Couples attended low-risk fraternity parties together, and men in the low-risk houses went out on dates more often. A man in one low-risk house said that about 70 percent of the members of his house were involved in relationships with women, including the pledges (who were sophomores).

Treatment of Women

Not all men held negative attitudes toward women that are typical of a rape culture, and not all social contexts promoted the negative treatment of women. When men were asked whether they treated the women on campus with respect, the most common response was "On an individual basis, yes, but when you have a group of men together, no." Men said that, when together in groups with other men, they sensed a pressure to be disrespectful toward women. A first-year man's perception of the treatment of women was that "they are treated with more respect to their faces, but behind closed doors, with a group of men present, respect for women is not an issue." One senior man stated, "In general, college-aged men don't treat women their age with respect because 90 percent of them think of women as merely a means to sex." Women reinforced this perception. A first-year woman stated, "Men here are more interested in hooking up and drinking beer than they are in getting to know women as real people." Another woman said, "Men here use and abuse women."

Characteristic of rape culture, a double standard of sexual behavior for men versus women was prevalent on this campus. As one Greek senior man stated, "Women who sleep around are sluts and get bad reputations; men who do are champions and get a pat on the back from their brothers." Women also supported a double standard for sexual behavior by criticizing sexually active women. A first-year woman spoke out against women who are sexually active: "I think some girls here make it difficult for the men to respect women as a whole."

One concrete example of demeaning sexually active women on this campus is the "walk of shame." Fraternity brothers come out on the porches of their houses the night after parties and heckle women walking by. It is assumed that these women spent the night at fraternity houses and that the men they were with did not care enough about them to drive them home. Although sororities now reside in former fraternity houses, this practice continues and sometimes the victims of hecklings are sorority women on their way to study in the library.

A junior man in a high-risk fraternity described another ritual of disrespect toward women called "chatter." When an unknown woman sleeps over at the house, the brothers yell degrading remarks out the window at her as she leaves the next morning such as "Fuck that bitch" and "Who is that slut?" He said that sometimes brothers harass the brothers whose girlfriends stay over instead of heckling those women.

Fraternity men most often mistreated women they did not know personally. Men and women alike reported incidents in which brothers observed other brothers having sex with unknown women or women they knew only casually. A sophomore woman's experience exemplifies this anonymous state: "I don't mind if 10 guys were watching or it was videotaped. That's expected on this campus. It's the fact that he didn't apologize or even offer to drive me home that really upset me." Descriptions of sexual encounters involved the satisfaction of men by nameless women. A brother in a high-risk fraternity, described a similar occurrence:

> A brother of mine was hooking up upstairs with an unattractive woman who had been pursuing him all night. He told some brothers to go outside the window and watch. Well, one thing led to another and they were almost

completely naked when the woman noticed the brothers outside. She was then unwilling to go any further, so the brother went outside and yelled at the other brothers and then closed the shades. I don't know if he scored or not, because the woman was pretty upset. But he did win the award for hooking up with the ugliest chick that weekend. . . .

DISCUSSION AND CONCLUSION

These findings describe the physical and normative aspects of one college campus as they relate to attitudes about and relations between men and women. Our findings suggest that an explanation emphasizing rape culture also must focus on those characteristics of the social setting that play a role in defining heterosexual relationships on college campuses (Kalof and Cargill 1991). The degradation of women as portrayed in rape culture was not found in all fraternities on this campus. Both group norms and individual behavior changed as students went from one place to another. Although individual men are the ones who rape, we found that some settings are more likely places for rape than are others. Our findings suggest that rape cannot be seen only as an isolated act and blamed on individual behavior and proclivities, whether it be alcohol consumption or attitudes. We also must consider characteristics of the settings that promote the behaviors that reinforce a rape culture.

REFERENCES

Boeringer S. B., C. L. Shehan, and R. L. Akers. 1991. "Social Contexts and Social Learning in Sexual Coercion and Aggression: Assessing the Contribution of Fraternity Membership." *Family Relations* 40:56–64.

Buchwald, E., P. R. Fletcher, and M. Roth (eds.). 1989. *Transforming a Rape Culture.* Minneapolis: Milkweed Editions.

Gwartney-Gibbs, P. and J. Stockard. 1989. "Courtship Aggression and Mixed-Age Groups." In *Violence in Dating Relationships,* edited by M. A. Pirog-Good and J. E. Stets. New York: Praeger.

Herman, D. 1984. "The Rape Culture." In *Women: A Feminist Perspective,* edited by J. Freeman. Mountain View, CA: Mayfield.

Kalof, L. and T. Cargill. 1991. "Fraternity and Sorority Membership and Gender Dominance Attitudes." *Sex Roles* 25:417–23.

Koss, M. P., T. E. Dinero, C. A. Seibel, and S. L. Cox. 1988. "Stranger and Acquaintance Rape: Are There Differences in the Victim's Experience? *Psychology of Women Quarterly* 12:1–24.

Koss, M. P., C. A. Gidycz, and N. Wisniewski. 1985. "The Scope of Rape: Incidence and Prevalence of Sexual Aggression in a National Sample of Higher Education Students." *Journal of Consulting and Clinical Psychology* 55:162–70.

Martin, P. Y. and R. Hummer. 1989. "Fraternities and Rape on Campus." *Gender & Society* 3:457–73.

O'Sullivan, C. 1993. "Fraternities and the Rape Culture." In *Transforming a Rape Culture,* edited by E. Buchwald, P. R. Fletcher, and M. Roth. Minneapolis: Milkweed Editions.

Sanday, P. R. 1990. *Fraternity Gang Rape.* New York: New York University Press.

19

Identity and Stigma of Women with STDs

ADINA NACK

In this article, Nack focuses her attention on a highly secretive group: college students with sexually transmitted diseases (STDs). She examines the way infected women deal with the deviance they have acquired. Employing a range of rich concepts from the interactionist and deviance literatures, she discusses the way women manage the stigma and identity associated with their sexually-diseased condition. People initially go through a stage of nonacceptance, or denial, of their deviance, where they try to distance themselves from the stigma by passing, covering, deceiving others, all the while feeling guilty about their deceptions. They then deflect their diseased identities onto others by transferring responsibility and blame for becoming infected to those they imagine have given it to them. After a time they begin to accept their deviant identities and make disclosures about their status to family members, friends, and sexual partners. We see the connection between stigma management behavior and self-concept as we follow the way the women's evolving strategies eventually lead them to change their vision of themselves, incorporating this deviant dimension into their core identity. Can you think of a secretive condition that you may have had that needed to be "managed" in the way that Nack describes? What is the nature of stigma in society? Why do some conditions or behaviors need to be hidden from others?

The HIV/AIDS epidemic has garnered the attention of researchers from a variety of academic disciplines. In contrast, the study of other sexually transmitted diseases (STDs) has attracted limited interest outside of epidemiology and public health. In the United States, an estimated three out of four sexually-active adults have HPV infections (human papillomavirus—the virus that can cause genital warts); one out of five have genital herpes infections (Ackerman 1998; CDC Server 1998). . . . This article focuses on how the sexual self-concept is transformed when the experience of living with a chronic STD casts a shadow of disease on the health and desirability of a woman's body, as well as on her perceived possibilities for future sexual experiences. . . .

STIGMA AND THE SEXUAL SELF

For all but one of the 28 women, their STD diagnoses radically altered the way that they saw themselves as sexual beings. Facing both a daunting medical and social reality, the women employed different strategies to manage their new stigma. Each stigma management strategy had ramifications for the transformation of their sexual selves. . . .

Stigma Nonacceptance

Goffman (1963) proposed that individuals at risk for a deviant stigma are either "the discredited" or "the discreditable." The discrediteds' stigma was known to others either because the individuals revealed the deviance or because the deviance was not concealable. In contrast, the discreditable were able to hide their deviant stigma. Goffman found that the majority of discreditables were "passing" as non-deviants by avoiding "stigma symbols," anything that would link them to their deviance, and by utilizing "disidentifiers," props or actions that would lead others to believe they had a non-deviant status. Goffman (1963) also noted that individuals bearing deviant stigma might eventually resort to "covering," one form of which he defined as telling deceptive stories. To remain discreditable in their everyday lives, nineteen of the women employed the individual stigma management strategies of passing and/or covering. In contrast, nine women revealed their health status to select friends and family members soon after receiving their diagnoses.

Passing The deviant stigma of women with STDs was essentially concealable, though revealed to the necessary inner circle of health care and health insurance providers. For the majority, passing was an effective means of hiding stigma from others, sometimes, even from themselves.

Hillary, a 22 year-old White senior in college, recalled the justifications she had used to distance herself from the reality of her HPV infection and facilitate passing strategies.

> At the time, I was in denial about it. I told myself that that wasn't what it was because my sister had had a similar thing happen, the dysplasia. So, I just kind of told myself that it was hereditary. That was kinda' funny because I asked the nurse that called if it could be hereditary, and she said "No, this is completely sexually transmitted." . . . I really didn't accept it until a few months after my cryosurgery.

Similarly, Gloria, a Chicana graduate student and mother of four, was not concerned about a previous case of gonorrhea she had cured with antibiotics or her chronic HPV "because the warts went away." Out of sight, out of her sex life: "I never told anybody about them because I figured they had gone away, and they weren't coming back. Even after I had another outbreak, I was still very promiscuous. It still hadn't registered that I needed to always have the guy use a condom."

When the women had temporarily convinced themselves that they did not have a contagious infection, it was common to conceal the health risk with partners because the women, themselves, did not perceive the risk as real. Kayla, a lower-middle class, White college senior, felt justified in passing as healthy with partners who used condoms, even though she knew that condoms could break. . . . Tasha, a White graduate student, found out that she might have inadvertently passed as healthy when her partner was diagnosed with chlamydia. "I freaked out—I was like, 'Oh my God! I gave you chlamydia. I am so sorry! I am so sorry!' I felt really horrible, and I felt really awful." Sara, a Jewish, upper-middle class 24-year-old, expressed a similar fear of having passed as healthy and exposed a partner to HPV. "Evan called me after we'd been broken up and told me he had genital warts. And, I was with another guy at the time, doing the kinda-sorta-condom-use thing. It was like, 'Oh, my gosh, am I giving this person something?' " Even if the passing is done unintentionally, it still brings guilt to the passer.

The women also tried to disidentify themselves from sexual disease in their attempts to pass as being sexually healthy. Rather than actively using a verbal or symbolic prop or action that would distance them from the stigma, the women took a passive approach. Some gave nonverbal agreement to put downs of other women who were known to have STDs. For example, Hillary recalled such an interaction. "It's funny being around people that don't know that I have an STD and how they make a comment like, 'That girl, she's such a slut. She's a walking STD.' And how that makes me feel when I'm confronted with that, and having them have no idea that they could be talking about me." Others kept silent about their status and tried to maintain the social status of being sexually healthy and morally pure. Kayla admitted to her charade: "I guess I wanted to come across as like really innocent and everything just so people wouldn't think that I was promiscuous, just because inside I felt like they could see it even though they didn't know about the STD." Putting up the facade of sexual purity, these women distanced themselves from any suspicion of sexual disease.

Covering When passing became too difficult, some women resorted to covering to deflect family and friends from the truth. Cleo summed up the rationale by comparing her behavior to what she had learned growing up with an alcoholic father. "They would lie, and it was obvious that it was a lie. But, I learned that's what you do. Like you don't tell people those things that you consider shameful, and then, if confronted, you know, you lie."

Hillary talked to her parents about her HPV surgery, but never as treatment for an STD. She portrayed her moderate cervical dysplasia as a pre-cancerous scare, unrelated to sex. "We never actually talked about it being a STD, and she kind of thought that it was the same thing that my sister had which wasn't sexually transmitted." When Tasha's sister helped her get a prescription for pubic lice, she actually provided the cover story for her embarrassed younger sister. "She totally took control, and made a personal inquiry: 'So, how did you get this? From a toilet seat.' And, I was like, 'a toilet seat,' and she believed me." When I asked Tasha why she confirmed her sister's misconception, she replied, "because I didn't want her to know that I had had sex." For Anne, a 28-year-old lower-middle class graduate student, a painful herpes outbreak almost outed her on a walk with a friend. She was so physically uncomfortable that she was actually "waddling." Noticing the strange behavior, her friend asked what was wrong. Anne told her that it was a hemorrhoid: that was only a partial truth because herpes was the primary cause of her pain. As Anne put it, telling her about the hemorrhoid "was embarrassing enough!"

Deception and Guilt The women who chose to deny, pass as normal, and use disidentifiers or cover stories shared more than the shame of having an STD—they had also told lies. With lying came guilt. Anne, who had used the hemorrhoid cover story, eventually felt extremely guilty. Her desire to conceal the truth was in conflict with her commitment to being an honest person. "I generally don't lie to my friends. And I'm generally very truthful with people and I felt like a sham lying to her." Deborah, a 32-year-old, White professional from the Midwest, only disclosed to her first sexual partner after she had been diagnosed with HPV: she passed as healthy with all other partners. Deborah reflected, "I think my choices not to disclose have hurt my sense of integrity." However, her guilt was resolved during her last gynecological exam when the nurse practitioner confirmed that after years of "clean" pap smear results Deborah was not being "medically unethical" by not disclosing to her partners. In other words, her immune system had probably dealt with the HPV in such a way that she might never have another outbreak or transmit the infection to sexual partners.

When Cleo passed as healthy with a sexual partner, she started, "feeling a little guilty about not having told." However, the consequences of passing as healthy were very severe for Cleo:

> No. I never disclosed it to any future partner. Then, one day, I was having sex with Josh, my current husband, before we were married, and we had been together for a few months, maybe, and I'm like looking at his penis, and I said, "Oh, my goodness! You have a wart on your penis! Ahhh!" All of a sudden, it comes back to me.

Cleo's decision to pass left her with both the guilt of deceiving and infecting her husband.

Surprisingly, those women who had *unintentionally* passed as being sexually healthy (i.e., they had no knowledge of their STD-status at the time) expressed a similar level of guilt as those who had been purposefully deceitful. Violet, a middle class, White 36-year-old, had inadvertently passed as healthy with her current partner. Even after she had preventively disclosed to him, she still had to deal with the guilt over possibly infecting him.

> It hurt so bad that morning when he was basically furious at me thinking I was the one he had gotten those red bumps from. It was the hour from hell! I

> felt really majorly dirty and stigmatized. I felt like "God, I've done the best I can: if this is really caused by the HPV I have, then I feel terrible."

When employing passing and covering techniques, the women strove to keep their stigma from tainting social interactions. They feared reactions that Lemert (1951) has labeled the *dynamics of exclusion:* rejection from their social circles of friends, family, and most importantly sexual partners. For most of the women, guilt surpassed fear and became the trigger to disclose. Those who had been deceitful in passing or covering had to assuage their guilt: their options were either to remain in nonacceptance, disclose, or transfer their guilt to somebody else.

Stigma Deflection

As the women struggled to manage their individual stigma of being sexually diseased, real and imaginary social interactions became the conduit for the contagious label of "damaged goods." Now that the unthinkable had happened to them, the women began to think of their past and present partners as infected, contagious, and potentially dangerous to themselves or other women. The combination of transferring stigma and assigning blame to others allowed the women to deflect the STD stigma away from themselves.

Stigma Transference I propose the concept of *stigma transference* to capture this element of stigma management that has not been addressed by other deviance theorists. Stigma transference is a specialized case of projection which . . . manifests as a clear expression of anger and fear. The women did not connect this strategy to a reduction in their levels of anxiety; in fact, several discussed it in relation to increased anxiety. . . .

Transference of stigma to a partner became more powerful when the woman felt betrayed by her partner. When Hillary spoke of the "whole trust issue" with her ex-partner, she firmly believed he had lied to her about his sexual health status and he would lie to others. Even though she had neither told him about her diagnosis nor had proof of him being infected, she fully transferred her stigma to him.

> He's the type of person who has no remorse for anything. Even if I did tell him, he wouldn't tell the people that he was dating. So it really seemed pretty pointless to me to let him know because he's not responsible enough to deal with it, and it's too bad knowing that he's out there spreading this to God knows how many other people.

Kayla also transferred the stigma of sexual disease to an ex-partner, never confronting him about whether or not he had tested positive for STDs. The auxiliary trait of promiscuity colored her view of him: "I don't know how sexually promiscuous he was, but I'm sure he had had a lot of partners." Robin, a 21-year-old White undergraduate, went so far as to tell her ex-partner that he needed to see a doctor and "do something about it." He doubted her ability to pinpoint contracting genital warts from him and called her a "slut." Robin believed that *he* was the one with the reputation for promiscuity and decided to "trash" him by telling her two friends who hung out with him. Robin hoped to spoil his sexual reputation and scare off his future partners. In the transference of stigma, the women ascribed the same auxiliary traits onto others that others had previously ascribed to them. . . .

In all cases, it was logical to assume that past and current sexual partners may have also been infected. However, the stigma of being sexually diseased had far-reaching consequences into the imaginations of the women. The traumatic impact on their sexual selves led most to infer that future, as yet unknown, partners were also sexually diseased. Kayla summed up this feeling: "After I was diagnosed, I was a lot more cautious and worried about giving it to other people or getting something else because somebody hadn't told me." They had already been damaged by at least one partner. Therefore, they expected that future partners, ones who had not yet come into their lives, held the threat of also being *damaged goods.*

For Hillary, romantic relationships held no appeal anymore. She had heard of others who also had STDs but stayed in nonacceptance and never changed their lifestyle of having casual, unprotected sex:

> I just didn't want to have anything to do with it. A lot of it was not trusting people. When we broke up, I decided that I was not having sex. Initially, it was because I wanted to get an HIV test. Then, I came to kind of a turning point in my life and realized that I didn't want to do the one-night-stand thing anymore. It just wasn't worth it. It wasn't fun. . . .

Blame The women's uses of stigma transference techniques were attempts to alleviate their emotional burdens. First, the finger of shame and guilt pointed inward, toward the women's core sexual selves. Their sexual selves became tainted, dirty, damaged. In turn, they directed the stigma outward to both real and fictional others. Blaming others was a way for all of the women to alleviate some of the internal pressure and turn the anger outward. This emotional component of the *damaged goods* stage externalized the pain of their stigma.

Francine recalled how she and her first husband dealt with the issue of genital warts. "We kind of both ended up blaming it on the whole fraternity situation. I just remember thinking that it was not so much that we weren't clean, but that he hadn't been at some point, but now he was." Francine's husband had likely contracted genital warts from his wild fraternity parties: "We really thought of it as, that woman who did the trains [serial sexual intercourse]. It was still a girl's fault kind-of-thing." By externalizing the blame to the promiscuous women at fraternity parties, Francine exonerated not only herself, but also her husband. . . .

For Violet, it was impossible to neatly deflect the blame away from both herself and her partner. "I remember at the time just thinking, 'Oh man! He gave it to me!' While, he was thinking, 'God, [Violet]! You gave this to me!' So, we kind of just did a truce in our minds. Like, OK, we don't know who gave it—just as likely both ways. So, let's just get treated. We just kind of dropped it." Clearly the impulse to place blame was strong even when there was no easy target.

Often, the easiest targets were men who exhibited the auxiliary traits of promiscuity and deception. Tasha wasn't sure which ex-partner had transmitted the STD. However, she rationalized blaming a particular guy. "He turned out to be kind of a huge liar, lied to me a lot about different stuff. And, so I blamed him.

All the other guys were, like, really nice people, really trustworthy." Likewise, when I asked Violet from whom she believed she had contracted chlamydia, she replied, "Dunno,' it could've been from one guy, because that guy had slept with some unsavory women, so therefore he was unsavory." Later, Violet contracted HPV, and the issue of blame contained more anger: "I don't remember that discussion much other than, being mad over who I got it from: 'oh it must have been Jess because he had been with all those women.' I was mad that he probably never got tested. I was OK before him." The actual guilt or innocence of these blame targets was secondary. What mattered to the women was that they could hold someone else responsible.

Stigma Acceptance

Eventually, every woman in the study stopped denying and deflecting the truth of her sexual health status by disclosing to loved ones. The women disclosed for either preventive or therapeutic reasons. That is, they were either motivated to reveal their STD status to prevent harm to themselves or others, or to gain the emotional support of confidants.

Preventive and Therapeutic Disclosures The decision to make a preventive disclosure was linked to whether or not the STD could be cured. Kayla explained, "Chlamydia went away, and I mean it was really bad to have that, but I mean it's not something that you have to tell people later 'cause you know, in case it comes back. Genital warts, you never know." Kayla knew that her parents would find out about the HPV infection because of insurance connections. Prior to her cryosurgery, Kayla decided to tell her mom about her condition. "I just told her what [the doctor] had diagnosed me with, and she knew my boyfriend and everything, so—it was kind of hard at first. But, she wasn't upset with me. Main thing, she was disappointed, but I think she blamed my boyfriend more than she blamed me." Sara's parents also reacted to her preventive disclosure by blaming her boyfriend: they were disappointed in their daughter, but angry with her boyfriend.

Preventive disclosures to sexual partners, past and present, were a more problematic situation. The women were choosing to put themselves in a position where they could face blame, disgust, and rejection. For those reasons, the women put off preventive disclosures to partners as long as possible. For example, Anne made it clear that she would not have disclosed her herpes to a female sexual partner had they not been, "about to have sex." After "agonizing weeks and weeks and weeks before trying to figure out how to tell," Diana, a 45-year-old African-American professional, finally shared her HPV and herpes status before her current relationship became sexual. Unfortunately, her boyfriend "had a negative reaction": "he certainly didn't want to touch me anywhere near my genitals." In Cleo's case, she told her partner about her HPV diagnosis because she wasn't going to be able to have sexual intercourse for a while after her cryosurgery. Violet described the thought process that lead up to her decision to disclose her HPV status to her current partner:

> That was really scary because once you have [HPV], you can't get rid of the virus. And then having to tell my new partner all this stuff. I just wanted to be totally up front with him: we could use condoms. Chances are he's probably totally clean. I'm like, "Oh my god, here I am tainted because I've been with, at this point, 50 guys, without condoms. Who knows what else I could have gotten (long pause, nervous laugh)?" So, that was tough. . . .

Many of the therapeutic disclosures were done to family members. The women wanted the support of those who had known them the longest. Finally willing to risk criticism and shame, they hoped for positive outcomes: acceptance, empathy, sympathy—any form of nonjudgmental support. Tasha disclosed to her mother right after she was diagnosed with chlamydia. "My family died—'Guess what, mom, I got chlamydia.' She's like, 'Chlamydia? How did you find out you got chlamydia?' I'm like, 'Well, my boyfriend got an eye infection.' (laughter) 'How'd he get it in his eye?' (laughter) So, it was the biggest joke in the family for the longest time!" In contrast, Rebecca, a White professional in her mid-fifties, shared her thought process behind *not* disclosing to her adult children. "I wanted to tell my younger one . . . I wanted very much for him to know that people could be asymptomatic carriers because I didn't want him to unjustly suspect somebody of cheating on him . . . and I don't believe I ever managed to do it . . . it's hard to bring something like that up." . . .

Consequences of Disclosure With both therapeutic and preventive disclosure, the women experienced some feelings of relief in being honest with loved ones. However, they still carried the intense shame of being sexually diseased women. The resulting emotion was anxiety over how their confidants would react: rejection, disgust, or betrayal. Francine was extremely anxious about disclosing to her husband. "That was really tough on us because I had to go home and tell Damon that I had this outbreak of herpes." When asked what sorts of feelings that brought up, she immediately answered. "Fear. You know I was really fearful—I didn't think that he would think I had recently had sex with somebody else . . . but, I was still really afraid of what it would do to our relationship." . . .

Overall, disclosing intensified the anxiety of having their secret leaked to others in whom they would have never chosen to confide. In addition, each disclosure brought with it the possibility of rejection and ridicule from the people whose opinions they valued most. For Gloria, disclosing was the right thing to do but had painful consequences when her partner's condom slipped off in the middle of sexual intercourse.

> I told him it doesn't feel right. "You'd better check." And, so he checked, and he just jumped off me and screamed, "Oh fuck!" And, I just thought, oh no, here we go. He just freaked and went to the bathroom and washed his penis with soap. I just felt so dirty. . . .

Disclosures were the interactional component of self-acceptance. The women became fully grounded in their new reality when they realized that the significant people in their lives were now viewing them through the discolored lenses of sexual disease.

CONCLUSION

The women with STDs went through an emotionally difficult process, testing out stigma management strategies, trying to control the impact of STDs on both their self-concepts and on their relationships with others. In keeping with Cooley's "looking glass self" (1902/1964), the women derived their sexual selves from the imagined and real reactions of others. Unable to immunize themselves from the physical wrath of disease, they focused on mediating the potentially harmful impacts of STDs on their sexual self-concepts and on their intimate relationships.

REFERENCES

Ackerman, S. J. 1998. "HPV: Who's Got It and Why They Don't Know." *HPV News* 8(2), Summer: 1, 5–6.

Centers for Disease Control and Prevention. 1998. "Genital Herpes." *National Center for HIV, STD, & TB Prevention.* Online. Netscape Communicator. 4 February.

Cooley, Charles H. [1902] 1964. *Human Nature and the Social Order.* New York: Schocken.

Goffman, Erving. 1963. *Stigma.* Englewood Cliffs, NJ: Prentice Hall.

Lemert, Edwin. 1951. *Social Pathology.* New York: McGraw-Hill.

PART III

Social Inequality

We live in a heterogeneous society, filled with many types of people. While it might be nice if people were different but equal, this is not usually the case. In our society, like most others, differences often translate into inequalities. The consequences of these inequalities range from apparent to subtle. Some are highly visible, and we can see them every day. Government policies favor some categories of people over others, social institutions privilege some groups of people over others, and individuals, in their daily functioning, treat some particular people better than others. While we can see the discrimination being applied in certain cases, in others the patterns of differential treatment are so deeply embedded in the criteria by which institutional decisions are made that we cannot easily recognize them. This is called *institutional discrimination*. The guidelines by which people are treated differently are complex and intersect according to several variables. They align people and the groups to which they belong into a *system of stratification,* with favored groups having greater access to the benefits of society such as money, privilege, prestige, and power. Thus, the ability to access wealth, to gain the respect of others, and to impose your will on others is unequally distributed in society.

The first way we see social inequality is in the system of **Social Class.** As William Domhoff notes in our first selection, a very small percentage of the population owns most of the wealth in our society, and has for many years. During the 1980s, however, the gap between the rich and the poor, which had narrowed during the 1960s and

'70s as the result of federal programs and the progressive tax system, began to widen steadily with government policies that favored the rich. America is a country with three distinct social classes: upper, middle, and lower (although some might argue that there is a distinction between the working poor, or working class, and the poor who do not work in the legitimate economy, or underclass). This contrasts significantly with many Latin American countries that are composed more exclusively of a very rich upper class and a very large peasant class, with little in the way of middle class. Our largest class, throughout most of our history, has been our middle class, and we have conceived of America as a country dominated by that ideology and way of life. But the 1980s brought, due to the economic policies of the Reagan and Bush administrations, an increasing divide between the rich and poor, accompanied by a shrinking of the middle class. Most middle-class families responded to this challenge by coming increasingly to rely on the income of the wife, the second wage earner, and this has placed greater burdens on other social institutions such as education and the family. Class inequality continues to be a growing problem in our society.

Our first selection in this area, by Domhoff, looks at the characteristics of America's social elite, the upper class. He sees this group, although it may have waned in influence or dominance slightly, as highly homogeneous and exclusive. He points out the distribution of wealth and income in our society (explaining the difference), and discusses the distinct social institutions that surround and socialize the upper class, from its prep schools and ivy league colleges to its interlocking elite social clubs. Next, Alan Wolfe follows with a fascinating analysis of the division of the American middle class into two separate and rather different groups of people. One is composed of traditional middle-class Americans, carried to their comfortable economic position by growing up here, following their parents into white-collar occupations, inheriting some money, and successfully pursuing the expected American Dream. The other consists of racial and ethnic minorities who worked their way up the ladder with the benefit of affirmative action, an expanding economy, permissive immigration policies, and dogged sacrifice for deferred gratification. Wolfe sets the stage for us to understand some of the differences between these groups. Finally, Mark Rank offers us a poignant examination of families struggling to survive at the lower end of the spectrum who have not taken recourse to the parallel (illicit) economy. We see these people's daily struggles as they exist in a situation that offers little hope of advancement.

A second dimension along which people are unequally arrayed is **Race and Ethnicity.** *Ethnicity* refers to the cultural characteristics of a group of people that are often derived from their heritage and geographical origin. *Race* is a more difficult and increasingly problematic concept that at one time referred to membership in a group characterized by different physical features including skin color and appearance. But the intermixing of population has grown so pervasive that we can no longer speak of "pure" racial groups, and the concept has lost its original value. Race now exists as a social construction that grounds its meaning in the way it is used in each society. In fact, even ethnicity is being eroded by intermarriage. Many Americans begin their experience in this country with a clearly defined heritage and ethnicity in the first generation (the group who moves here) and then begin to assimilate in the second generation as young members move through the public schools, speak the language, and acquire American cultural values. People of different racial/ethnic groups are conceived and treated according to common stereotypes in this country, with some groups, notably Asians, benefitting from the "model minority" perception, and others, notably blacks and Hispanics, still suffering profound prejudice that results in blocked upward mobility and social acceptance.

Americans have conceived of ourselves as an immigrant-formed country, as most of our population derives its roots elsewhere. Large waves of immigrants populated our land in the eighteenth, nineteenth, and early twentieth centuries. We considered ourselves a *melting pot* where people from all countries came together and formed into one, ethnic differences evaporating as new members were exposed to the dominant culture and mass education. But as time passed and distinctions remained, some groups still failing to attain parity, we began to recognize the relevance of a *cultural pluralist* model, our society consisting of many different groups living together. Recent increasing waves of immigration have raised issues of race and ethnicity to the fore once again, as our country struggles with its pluralistic composition. Troubling problems include segregation, the difficulties still faced by American blacks, the large influx of Asians and Hispanics, and the "browning" of America. Our first selection, by Joel Perlmann, addresses the issue of racial intermarriage, mixed identity, and the future composition of the population. We offer two selections on new immigrant groups, as this topic is currently critical. Two very contrasting portraits are presented in Cecilia Garza's tale of Mexican American immigrants who straddle the border and live a life of turmoil, separation, and exploitation, and Ilsoo Kim's investigation of family- and business-centered Korean American immigrants who are finding a social and economic niche that offers the promise of success. We close with Edward Shapiro's selection on the loss of ethnicity, as we see how successful assimilation is endangering the very existence of Jewish Americans.

Gendered inequality is our last focus in this Part, with men attaining a greater share of the world's valued resources than women. In many countries the gap between men and women is wider than in the United States. India and China look to male children to support the family while girls represent an economic drain and social loss, moving in with in-laws and having to be supported by a dowry. These countries struggle with difficult problems such as female infanticide, fetal (female) sex-screening and abortion, and the malnourishment and premature death of female children. Women in many parts of the Arab world remain secluded and veiled, symbolic property of their male relatives, unable to drive, hold jobs, or travel freely in public. The United States, along with some of the socialist democracies in Western Europe, has the most liberated and powerful women in the world. But even in America we live in a *patriarchy*, a system of male dominance embedded in the culture and social structure. Unequal legal rights, opportunities, and treatment of women foster their reduced power economically, socially, politically, occupationally, and interpersonally.

The role of women has undergone some profound changes since the beginning of the women's movement in the 1970s. Women have made some moves out of traditional gender roles and pursuits, with resulting changes in their interactional patterns. Avenues of opportunity for gender parity have opened to them that were closed before, but many blocks remain. Men's lives have not been transformed as much over this period, with men remaining more trapped in traditional gender obligations and outlooks. We see the evidence of this in our first selection, where we share some of our own research into preadolescent children's gender socialization. By examining boys' and girls' popularity variables, you can see that girls, while still following traditional roles, have more options open to them than boys, whose choices have increased less. We then move up the life cycle to adult men and women and examine the way they communicate with each other. Deborah Tannen shows us the gendered characteristics of men's and women's concerns and the way they talk to each other. Finally, we turn to the world of work, where women are still earning approximately 75 cents on the dollar compared to the same labor of men. Barbara Reskin and Irene Padavic share with us some of the occupational dilemmas in which women are caught, although they are necessarily limited, by space, from discussing other relevant issues such as tokenism (promoting one or two women to highly visible positions while keeping the rest down), mommy tracks (separate and limited advancement ladders for women as the result of their child-rearing obligations), and the role stereotyping of women at work into categories such as the mother, the helpless girl, the flirt, the iron maiden, and the bitch.

20

Who Rules America? The Corporate Community and the Upper Class

G. WILLIAM DOMHOFF

In this recently updated study, Domhoff offers us a glimpse into the position and practices of the upper class, that group of interlocking families that holds itself aloof from the rest of society, owns and controls a disproportionate share of American wealth, and seeks to retain its elite status atop the social world through inbreeding and exclusivity. While some dilution of this population has occurred over time, Domhoff suggests that these old-line WASPs (White, Anglo-Saxon Protestants) still retain family-generated control over significant portions of corporate America, especially those companies with private stock ownership. Domhoff offers us figures that trace the breakdown of wealth and income in America over the century, showing that despite some groups' movement into high-income brackets, wealth ownership has not significantly shifted. He then gives a glimpse into the social institutions, particularly the schools and social clubs, within which the upper class are educated, socialized, and circulate. He compares these to respectable versions of "clan" establishments, in that they offer a protected enclave where members can retreat in private to enjoy their leisure, away from contact with lesser or dissimilar others. Do you know any members of this group described by Domhoff? Why or why not? How does social class influence social behavior? Even though Domhoff shows that members of the overclass come from common backgrounds and emphasize inbreeding, what differences characterize them? What is the role of their status-consciousness and exclusivity, and has it become archaic in contemporary society?

From *Who Rules America? Power and Politics in the Year 2000* by G. William Domhoff.

Most Americans do not like the idea that there are social classes. Classes imply that people have relatively fixed stations in life. They fly in the face of beliefs about equality of opportunity and seem to ignore the evidence of upward social mobility. Even more, Americans tend to deny that social classes are based in wealth and occupational roles but then belie that denial through a fascination with rags-to-riches stories and the trappings of wealth. . . .

If there is an American upper class, it must exist not merely as a collection of families who feel comfortable with each other and tend to exclude outsiders from their social activities. It must exist as a set of interrelated social institutions. That is, there must be patterned ways of organizing the lives of its members from infancy to old age that create a relatively unique style of life, and there must be mechanisms for socializing both the younger generation and new adult members who have risen from lower social levels. If the class is a reality, the names and faces may change somewhat over the years, but the social institutions that underlie the upper class must persist with remarkably little change over several generations. This emphasis on the institutionalized nature of the upper class, which reflects a long-standing empirical tradition in studies of it, is compatible with the theoretical focus of the "new institutionalists" within sociology and political science. . . .

WEALTH AND POWER: WHO BENEFITS?

In considering the distribution of wealth and income in the United States, it must be stressed that they are two separate issues. Wealth distribution has to do with the concentration of ownership of marketable assets, which may include tangible things such as land, machinery, and animals, and intangibles such as stocks, bonds, and copyrights, but also insurance policies, houses, cars, and furniture. Income distribution, on the other hand, has to do with the percentage of wages, dividends, interest, and rents paid out each year to individuals or families at various income levels. In theory, those who own a great deal may or may not have high incomes—depending on the returns they receive from their wealth—but in reality, those at the very top of the wealth distribution also tend to have the highest incomes, mostly from dividends and interest. . . .

The most important focus of wealth and income studies is on the highest levels of wealth distribution and the percentage of overall income that is derived from that wealth. Numerous studies show that wealth distribution is extremely concentrated and that it has been very stable over the course of the twentieth century, although there was a temporary decline in wealth concentration in the 1970s (in good part due to a decline in stock prices). By the late 1980s, however, wealth distribution was as concentrated as it had been in 1929, when the top 1 percent had 36.3 percent of all wealth. The percentage of yearly income received by the highest 1 percent of wealthholders also remained constant within the context of some mild fluctuations. In 1958, for example, the top 1.5 percent of wealthholders received 13 percent of yearly income; in 1992, the top 1 percent received 15.7 percent.[1] Table 1 presents figures for 1983, 1989, and 1992 for net

Table 1 The Distribution of Net Worth, Financial Wealth, and Income for 1983, 1989, and 1992

	Net Worth			Financial Wealth			Income		
	1983	1989	1992	1983	1989	1992	1983	1989	1992
Top 1%	33.8	39.0	37.2	42.9	48.3	45.6	12.8	16.4	15.7
Next 19%	47.6	45.6	46.6	48.4	45.8	46.7	39.0	39.0	40.7
Bottom 80%	18.7	15.4	16.3	8.7	6.1	7.8	48.1	44.5	43.7

SOURCE: Edward Wolff, *Top Heavy* (New York: Twentieth Century Fund, 1996), p. 67. Reprinted with permission from the Twentieth Century Fund, New York.

Net Worth = Total assets minus debts.

Financial Wealth = Net worth minus net equity in owner-occupied housing (a good indication of liquidity).

worth, financial wealth, and income for the wealthiest 1 percent, the next 19 percent, and the bottom 80 percent. . . .

There are newly rich people who are not yet assimilated into the upper class, and there are highly paid professionals, entertainers, and athletes who for a few years make more in a year than many members of the upper class. However, for the most part it is safe to conclude that the people of greatest wealth and highest income are part of—or are becoming part of—the upper class.

Without a doubt, then, the .5 to 1 percent of the population that makes up the upper class is also the .5 to 1 percent who owned 45.6 percent of the financial wealth in 1992. In terms of the "Who benefits?" indicator of power, the upper class is far and away the most powerful group in society. . . .

Prepping for Power

From infancy through young adulthood, members of the upper class receive a distinctive education. This education begins early in life in preschools that frequently are attached to a neighborhood church of high social status. Schooling continues during the elementary years at a local private school called a day school. During the adolescent years the student may remain at day school, but there is a strong chance that at least one or two years will be spent away from home at a boarding school in a quiet rural setting. Higher education will take place at one of a small number of heavily endowed private colleges and universities. Large and well-known Ivy League schools in the East and Stanford in the West head the list, followed by smaller Ivy League schools in the East and a handful of other small private schools in other parts of the country. Although some upper-class children may attend public high school if they live in a secluded suburban setting, or go to a state university if there is one of great esteem and tradition in their home state, the system of formal schooling is so insulated that many upper-class students never see the inside of a public school in all their years of education.

This separate educational system is important evidence for the distinctiveness of the mentality and life-style that exists within the upper class because schools play a large role in transmitting the class structure to their students. Surveying and summarizing a great many studies on schools in general, sociologist Randall Collins concludes: "Schools primarily teach vocabulary and inflection, styles of dress, aesthetic tastes, values and manners."[2] His statement takes on greater significance for studies of the upper class when it is added that only one percent of American teenagers attend independent private high schools of an upper-class nature.[3] . . .

The linchpins in the upper-class educational system are the dozens of boarding schools founded in the last half of the nineteenth and the early part of the twentieth centuries. Baltzell concludes that these schools became "surrogate families" that played a major role "in creating an upper-class subculture on almost a national scale in America."[4] The role of boarding schools in providing connections to other upper-class social institutions is also important. As one informant explained to Ostrander in her interview study of upper-class women: "Where I went to boarding school, there were girls from all over the country, so I know people from all over. It's helpful when you move to a new city and want to get invited into the local social club."[5]

It is within these few hundred schools that are consciously modeled after their older and more austere British counterparts that a distinctive style of life is inculcated through such traditions as the initiatory hazing of beginning students, the wearing of school blazers or ties, compulsory attendance at chapel services, and participation in esoteric sports such as squash and crew. Even a different terminology is adopted to distinguish these schools from public schools. The principal is a headmaster or rector, the teachers are sometimes called masters, and the students are in forms, not grades. Great emphasis is placed on the building of "character." The role of the school in preparing the future leaders of America is emphasized through the speeches of the headmaster and the frequent mention of successful alumni. Thus, boarding schools are in many ways the kind of highly effective socializing agent that sociologist Erving Goffman calls "total institutions," isolating their members from the outside world and providing them with a set of routines and traditions that encompass most of their waking hours.[6] The end result is a feeling of separateness and superiority that comes from having survived a rigorous education. As a retired business leader told one of my research assistants: "At school we were made to feel somewhat better [than other people] because of our class. That existed, and I've always disliked it intensely. Unfortunately, I'm afraid some of these things rub off on one."[7]

Almost all graduates of private secondary schools go on to college, and almost all do so at prestigious universities. Graduates of the New England boarding schools, for example, historically found themselves at one of four large Ivy League universities: Harvard, Yale, Princeton, and Columbia. . . . Now many upper-class students attend a select handful of smaller private liberal arts colleges, most of which are in the East, but there are a few in the South and West as well.

Graduates of private schools outside of New England most frequently attend a prominent state university in their area, but a significant minority go to Eastern Ivy League and top private universities in other parts of the country. . . . A majority of private school graduates pursue careers in business, finance, or corporate law. For example, a classification of the occupations of a sample of the graduates of four private schools—St. Mark's, Groton, Hotchkiss, and Andover—

showed that the most frequent occupation for all but the Andover graduates was some facet of finance and banking. Others became presidents of medium-size businesses or were partners in large corporate law firms. A small handful went to work as executives for major national corporations[8]. . . .

Social Clubs

Just as private schools are a pervasive feature in the lives of upper-class children, so, too, are private social clubs a major point of orientation in the lives of upper-class adults. These clubs also play a role in differentiating members of the upper class from other members of society. According to Baltzell, "the club serves to place the adult members of society and their families within the social hierarchy." He quotes with approval the suggestion by historian Crane Brinton that the club "may perhaps be regarded as taking the place of those extensions of the family, such as the clan and the brotherhood, which have disappeared from advanced societies."[9] Conclusions similar to Baltzell's resulted from an interview study in Kansas City: "Ultimately, say upper-class Kansas Citians, social standing in their world reduces to one issue: where does an individual or family rank on the scale of private club memberships and informal cliques?"[10]

The clubs of the upper class are many and varied, ranging from family-oriented country clubs and downtown men's and women's clubs to highly specialized clubs for yacht owners, gardening enthusiasts, and fox hunters. Many families have memberships in several different types of clubs, but the days when most of the men by themselves were in a half dozen or more clubs faded before World War II. Downtown men's clubs originally were places for having lunch and dinner, and occasionally for attending an evening performance or a weekend party. But as upper-class families deserted the city for large suburban estates, a new kind of club, the country club, gradually took over some of these functions. The downtown club became almost entirely a luncheon club, a site to hold meetings, or a place to relax on a free afternoon. The country club, by contrast, became a haven for all members of the family. It offered social and sporting activities ranging from dances, parties, and banquets to golf, swimming, and tennis. Special group dinners were often arranged for all members on Thursday night—the traditional maid's night off across the United States. . . .

Initiation fees, annual dues, and expenses vary from a few thousand dollars in downtown clubs to tens of thousands of dollars in some country clubs, but money is not the primary barrier in gaining membership to a club. Each club has a very rigorous screening process before accepting new members. Most require nomination by one or more active members, letters of recommendation from three to six members, and interviews with at least some members of the membership committee. Names of prospective members are sometimes posted in the clubhouse, so all members have an opportunity to make their feelings known to the membership committee. Negative votes by two or three members of what is typically a ten- to twenty-person committee often are enough to deny admission to the candidate. The carefulness with which new members are selected extends to a guarding of club membership lists, which are usually available only to club members. Older membership lists are sometimes given to libraries by members or their surviving spouses, but for most clubs there are no membership lists in the public domain. . . .

CONCLUSION

This chapter establishes the existence of a social upper class that is nationwide in scope through private schools, clubs, summer resorts, retreats, and other social institutions, all of which transcend the presence or absence of any given person or family. Families can rise and fall in the class structure, but the institutions of the upper class persist. This upper class makes up from .5 to 1 percent of the population, a rough estimate based on the number of students attending independent private schools, the number of listings in past *Social Registers* for several cities, and detailed interview studies in Kansas City and Boston. The disproportionate share of wealth and income controlled by members of the upper class is evidence for a class-domination theory in terms of the "Who benefits?" indicator of power.

Not everyone in this nationwide upper class knows everyone else, but everybody knows somebody who knows someone in other areas of the country—thanks to a common school experience, a summer at the same resort, or membership in the same social club. With the social institutions described in this chapter as the undergirding, the upper class at any given historical moment consists of a complex network of overlapping social circles knit together by the members they have in common and by the numerous signs of equal social status that emerge from a similar life-style. Viewed from the standpoint of social psychology, the upper class is made up of innumerable face-to-face small groups that are constantly changing in their composition as people move from one social setting to another.

NOTES

1. James Smith, "An Estimate of the Income of the Very Rich," *Papers in Quantitative Economics* (Lawrence: University of Kansas Press, 1968); Edward Wolff, *Top Heavy* (New York: Twentieth Century Fund, 1996).
2. Randall Collins, "Functional and Conflict Theories of Educational Stratification," *American Sociological Review* 36(1971): 1010.
3. "Private Schools Search for a New Role," *National Observer* (August 26, 1968), p. 5. For an excellent account of major boarding schools, see Peter Cookson and Caroline Hodge Persell, *Preparing for Power: America's Elite Boarding Schools* (New York: Basic Books, 1985).
4. E. Digby Baltzell, *Philadelphia Gentlemen: The Making of a National Upper Class* (Glencoe, IL: Free Press, 1958), p. 339.
5. Susan Ostrander, *Women of the Upper Class* (Philadelphia: Temple University Press, 1984), p. 85.
6. Erving Goffman, *Asylums* (Chicago: Aldine, 1961).
7. Interview conducted for G. William Domhoff by research assistant Deborah Samuels, February 1975; see also Gary Tamkins, "Being Special: A Study of the Upper Class" (Ph.D. Dissertation, Northwestern University, 1974).
8. Steven Levine, "The Rise of the American Boarding Schools" (Senior Honors Thesis, Harvard University, 1975), pp. 128–30.
9. Baltzell, *Philadelphia Gentlemen,* p. 373.
10. Richard P. Coleman and Lee Rainwater, *Social Standing in America* (New York: Basic Books, 1978), p. 144.

21

The Duality of the Middle Class

ALAN WOLFE

Wolfe illuminates our image of mainstream contemporary America with his analysis of our largest class group, how it has evolved, and its division into two groups that are in many ways at odds with each other. Although Americans have thought of our country as filled predominantly by the middle class for a long time, we have not exactly examined what we meant by this. Wolfe lodges his definition in a style of living, a comfort zone constituting the suburban, "American Dream." Yet by tracing the economics of our country, he shows that getting there is not as assured as it used to be, something we may have noticed as many industries (like steel and automobile manufacturing) suffered cutbacks and laid off their workers. Some of these people slipped into the lower class, losing their jobs and homes, while at the same time, other entrepreneurs worked their way up into the upper class. We saw this in Domhoff's selection, as we noticed that the highest strata of our country is not filled as exclusively by a dominant WASP establishment that is homogeneous and elite. Those who managed to hold on to their middle-class position represent one group, while they were joined by another who earned their way into the middle class by hard work. Wolfe offers some insights into who forms these two groups and how their ideologies conflict. As you read it, you may want to trace the position of your ancestors and see what they were doing over the time period he describes. How do the occupational histories and geographic residencies of your family members compare to the groups in this selection? Which morality and set of attitudes toward the upper and lower classes best describes your family and you?

Middle-class anxieties about the economy, crime, and social issues seem certain to dominate American politics for years to come. Yet it has become very difficult to define clearly what it means to be middle class. The nation's images of bourgeois life are increasingly obsolete: yeoman farmer, small-town merchant, independent entrepreneur, male breadwinner, stay-at-home mom, well-paid factory worker, hard-working school teacher, self-employed lawyer, family physician. Is Zöe Baird, whose name was never mentioned without note of her $500,000 income, middle class? Are the mostly blue-collar Reagan Democrats? Is a former executive who is struggling to start a new business but in reality living on his wife's income as a social worker? Is anyone without health insurance, whatever his or her income? Are blacks who have made it to the suburbs? Korean grocers? Divorced mothers of small children? An assistant professor of anything? As we watch more Americans fall from the middle class, we ought to know at what point we should begin to roll out the nation's safety net. But even spelling out a formula in dollars and cents is nearly impossible. We cannot even decide at what point we consider people rich. Candidate Bill Clinton pledged to make the rich pay a larger share of the nation's taxes, but the definition of rich has bounced around. President Clinton's tax plan now calls for higher income taxes on couples earning more than $140,000, and a special "millionaires'" surcharge on those earning more than $250,000.

It may be hard to determine where the economic boundary lines of middle-class life should be drawn, but it is not that difficult to figure out what has happened to the core of the middle class during the 1980s and '90s. Most sensibly defined, a middle-class job is one that makes it possible to afford certain basics: a home of one's own, a car or two, and some child care. By this definition, middle-class jobs have most definitely disappeared over the past 15 years. There is much truth to the notion that the middle class, as economists Frank Levy and Richard J. Murnane write, has been "hollowed out": More people have moved to points where the middle class blends into the class above or the class below.

This change has its roots in the economic turmoil of the 1970s. In 1973, the year the first oil crisis began, the country entered an era of slower economic growth, and in 1979 income inequality began a comparatively rapid increase. Because of this relatively clear turning point in time, one can picture two middle classes in America: one that rose to its status when economic growth was assumed and opportunity abundant, and one that achieved its status at a time when very little could be taken for granted. What divides these two groups is not how much money their members make but the different degrees of effort involved in making it. So different are the experiences of these two middle classes that, for all their economic similarity, they have little in common culturally or morally. There is no longer one thing called "the middle class" in America, and there is no longer a single middle-class morality. It is far more accurate to say that we have at least two middle-class moralities, each defined by different opportunities, expectations, and outlooks.

For those whose income and status began to rise in the 1950s, passage into the middle class was nearly as automatic as the progress through the seven ages of man. Each step seemed preordained: the breadwinner's income rose, the family moved to a larger apartment, then bought its first house, along with a car, a television, and a few other accouterments of the good life. The children were sent off to college, perhaps the first in their families to go, and the parents could look forward to spending their retirement years in Florida or Arizona. Dad might have been a middle manager with Prudential, the owner of a small business, a salesman, or a shopkeeper with an expanding clientele. He might have worked incredibly hard or he might have worked nine to five, but the robust economy guaranteed at least minimal affluence. Mom stayed home, though after the kids were grown she might have taken a job as a receptionist or gone back to school. Many people in this generation became middle class simply by being there. To be sure, one had to be of the right race. At least some initiative and hard work were needed—everyone knew people who were left behind. But for more Americans than ever before, the goal was in reach, and never before had so many reached it.

Money, for this generation, was always an awkward proposition. With the Great Depression never far from consciousness, income was something to be saved, not spent. Yet this generation was willing to share with those left behind some of the surplus generated by the economy. The Democrats did rather well during the go-go years of the 1960s, in part because middle-class prosperity was compatible with, if not fueled by, activist government. In neither lifestyle nor politics did this generation flaunt its good fortune, understanding very well how unreal its prosperity was. Anything won with so little effort could be lost with even less. *Security* became the watchword for the first postwar middle class, as if the right combination of public policy and private behavior could make permanent what was too good to be true.

The postwar generation maintained its liberalism through old age; the elderly living in Florida still vote on the basis of who will best protect the government programs that will guarantee them economic security until they die. At the same time, this generation passed on some aspects of its liberalism to its children. Although all of America turned more conservative in the 1980s, young urban professionals—those whose privileged educations or first home purchases were made possible by the advantages of their parents—remained culturally liberal. More tolerant than their parents—they came of age, after all, in the 1960s and after—the children of the immediate postwar bourgeoisie reacted against what they saw as the overly materialistic concerns of their parents' generation. What eventually became support for multicultural education, instinctive identification with feminism, and tolerance of diverse lifestyles had its origins in the committed cultural libertarianism of the 1960s. It was not that the younger generation's views on religion, the family, or love of country were well thought out. It was more that to many of its members, these were issues that never arose. One of the things that made being middle class so delightful during the 1960s was that you never had to think much about the obligations of community or the need to contain the libido for the sake of civilization.

The long national economic downturn that began around 1973 did not destroy the middle class, but it did halt the postwar escalator that automatically carried millions of fortunate Americans upward into affluence. Everyone knew someone who was no longer assured of the house in the suburbs, the new car, the good schools. Downward mobility was no longer merely a term in sociology textbooks. But just as large numbers of people saw the American Dream slip away, a surprising number of newcomers grabbed onto it. Some were urban white ethnics—policemen, civil servants, unionized blue-collar workers whose jobs were spared—who were driven from the cities by crime and who, with the aid of a second paycheck from the wife's new job, moved out beyond the established suburbs in search of a middle-class lifestyle they could afford. Others were freshly minted graduates of the state universities and community colleges—vastly expanded during the good years—who took jobs in engineering, insurance, and other flourishing service industries. An unprecedented number of African Americans joined the middle class. The tide of upward mobility was powerful enough to transform neighborhoods and regions. In New York City, Asians pushed out into urban neighborhoods beyond Manhattan, bringing new vibrancy to once-thriving Jewish neighborhoods such as Flushing. The middle-class accent of Miami became Spanish, while Iranians installed themselves in the tonier sections of Beverly Hills. The second postwar middle class, though smaller than the first, was certainly more diverse.

Middle-class status, when no longer automatic, became more of a commodity, something one purchased through hard work and sacrifice. The new arrivals came to see merit, rather than position on a growth curve, as the prerequisite for a middle-class lifestyle. Under such competitive conditions, money moved to the center of people's consciousness. A class once known for saving began to spend. Often there was little choice. Even with the two (often rather high) incomes needed just to purchase a house with access to decent schools, there was little left over to put away for the future. In some major cities, even people with six-figure incomes and boasting only the normal trappings of suburban life learned to live with a certain sense of precariousness about their existence. With less of the security that comes from having money in the bank, the middle class became much more wary of government-led altruism. The tax revolts and attacks on waste in government that began in the late 1970s were symptoms of a new politics of increased self-concern. It had taken some time, as well as a shift in generations, but finally the middle class was living up to the cliché that money breeds increased conservatism.

This second middle class, like its predecessor, is moved by considerations of security, but its concerns are more psychological than economic. They try to save moral capital rather than economic capital. Uncertain that they can maintain their economic privileges, these newcomers to the American Dream are determined to hold on to their social and cultural ones. They look to government not to intervene in the economy to help workers and minorities get ahead but to reinforce the rules of civil order. The control of crime becomes more important than the control of business. Government, they believe, ought to regulate sexuality (teen access to abortion, for example) and the display of dirty pictures, and it ought to keep its own house in order as well. Even if families have trouble balancing their budgets, government should balance its own, and politicians had better not get the idea that they are better than the people who elected them or they will be humbled.

For those who achieved middle-class status the hard way, the cultural enemy is the old middle class already encamped in the tonier inner suburbs, and especially those of its descendants in the baby-boom generation who have embraced far more liberal and culturally libertarian views: the "new class" of attorneys, journalists, managers, and other professionals who make their living by manufacturing and manipulating information. For its part, this more cosmopolitan middle class looks down its collective nose at the tastes and sensibilities of the newcomers in the tract homes and townhouses on the fringes of suburbia.

Hence ariseth the new class war. . . .

Work is not only a way of making things but a way of making meaning. At least since the early 19th century, but probably originating some time before that, Americans have been attracted to ideologies of production as much as to production itself. In making things, they came to believe, people made themselves.

Classical republican ideals about production are the heart of the moral worldview of the more newly arrived middle class in America, an ideal strengthened, rather than weakened, by the increasing difficulty its members have in finding productive work. For those who believe in the sanctity of work, morality is defined by the perception that those who do not make things—lawyers, stockbrokers, "bureaucrats"—deserve a lower place in the moral hierarchy. This is as it has always been, but with one crucial difference: For over a century, the foil that helped define middle-class ideas about the importance of work was the idle rich, with their coupon-clipping frivolity and conspicuous consumption. Now that high society has all but disappeared from America's consciousness, the urban underclass increasingly bears the burden of comparison. There, bourgeois propriety finds the same defining symbols: uncontrolled sexuality, flamboyant spending, money without work, and the appearance of government protection. Nothing is more certain to arouse the fury of the new middle class than the "welfare mother," whose seemingly irresponsible behavior not only goes unpunished but is in fact rewarded with money taken from the pockets of hard-working taxpayers like themselves.

If one middle class believes in work, the other believes in career. These contrasting beliefs also imply different ways of thinking about time and space. Because work involves producing things, it takes place within boundaries. Not only is it tied to a specific neighborhood, employer, or industrial quarter, it is time-bound and regulated by hours or weeks. Careers, by contrast, tend to be loosened from the constraints of space and time. People who have careers are prepared to move anywhere in search of the next stage, either within the firm or within the country. They are not, however, prepared to punch a clock. Process, not output, counts as the measure of success. Those who follow careers manage rather than produce. Indeed, one of the things they devote a great deal of time to managing is the transition to an economy that produces less.

Career-followers tend to view those bound to specific hours and places as slow-moving and backward, "time-servers" lacking in cosmopolitan sophistication. They work at jobs that pollute the environment and belong to hidebound unions that are bastions of conservatism and special privilege. Working people vote against the higher taxes needed to keep the local schools in the right loops for the right colleges. From the perspective of the wealthier middle class, Americans who produce things put tacky sculptures on their front lawns, ice cubes in their (sweet) white wine, pictures of their children on their walls, sugar in their (disgustingly weak) coffee, cigarettes in their ashtrays, and dirt bikes in their driveways. The career-followers are undisturbed by the decline of industrial America—old factories can be converted into attractive shopping malls and offices, after all—and tend to believe that given a choice the country would turn every industrial community into a Silicon Valley. Visions of postindustrial society may no longer preoccupy social scientists, but they lie behind the dreams of the older, more entrenched, middle class.

Unappreciative of productive work, this middle class is hardly prepared to insist that the underclass be required to submit to its rigors. Unlike the more recently arrived middle class, which tends to move to the outer suburbs, the older middle class lives closer to the city and even, on occasion, "gentrifies" urban neighborhoods in the city itself. From this position of greater proximity to the poor, being unproductive is seen not as a sin but as a condition. It can even, in more sophisticated understandings, be seen as a kind of career. Youngsters in the gang business, for example, work pretty hard at what they do. They, too, are liberated from the constraints of space and time—they certainly keep irregular hours— and often possess an entrepreneurial flair. Even welfare can be understood as a career. Welfare recipients, like many urban professionals, are creatures of the bureaucracy. And while they may not be producing anything at the moment, welfare is something like a career interlude, necessary before work can be resumed.

The virtue of productivity, once a crucial American symbol, is now contested. For those wishing no more than to say good-bye to all that, unproductive behavior, while not necessarily appealing, is also not especially threatening. But to those who labor in traditional jobs, urban loitering, always unforgivable, approaches anathema. The more Americans are forced to compete for a diminishing number of good jobs, the more they will also differ over the meaning of jobs themselves.

SOCIAL CLASS:
Underclass

22

Welfare Recipients Living on the Edge

MARK RANK

In this very rich selection filled with poignant quotes from welfare recipients, Rank paints a vivid picture of the difficulties of day-to-day living for people in the lowest economic stratum of society. He shows us how some of these men and women ended up on public assistance and their feelings about it, their relations to the world of work, and the problems they have making ends meet, especially as they invariably run out of money toward the end of the month. Readers will find the consequences and effects of a public assistance life, with its poverty, disease, and hunger, particularly stirring. Like the last article, do you know anybody on welfare? Why or why not? How are the everyday life experiences of these welfare recipients different from or similar to the life you and your parents live? What do you think the opportunities are for people on welfare? How many of them do you think have had a chance for a college education such as the one you are receiving? How might this affect their futures?

There's lots of nights that I go to bed and I lay awake and wonder how I'm gonna pay the bills. Because gas and electric is high. The telephone is high. And everything's going up instead of down. They keep everything up so high. It makes you wonder if you can make it through the month. When I take my check to the bank, I feel this way—I'm putting this check in the bank but it isn't gonna be my money. The bills are gonna have that money, not me.

—ELDERLY WOMAN LIVING ON FOOD STAMPS AND SOCIAL SECURITY

Public assistance programs provide a bare minimum on which to live. They are not intended to raise a family's standard of living above the poverty line, only to provide limited assistance to particular categories of people. What is it like to live on welfare and in poverty on a day-to-day basis?

THE CONSTANT ECONOMIC STRUGGLE

Perhaps most apparent when one listens to welfare recipients describe their daily lives and routines is the constant economic struggle that they face. This includes difficulties paying monthly bills, not having enough food, worrying about health care costs, and so on. The amount of income received each month is simply insufficient to cover all these necessary expenses. Having talked with dozens of families, having seen the daily hardships of recipients, having felt their frustrations and pain, I have no doubt that these families are indeed living on the edge.

This economic struggle is typified by the experience of Mary Summers. A fifty-one-year-old divorced mother, Mary and her two teenage daughters have been on public assistance for eleven months. She receives $544 a month from AFDC and $106 a month worth of Food Stamps. After paying $280 for rent (which includes heat and electricity), she and her daughters are left with $370 a month (including Food Stamps) to live on. This comes to approximately $12 a day, or $4 per family member. While this may seem like an implausibly small income for any household to survive on, it is quite typical of the assistance that those on welfare receive.

Mary turned to the welfare system because she had been unable to find work for two years (this in spite of a rigorous search for a bookkeeping or accountant's position, jobs she has held in the past). She comments, "This is probably about the lowest point in my life, and I hope I never reach it again. Because this is where you're just up against a wall. You can't make a move. You can't buy anything that you want for your home. You can't go on a vacation. You can't take a weekend off and go and see things because it costs too much. And it's just such a waste of life."

I asked Carol Richardson to describe what her day-to-day problems were. Having lived in poverty for most of her forty-five years—Carol and her five children have received welfare on and off for twenty years—she can be considered an expert on the subject: "Making ends meet. Period. Coming up with the rent on time. Coming up with the telephone bill on time. Having food in the house. It looks like we've got enough now. I got Food Stamps last month. Otherwise we would be down and out by now. It's just keepin' goin' from day to day. Carfare, busfare, gas money. . . ."

Joyce Mills described her feelings about applying for public assistance. At the time of the interview, she was no longer receiving AFDC. I asked her about her most pressing problems. Although she was working full-time in a clerical position, and although she was receiving eighty-one dollars a month from the Food Stamp program, she was clearly having a difficult time surviving economically:

> I think it's just trying to make ends meet. I don't think I go out and spend a whole lot of money, not like I'd like to. I'd like to be able to fix the kids' rooms up nice and I haven't been able to. And if I do have the money, it's always something else that comes up. It's really frustrating. Then when I think, if my husband were not incarcerated and working with the job that he had, God, we could, both of us, be raking in very close to a thousand dollars a month. And it just makes me so mad. I've been juggling bills around trying to make ends meet. Now my muffler is off of the car and I gotta get that fixed.
>
> I've been using that car very little since this happened this week. If it wasn't for this car, I would have had the money for Tommy's violin. And last night it just comes to a head, and I just released my tensions through tears. And thank God the kids were sleeping. It's so frustrating when you're trying to lead a normal life, but you can't do it. And then I was trying to think, "Well, if I get this in this month. If I can take so much out of this from the savings account. . . ." There was a time where I could have done that, but now I'm down to nothing. And this past check, it was down so low that I didn't have anything. I got paid and that week it was gone. It was all gone. So it was hard. I had to resort to borrowing.

Perhaps most revealing is her remark "It's so frustrating when you're trying to lead a normal life, but you can't do it." For many recipients, particularly those with children, this expresses in a nutshell their economic frustrations and the dilemmas they face.

Trying to lead a "normal" life under the conditions of poverty is extremely difficult. For example, Marta Green describes how not having a car for transportation severely restricts the kinds of activities she can do with her children, particularly during the wintertime.

> In the winter we don't go anywhere. Because it's very hard without a car. I always had a car until last winter. It was very hard. Because we had to wait there sometimes twenty minutes for the bus. And with the kids and the very cold days, it's very hard. I only took them out last winter once, besides the Saturday afternoons that we go to church.
>
> One Sunday, they had some free tickets to go to the circus. And I only had to buy my ticket. They both had theirs free. The circus was at seven. And it was done by ten. And we were waiting for the bus until eleven thirty that Sunday night. In the middle of the winter. And then finally we started walkin' home. We walked all the way home. We made it home by twelve thirty. They were tired and almost frozen. And then I thought, this is it, no more. So we really don't go very much anywhere.

As a result of daily economic constraints, the diversions open to welfare recipients must be quite simple and inexpensive. Going for walks, visiting friends or relatives, playing with their children, watching television, and reading are the types of activities available to provide some enjoyment and diversion for recipients during their otherwise hard times.

Ironically, those who have the fewest economic resources often pay the most for basic necessities, frequently of inferior quality. Take Marta Green as an example. Without access to an automobile, Marta often shops at a small neighborhood grocery store that charges higher prices for food than do larger supermarkets. In addition, the selection and quality of food items are generally worse than the offerings of the large chains. Without a bank account, Marta has to pay to cash her government checks and to make other monetary transactions. Her winter utility bills also run higher because she cannot afford better-quality housing with improved insulation. This pattern has been found repeatedly among the poverty-stricken.

The hard times and daily problems are succinctly summarized by forty-six-year-old Alice Waters, who is on SSI as a result of cancer and whose husband died of emphysema: "Day-to-day problems? No money, that's number one. No money. Bein' poor. That's number one—no money."

THE END OF THE MONTH

Food Stamp and AFDC benefits, received monthly, are usually not enough to provide adequately throughout a particular month. Many recipients find that their Food Stamps routinely run out by the end of the third week. Even with the budgeting and stretching of resources that recipients try to do, there is simply not enough left. Tammy and Jack Collins describe the process:

Tammy: Mainly it's towards the end of the month, and you run out of Food Stamps and gotta pay rent. Tryin' to find enough money to buy groceries. It's the main one.

Q: When that comes up, do you turn to somebody to borrow money, or do you just try to stretch what you've got?

Tammy: I try to stretch. And sometimes his ma will pay him for doin' things on the weekend for her, which will help out. She knows we need the money.

Jack: We collect aluminum cans and we got a crusher in the basement and we sell them.

Q: What happens when you run out of food?

Tammy: That's when his ma helps us.

Recipients often rely on some kind of emergency assistance, such as food pantries or family and friends, to help them through. Such networks provide an important source of support. Of the food pantries I visited, all reported that the numbers of people coming in for emergency food supplies increased dramatically during the last ten days of each month.

I asked Carol Richardson if she ran out of food, particularly at the end of the month.

Carol: Yeah. All the time.

Q: How do you manage?

Carol: We've got a food pantry up here that they allow you to go to two times a month. They give you a little card. And in between those times, we find other food pantries that we can get to. We've gone to different churches and asked for help all the time. And we get commodities at the end of the month. Cheese and butter. And then we usually get one item out of it, which helps an awful lot.

Borrowing from friends or relatives is another end-of-the-month strategy. We asked Rosa and Alejandro Martinez, an elderly married couple, about this. Rosa responded, "Sometimes we're short of money to pay for everything. We have to pay life insurance, mine and his. We have to pay for the car. We have to pay the light. And we have to pay the telephone. And there are a lot of expenses that we have to pay. And sometimes we can't meet them all. And between the month, I borrow, but I borrow from my friends to make ends meet."

Others deal with the financial squeeze at the end of the month differently. For example, Clarissa and John Wilson, a married couple in their thirties, rely on extra money from a blood plasma center to help them through. I asked them if they had bills which they are unable to pay.

Clarissa: Yeah, that's just frustrating. When you know that you can't pay your bills, and where are you gonna get the money to pay them. If you don't pay 'em, they're gonna always be with you. And that's just frustrating.

John: That's the main reason that we're always going to University Plasma. That's the main reason. To keep up . . . at least try to keep up, some of the bills. If the bills weren't a problem, we wouldn't go every week like we do.

Clarissa: Sometimes I don't see why we're gettin' the aid check because it still doesn't meet our needs—half of our needs. Maybe forty percent of it, I'd say.

In short, at the end of the month, the economic struggles that recipients face loom even larger. Even the basic necessities may be hard to come by.

A SET OF DOMINOES

For welfare recipients, there is little financial leeway should any unanticipated expenses occur. When nothing out of the ordinary happens, recipients may be able to scrape by. However, when the unexpected occurs (as it often does), it can set in motion a domino effect touching every other aspect of recipients' lives. One unanticipated expense can cause a shortage of money for food, rent, utilities, or other necessary items. The dominoes begin to fall one by one.

Unanticipated expenses include items such as medical costs and needed repairs on a major appliance or an automobile. During these crises, households must make difficult decisions regarding other necessities in their lives. I asked

Cindy and Jeff Franklin to describe how they deal with these kinds of problems.

Cindy: Well, I think it's running out of money. (*Sighs.*) If something comes up—a car repair or (*pause*) our refrigerator's on the fritz. . . . We have enough money for a nice, adequate, simple lifestyle as long as nothing happens. If something happens, then we really get thrown in a tizzy. And I'd say that's the worst—that's the worst.

Jeff: Yeah, 'cause just recently, in the last month, the car that we had was about to rust apart. Sort of literally. And so we had to switch cars. And my parents had this car that we've got now, sitting around. They gave it to us for free, but we had to put about two hundred dollars into it just to get it in safe enough condition so that we don't have to constantly be wondering if something's gonna break on it.

Cindy: I think that sense of having to choose—the car is a real good example of it—having to choose between letting things go—in a situation that's unsafe, or destituting ourselves in order to fix it. Having to make that kind of choice is really hard.

When welfare recipients must make these types of choices, it is seldom because they have budgeted their finances improperly. Rather, they simply do not have enough money to begin with, often not enough to cover even the basic monthly expenses. Among the poverty-stricken, this is a major recurring problem. Public assistance programs help, but they just do not provide enough. Households are forced routinely to make hard choices among necessities.

One particularly difficult choice facing some households in wintertime is choosing between heat or food—the "heat-or-eat" dilemma. A three-year study done by the Boston City Hospital showed that the number of emergency room visits by underweight children increased by 30 percent after the coldest months of the year. In explaining the results, Deborah Frank, who led the study team, noted, "Parents well know that children freeze before they starve and in winter some families have to divert their already inadequate food budget to buy fuel to keep the children warm" (*New York Times,* September 9, 1992). An underweight child's ability to fight infection and disease becomes even more impaired when that child is also malnourished.

Tammy and Jack Collins described earlier how they are pushed to stretch their food supplies and to juggle their bills, particularly at the end of each month. When asked about her children's needs, Tammy responded, "Their teeth need fixing. That's the worst. See, our insurance doesn't cover that. And we've got one that needs an eye checkup that we can't get in to get done because the insurance don't cover that." If they decide to get their children's teeth fixed or eyes examined, the Collinses know they will have to make some hard choices as to which bills not to pay or which necessities to forgo. The dominoes thus begin to fall.

CONSEQUENCES AND EFFECTS

Much has been written about the negative consequences of poverty. The poor suffer from higher rates of disease and crime, experience more chronic and acute health problems, pay more for particular goods and services, have higher infant mortality rates, encounter more dangerous environmental effects, face a greater probability of undernourishment, and have higher levels of psychological stress. . . .

As mentioned earlier, hunger is a real consequence of poverty. Many of the families I talked to admitted that there were times when they and their children were forced to go hungry and/or significantly alter their diets. Running out of food is not uncommon among those who rely on public assistance.

A widow in her sixties, Edith Mathews lives in a working-class, elderly neighborhood. When she received forty-five dollars' worth of Food Stamps a month, they were not enough to provide an adequate diet (she was subsequently terminated from the program for not providing sufficient documentation). Edith suffers from several serious health problems, including diabetes and high blood pressure. The fact that she cannot afford a balanced diet compounds her health problems. She explains:

> Toward the end of the month, we just live on toast and stuff. Toast and eggs or something like that. I'm supposed to eat green vegetables. I'm supposed to be on a special diet because I'm a diabetic. But there's a lotta things that I'm supposed to eat that I can't afford. Because the fruit and vegetables are terribly high in the store. It's ridiculous! I was out to Cedar's grocery, they're charging fifty-nine cents for one grapefruit. I'm supposed to eat grapefruit, but who's gonna pay fifty-nine cents for one grapefruit when you don't have much money? But my doctor says that that's one thing that's important, is to eat the right foods when you're a diabetic. But I eat what I can afford. And if I can't afford it, I can't eat it. So that's why my blood sugar's high because lots of times I should have certain things to eat and I just can't pay. I can't afford it.

Similarly, Edith is often forced to reuse hypodermic needles to inject insulin. While she is aware that this could be dangerous, she feels she has little choice: "And then those needles that I buy, they cost plenty too. Twelve dollars and something for needles. For a box of needles. And you're only supposed to use 'em once and throw 'em away. But who could afford that? I use 'em over, but they said you shouldn't do that. Sometimes they get dull and I can't hardly use 'em. But I just can't afford to be buying 'em all the time. It's outrageous the way they charge for things."

Nancy Jordon was asked about not having enough food for her three children. In her mid-thirties, Nancy had been receiving public assistance for two months. Her income from working as a cosmetologist was simply too low to survive on as a single parent. She explains that not having enough money for food has had physical consequences not only upon her children but upon her as well:

Nancy: Well, as long as I got money. If not, I have to resort to other measures. It's a sad thing but a woman should never be broke because if she's got a mind, and knows how to use it, you can go out in the streets. Which is the ultimate LAST resort is to go to the streets. But at a point in a woman's life, if she cares anything about her children, if she cares anything about their lifestyle, they'll go. Matter of fact, some would go to the streets before they would go to aid.

Q: Have you had to do that, in the past, to feed your kids?

Nancy: A couple of times yes.

In addition to suffering from hunger, recipients may let health problems go unattended until they became serious, live in undesirable and dangerous neighborhoods, or face various types of discrimination as a result of being poor. But perhaps the most ubiquitous consequence of living in poverty and on public assistance is the sheer difficulty of accomplishing various tasks most of us take for granted: not being able to shop at larger and cheaper food stores because of lack of transportation and so paying more for groceries; having to take one's dirty clothing on the bus to the nearest laundromat with three children in tow; being unable to afford to go to the dentist even though the pain is excruciating; not purchasing a simple meal at a restaurant for fear it will disrupt the budget; never being able to go to a movie; having no credit, which in turn makes getting a future credit rating difficult; lacking a typewriter or personal computer on which to improve secretarial skills for a job interview. The list could go on and on.

These are the types of constraints experienced day in and day out by most of the people I saw and spoke with. The barriers facing the poor are often severe, apparent, and ongoing. They represent an all-pervasive consequence of living in poverty. Even the little things in life can become large. Recall Marta Green's difficulties when she took her children to the circus. Or consider the time the utility company shut off the Collins's gas: Tammy and Jack found that simply giving their children baths became a formidable task. Tammy explains: "They used to shut our gas off during the summer. And then I had no hot water for bathin' the kids. I had to heat it on the stove downstairs and carry it up to the bathtub. That was the worst times. I had an electric stove. I used to sit a big pan on top of it, and heat it."

For Cindy and Jeff Franklin, economic problems have turned what many of us would consider insignificant decisions into major quandaries.

Jeff: I mean there's times when I'm taking the bus, and I have to wonder can we really afford to take the bus this morning?

Cindy: Right now, today, we've got just over a dollar. And three dollars in food stamps. And that's all we've got till the end of the month, till the end of the month. That's fairly unusual because the car ate up so much money this month. Usually we manage our money better than that. But we really have to think. I mean, it's not just thinking about whether or not you can afford to go to a movie,

> but you have to think about can the kids and I stop and get a soda if we've been out running errands. It's a big decision, 'cause we just don't have much spending money.

Living in poverty and on public assistance is a harsh, ongoing economic struggle that becomes more acute by the end of the month. There is very little slack in the rope. When an unanticipated event leads to economic stress, entire lives suffer, domino fashion. Everyday life is complicated by a variety of negative physical consequences, including the difficulty of carrying out the most basic tasks. As Mary Summers commented at the beginning of this chapter, "This is where you're just up against a wall. You can't make a move."

23

Multiracials, Intermarriage, Ethnicity

JOEL PERLMANN

One of the hottest trends in the study of ethnicity involves its definition. With intermarriage eroding classic definitions of racial categories, Perlmann suggests that we may be entering a new era of American race and ethnicity relations, moving us beyond the melting pot of ethnic assimilation and the subsequent backlash that led to ethnic reidentification. Despite prejudice and discrimination, American society is witnessing growing rates of ethnic and racial intermarriage, leading to an ambiguity of demographic categorization, most clearly seen in the furor over census classifications. This trend, he argues, is occurring in most pronounced fashion among white ethnics and between whites and Asians, but has also spread into the Hispanic communities and last to the most stigmatized of intermarriage combinations: black-white. Perlmann proposes that our society in the future will be more characterized by racial intermixing, and that people of multiracial composition will comprise a greater segment of the population. Do you know anyone who has married someone of a different racial or ethnic group? How did their relatives react to this? Now, think back to the America in which your grandparents lived. How would the reaction to intermarriage be the same or different from today? Do you think that the rising rates of intermarriage are a positive or negative sign for the future of society?

If a child has a white mother and a black father, then the child is racially . . . what? Deciding how the next census, to be held in the year 2000, should handle the "multiracial" child is a hot topic this year—from front page stories in the *New York Times,* to disquisitions on the origins of Tiger Woods, to congressional hearings. There are two bizarre features to the way the issue is being treated, and they tell us much about the state of American thinking about race. First, there has been virtually no discussion of the fact that racial intermarriage is a form of ethnic intermarriage—intermarriage across those ethnic lines we choose to call race lines. And the Census Bureau knows how to count the offspring of ethnic intermarriages: It has done so for over a century. Second, neither the Bureau nor the media have been drawing the important connection between this topic (how to count the "multiracial" child) and another topic that has also been receiving front-page attention for several years, namely, the Census Bureau's regularly updated projections of the future racial composition of the American people. A recent projection, for example, was reported in the *New York Times* on March 14, 1996, also on page 1. These projections of racial composition pivot on expectations for intermarriage in the future, and the expectations are strange indeed.

How "multiracial" children will be counted in the next census is receiving attention now because the decision is needed soon, because such children have found a voice and because the decision might affect race-based laws and politics. Specifically, many people concerned with civil rights enforcement or with the power of racial minorities do not want to add messy complexities to the struggle for racial equality. It is hard enough to achieve programs against discrimination (or in favor of affirmative action) today without considering how such programs should treat mixed-race individuals. Similarly, the future racial composition of the country is a hot-button issue. Telling Americans that their country will be more than half nonwhite by the middle of the next century—what *Time* called "the Browning of America" some years ago—stimulates different reactions. To some, the demographic projection says America had better wake up to the needs of its "minorities": They are soon to be its majority. To others, the projection says America had better restrict immigration to avoid reaching a "nonwhite majority." In fact, by the middle of the next century America will very likely be "more than 50 percent nonwhite" and also "more than 50 percent white" by the Bureau's logic.

ETHNIC INTERMARRIAGE—AMERICAN AS APPLE PIE

American history would be unrecognizable without ethnic intermarriage. From colonial times to the present, immigrants typically married their own, the second generation did so much less consistently, and the third generation still less consistently; probably a majority of third-generation members have married people of other ethnic origins than their own. And by the fourth and fifth generations—who even kept track? The evidence for the history of ethnic intermarriage is about as overwhelming and unambiguous as any generalization about the American population can be: from de Crèvecoeur in the eighteenth century noticing "this new man, the American" arising out of various European immigrant stocks to the data from census after census in the twentieth century.

Intermarriage operated most fully among the descendants of European groups (Italians, Poles, Irish, English, Germans, Scandinavians, etc.); it was crucial to the making of "Americans" out of the descendants of "hyphenated Americans." Intermarriage was least prevalent, perhaps, across the black-white divide—not least because such intermarriage was so often illegal. Notwithstanding the lower prevalence of black-white intermarriage, there has been a surprisingly high degree of intermarriage across lines one might not have expected: between "white" and "red" for example. Huge proportions of people descended from "American Indians" tell us that they are part "American Indian" (or "Native American") and part "white."

Today black-white intermarriage is growing in importance—especially among younger, middle-class, college-educated blacks; nevertheless, it remains the one divide across which intermarriage has remained strikingly less common than elsewhere. If "multiracial" today meant only the offspring of black-white marriages, it would be less of an issue (although a growing issue nonetheless, given changing consciousness). However, "multiracials" also include rapidly growing numbers of children in marriages of Asians (immigrants or their children, for example) and non-Asians; and the way the counting goes, it can also be interpreted to include Hispanics and non-Hispanics.

Well, then, how has the Census Bureau handled the offspring of other ethnic intermarriages for the past century—those ethnic intermarriages that have not crossed racial lines? When the Census Bureau asks a native-born child of foreign-born parents where the parents were born, the answer is often that they were born in two different countries. No problem; the bureau records two countries of origin. Both parents born in Italy? Fine. One born in Italy, one in Poland? Also fine (the question about parents' birthplaces was dropped in the most recent censuses but will probably reappear by census 2000).

There is an even stronger model for the Bureau's "multiracial problem" in the census ancestry question. Lately, in the 1980 and 1990 censuses, the Bureau has also asked each individual with what ethnic ancestry he or she identified—and ethnic ancestry means a place on the globe: Italy, Poland, etc. The Bureau introduced this ethnic ancestry question in order to allow Americans to state an ethnic affiliation even if one's immigrant ancestors had come to the United States from the country of ancestry many generations back. Now the crucial point is that on the ancestry question, the Census Bureau instructs Americans that they can identify themselves with more than one ethnic ancestry. Many millions of Americans have taken the trouble to list two ethnic ancestries, millions more list three ancestries. Yet even these multiple listings are only a partial indicator of the real complexity of American ancestry, since many people are descended from far too many different ancestries to list them all (or even to know them all). In 1980, "English" was listed before "German" in the Bureau's examples of ancestry; in 1990 the ordering was reversed. As a result, the percentage claiming identity with English ancestry declined by a large fraction, and the number claiming identity with German ancestry rose by a comparable amount. There are many such examples, but even these examples of confusion show us something important, namely, the long-term results of population mixing and the attenuation of connections with the origins of ancestors (whom one never knew). In short, keeping track of American ancestries at the Bureau eventually gets messy because of intermarriage patterns—and that is as it should be. A simple answer would be a false answer.

RACIAL INTERMARRIAGE

What has all this got to do with racial mixing today and with the American racial profile tomorrow? It has everything to do with those topics—once we understand racial intermarriage as a variant of ethnic intermarriage. Whatever meaning race has for anthropologists and biologists today—and that is probably no meaning—race is thought of as different from ethnicity by most Americans. And in some sense that is a good thing. One useful way to think about racial differences is simply that these are ethnic differences that were treated in very distinct ways in the American past (and to some extent are still treated so today). For that reason, if we want to understand problems like American economic inequality we cannot ignore people's racial origins; to blithely throw out race classifications in our present censuses would not be smart or fair.

Nevertheless, as barriers begin to crumble, ethnic groups that were especially stigmatized in the past are increasingly able to partake of the American mainstream. Among other things, this change means that they intermarry more often with members of other ethnic groups. We see this happening on a large scale among Asian immigrants today, among American Indians and Hispanics past and present, and finally, although still to a lesser extent, even among blacks today. Will intermarriage characterize the descendants of "new" immigrant groups, that is, Hispanics and Asian immigrants, as fully as (or more fully than) it has characterized the descendants of Europeans? And how much will black-white intermarriage increase? We need not know the answers to these questions in order to notice that intermarriage rates for the children of new immigrants (mostly, as *Time* had it, "brown"), are already high and cannot be ignored, and that black-white intermarriage is now a meaningful social reality.

SO HOW TO COUNT?

We must think about the children of interracial marriages in the same way that we think about the children of other interethnic marriages. The easiest way to do that is to let people identify themselves in terms of race just as we let them identify themselves in terms of ethnic ancestry. At the moment, the census ancestry question encourages respondents to list more than one ancestry, and the census race question tells Americans that they cannot belong to more than one race; but they can and they do. This contradictory position must be changed: People must be told explicitly that they can list themselves as members of more than one race. To create a new category, "multiracial"—a key proposal being debated today—is better than no change, but it is surely the wrong solution. The creation of a "multiracial" racial category was sensibly rejected by the government's task force on racial classification, and the rejection should not be overturned. Such a category would add nothing to the information conveyed when a person declares his or her origins to be, for example, Asian and white. It would not only be redundant to have the category, it would be inappropriate in implying that somehow being multiracial deserves to be treated differently than being multiethnic. A person reports as, for example, Polish and Italian in terms of ancestries (or parents' birthplaces)—not as multiethnic.

If we do let people identify with more than one race, counts will surely become more complicated than they are now, but they will be reasonably straightforward for as long as they are worth having, and if they become as confused, on occasion, as the ethnic ancestry data sometimes becomes, that will be good to know too. How such counts of multiracial people will be used in connection with enforcing civil-rights legislation (including affirmative action and voting rights laws) will have to be worked out (and the counts issue can be worked out). But counting the racially mixed in connection with laws that rely on race counts is going to be an issue in the coming years no matter what the Census Bureau does about the multirace category. Keeping track of the magnitude of groups over time will become messier and will cause technical complexities. But those complexities reflect the experience of the groups; to avoid those complexities is to deny that a fundamental force in American life operates among "racial minorities" as well as among other Americans.

24

Mexican American Domestic Workers

CECILIA GARZA

American society is once again witnessing a wave of immigration and dealing with it in diverse ways. Severe political resistance has arisen to the influx of Mexican Americans who smuggle themselves across the border, although Garza argues that this is heavily influenced by prevailing economic conditions, with quiet acceptance predominating in periods where cheap, foreign labor is highly needed, and loud argument arising when native jobs are perceived as threatened. Yet, Garza notes, the influx of undocumented illegal aliens could not flourish without Americans supporting and employing these workers. Garza focuses on one segment of this population: women who cross into the country and obtain employment as domestic workers. She uses the case study approach to excellent effect, offering us portraits of several women, in different situations, who are picked up, networked, housed, employed, and often exploited by American men and women who need inexpensive housecleaning, child care, and older adult care to maintain their own standard of living and positions in the higher-strata labor force. Yet, interestingly, these household workers often leave families back home in Mexico, send most of their earnings back there, and repeatedly cross the border to visit, maintaining strong roots in their countries of origin. This study offers us a compelling look into the lives of a growing segment of Americans who are officially rejected, poorly treated, self-sacrificing, yet highly connected to their fragmented kinship networks. Do you know anyone who either has worked as an undocumented domestic worker or who has hired such a person? What do you think are the motivations for hiring these people? What do you know about the conditions under which they work? Does the picture of these people painted by Garza coincide with your own previous images of them?

This paper presents the situational context for undocumented household workers in the border city of Laredo, Texas. I will highlight a few case histories to illustrate the effect of economic need. My thesis is that the occupation of domestics along the U.S.-Mexico border is entrenched in a larger system of mutual self-interest where individuals and corporations can benefit, while others may become its victim. . . .

BACKGROUND ON IMMIGRATION POLICIES

The pattern of immigration policies affecting Mexican Americans depicts a pattern of self-serving interests. When workers are needed, the U.S. immigration policies have been lenient. During difficult economic times, the policies are strict and the enforcement severe and punitive. The Border Patrol was established in 1924 and during the peak years of the depression, enforcement restrictions were strengthened. In the process, many Mexican Americans were repatriated along with the illegal immigrants. During World War II when workers were needed, the Bracero Program was legislated to admit mostly agricultural workers. After the war, the INS's Operation Wetback embarked upon a massive program to deport illegal Mexican immigrants. Mexican Americans, mistaken to be Mexican nationals, were also deported.

According to recent media reports, the United States is increasingly experiencing anti-immigrant sentiment. Rhetoric espousing strict enforcement of the border has been politically advantageous this election year. With the passage of the Immigration Reform and Control Act in 1986, amnesty was available to immigrants who could prove they had lived in the U.S. since January 1, 1982. Those who qualified were permitted to obtain their "green card" (not necessarily green) legalizing their status. Those who did not qualify continue to survive in an underground and unregulated economy.

Many immigrants did not qualify because they could not show documents proving their residence before 1982. Ironically, their illegal existence in the U.S. is precisely what they were effectively hiding, and now they were asked to present documents demonstrating what they had been successfully hiding. . . .

THEORETICAL ORIENTATION AND PROFILE OF DOMESTIC WORKERS

The global economy plays a role in the shaping of the climate within which domestic servitude exists. Multinational corporations establish assembly plants known as maquiladoras along the Mexican side of the border in search of cheap labor. The arrangement proves convenient for the managerial executives who live on the American side of the border. These plants have attracted thousands of individuals and families from the interior of Mexico (Martin 1993). Unfortunately, there are many more job applicants than jobs and the overflow creates a surplus of labor willing and pressured to work for low wages. Much of this surplus labor turns to domestic work. In turn, the distance between the social classes increases, such as when many middle and lower-middle class families can afford the otherwise luxury of maid service.

Undocumented domestic workers are not the typical illegal aliens depicted in the media as the young, Mexican male, "wetback" who swam across the Rio Grande. One woman I interviewed, however, did swim her way across to the U.S. Illegal immigrants are increasingly a heterogeneous population (Hondagneu-Sotelo 1996). The undocumented household worker functioning in an unregulated economy is profiled as Latina between the ages of fifteen and thirty-nine (Passel 1986). These workers are often women doing domestic and child care work or caring for the elderly and disabled, a growing and significant sector of the population.

Feminist theory argues that precisely because domesticity is primarily a female occupation, it does not receive recognition under public policies (Fine 1988; Davis 1993). Romero (1994, 135) argues "domestic service challenges feminist notions of sisterhood . . . it accentuates contradictions of ethnicity, race, and class in feminism; privileged women of one class use the labor of other women to escape aspects of sexism."

Domestic work is not recognized as a category of essential labor as other technical and specialized occupations are. Yet, domestics, whether legal or undocumented, provide a valuable resource to society and a savings to the federal government. To illustrate, there are numerous household workers who care for the elderly in their own homes preventing costly institutionalization. Household workers also facilitate the job entry for dual-career households. Domestics supervise child care in a home environment providing a savings in tax deductions for child care expenses. The dollar value of these services merits study and occupational recognition for immigration policy consideration. . . .

THE UNDOCUMENTED WORKERS

The supply of household workers is plentiful, but the network is intricate. There are two major groups of undocumented workers. There are those who cross the border daily and those from the interior of Mexico who are live-in domestics. The latter cross and stay from five or six weeks to several months at a time, and return to visit their children, parents, spouses, etc. for three or four weeks.

The following is a synopsis of some of the case histories, depicting a microscopic view of the human life hidden behind the data:

Norma. Norma is a petite, 33 year-old female. She is employed by an upper-class family where both spouses work. The family has one daughter in college and a son in boarding school. Norma is the second child of eleven. There would have been fifteen children, but four died, two were aborted unwillingly (miscarriages). Norma's mother was doing heavy labor when she miscarried; two others died as infants, one dehydrated, and the other died from bronchitis. There was no talk of family planning or the means with which to pay for it. Norma's father worked as a miner. He was disabled early in his otherwise productive lifetime because of the effect of the mine pollution which was destroying his lungs. The doctor said he must quit or risk an early death. The mother worked at home raising the children and tending a small amount of crops, cattle and chickens. They lived in an ejido which they presumed to be theirs, but could be taken from them at any time. They paid taxes on any type of income producing item, including each head of cattle.

Norma stayed home without schooling until age 11 when she joined a convent. There she received elementary education and the first year of middle school. Her hopes were to remain in the convent, but her father did not allow it. She must work to help the family, and the convent life was not a productive enterprise. At the convent she learned many of the skills which would later help her. She studied home management, cooking, sewing, first aid, and care of the sick. More importantly, she was no longer illiterate. Leaving the convent has been her biggest disappointment in life.

Jobs are scarce throughout Mexico. The best job one can reasonably hope for is a factory job paying 150 pesos per week, or the equivalent of about $25. Her biggest hope, like that of thousands of other Mexican citizens, was to migrate to the U.S., find work and send money home.

Norma began the process of moving to the U.S. when she was 19 years old. She first travelled to Nuevo Laredo from San Luis Potosi. There she lived with a family known to a family from back home. Within a week she had convinced a family member to help her cross the river to Laredo. He was reluctant because he had never done this before, and there were many risks to their safety and lives. Nevertheless, Norma knew she could not stay with this family forever, and she must take the risk. With indescribable fear she swam across during the day. Once on the U.S. side, she tried to mingle with other people and act inconspicuous. She had heard stories of rapes and beatings committed by hoods and even immigration officials.

Before crossing she had made arrangements to be taken to the house where a sister worked. It cost her $200 for the service of the lookout-contact person. With the permission of the sister's employer, she stayed there for fifteen days until she found employment as a domestic. She worked temporarily for a few families returning to stay with her sister until 1981 when she found more permanent employment. She worked for a prominent family from 1981 to 1985. The family also allowed her to work for other families during the weekends. Since she could not go out in public and risk deportation, her only outside activities were to work in other homes. There is an underground support network among families who share or trade their housekeepers. The intermittent family provides roundtrip transportation which helps protect the safety of the housekeeper.

This system of exchange works favorably, but it does have its desavenencias, or points of discord which may alienate families. For example, the housekeepers communicate among themselves and share their experiences with their respective employers. They compare working conditions such as hours, pay, free time (if any), leave, and special considerations. They are permitted only limited use of the telephone and access to food, thus making these "desirable treats." When the differences in working conditions are too disparate, they may proceed to transfer to a household which offers better working conditions.

There are serious risks to leaving the employ of a family. The losing family may cut relations with the receiving family or worse yet, may report the housekeeper to INS. This happened to Norma once when she found another home for a peer. However, she was able to talk herself into convincing the losing employer that she was legal.

In 1985, six years after she left her home, she went to visit her parents. She saved money for the trip and for her parents. During the six years she had also been mailing money to them, but sometimes the money was lost. Now she is assured they will receive the money because two brothers who now live in Houston take them money. These two brothers filed for legalization during the amnesty period and now can come back and forth freely.

When she returned from the latest trip, Norma did not feel the need to swim to get across. Instead, she used the bridge and mingled with people she could recognize as locals. With subsequent returns, she waits for a large crowd to cross believing this will give her a better chance to blend in and get in undetected. Sometimes this does not work, and she is sent back.

Norma plans another trip this year, but she is always in fear, especially now that her sister is with her. Her sister suffers from epilepsy which complicates matters even more. She said she agreed to bring her sister with her to give her parents a break from taking care of her. Neither of them have health insurance. Their employer has arranged for them to be seen by local physicians, but they pay out of their own earnings. The psychologist who is treating her sister, for example, is charging them $200 per visit. Norma is desperate to find a way to help her sister.

Norma says she has been fortunate to have fine people as employers. They have allowed her to work in other homes, visit her parents, and help other peers. Norma was able to bring her sister with her and even take English classes. Her greatest aspiration is to legalize her status and to study nursing. She says any document can be obtained for the right amount of money. However, she wants to proceed according to the law and live a free person, like any other human being. She is currently studying English and hopes to get her GED and proceed from there. She lost the chance to apply for amnesty because of an ineffective attorney. He was incompetent, but claimed he simply could not help her.

Norma is a brave, intelligent, and determined young woman. She has sacrificed her personal happiness for the sake of her family. It has now been fifteen years since she left her parents' home. She hopes to have a family of her own someday. Norma has been making payments on a piece of land in Laredo, and it is almost paid for. She has been offered marriage in exchange for legalization, but she says she could never do that. It would not be morally correct, and in her view she never wants to lose her sense of morality. She wants to proceed legally and live to see the day when she can breathe freely without feeling fearful palpitations and without constantly looking behind her shoulder.

Nora. Nora is a petite, frail fifteen-year-old female. She was unemployed when I interviewed her and temporarily staying with a family who knew her sister. Nora came to Texas from southernmost Mexico. She first came to the Rio Grande Valley to work with a family whose business, it turned out, was to exploit undocumented domestics. After working a few weeks with an American family, the employer threatened to call the Border Patrol unless she paid $300. In fear of being deported, Nora contacted her sister in Florida who sent her the money. The sister also called a former employer in Laredo asking her to receive her sister at the bus station. The sister begged her former employer to allow Nora to stay at her former employer's house until she arranged transportation for Nora to join her in Florida. Nora stayed in Laredo four weeks until a man known to transport people returned from a trip to Georgia where he had been making such deliveries. Nora eventually left Laredo and joined her sister in Florida.

Mari. Mari is a 41-year-old female. She works for an upper-middle-class family with two teenage children and three adults, one of whom is elderly and disabled. Mari looks much older than her biological age. She comes from a small town near Ciudad Victoria. She is married and has two children, a daughter, 15, and a son, 12. Her mother looks after her children. Her father is working as a night watchman, but is planning to retire soon since he has recently developed diabetes. Her mother is 61 years old and in fairly good health. Mari is planning to arrange for her son to move to Mexico City to live with her sister. She wants him to begin to know his way around a big city and prepare for work and hopefully a career. By living with her sister's family, Mari's son could also have a male role model in his life. Mari's husband is currently living in Houston where he has worked for the past five years. He works as a roofer, but he has worked mostly in agriculture.

Mari was employed as a secretary for nine years before she was laid off or as she called it, "Me congelaron mi plaza," or froze my job. She worked for the government earning the equivalent of 200 U.S. dollars per month. According to Mari, her supervisor insisted that she go out with him, and she refused him. She was denied employment, unemployment benefits and lost all other benefits, including retirement. This was a very difficult time for her as she was unable to find work elsewhere. She said she came close to a serious breakdown and wanted to die. She asked her husband to come see her from Houston because she was in such a bad state of desperation. This period of unemployment lasted eight months. At that point in time, she decided to leave for the U.S. to find employment.

The most painful scene was to leave her children behind. The emotional trauma was indescribable, but she saw no other way. Her parents supported her decision and agreed to look after her children.

Mari first worked in Mission, Texas, for her sister who was already in the U.S. and had legally immigrated. She then went to Laredo with a friend of the family who was a legal resident. She first found employment as a housekeeper with a widower who lived with his son. There she was to earn $60 per week. She did not last an entire week. He was very abusive and became exasperated when she could not find utensils or dishes in his kitchen. She could not take his profanity and called her friend to come pick her up. The man kicked her out of the house and made her wait outside the house at 11:00 P.M. until her friend came for her. Mari said she was very scared but felt somehow lucky that at least she had someone to call.

Later Mari found employment with her current family where she has been working as a housekeeper for a year. This family has been kind to her, treating her almost as a family member. They have been considerate in allowing her to visit her family. She does this about every two months and stays usually a week at a time. During these trips her husband also tries to meet her so the family can be reunited for a few days. The last time they celebrated her daughter's fifteenth birthday. Mari says that she was able to afford to give her daughter a party because both she and her husband are working. The fifteenth birthday is of particular importance to many Mexican and Mexican-American families because it represents a rite of passage into adulthood. Mari's income is for her children and their education. Unlike other domestics, Mari does not pay for food or shelter. This helps her save nearly all of her income, which she either sends to them or takes with her when she visits.

Mari has a Mexican passport to get back and forth into the country. She does not have a permit to work. Once she crosses the border she calls her employer who makes arrangements to pick her up at a certain location not frequented by other domestic workers. She usually chooses a very public place like a mall, where it would be perfectly legitimate for her to be. The passport entitles her to visit and shop, but not to work. She was able to obtain her passport while she was employed. Today, it is very difficult to get a passport, if the government thinks it is going to be used to leave the country to work abroad. To get a passport, applicants must be working and show financial security with bank statements and property ownership.

While she is working in the U.S., Mari rarely leaves her employer's house. She is afraid of being detained by INS officials and either sent back, have her passport picked up, or worse yet, be imprisoned. Thus, she stays inside the house the entire time, until she leaves to go back to visit her family. She thinks this routine will last as long as it takes for her children to be educated and independent of their parents' support. For now, Mari hopes to stay healthy and maintain employment.

Chabela. Chabela is a 40-year-old attractive female. She also looks older than forty. She lives in Nuevo Laredo with her fourteen-year-old daughter, her seventy-year-old mother, and lately her brother who is divorced. Her brother is an albanil, a mason, who drinks too much, like her father did. Her mother left her father for the same reason.

Chabela works as a domestic and earns $25 per day. She works four days with one family and one or two days with another family. At the house in which I interviewed her, she irons clothing and also earns an extra five dollars for bathing an elderly lady. There is an unwritten code, a type of collective conscience, among domestics that they do not take work from each other if it means someone else can earn some money. Mari, the full-time domestic, said she could bathe the lady, but it would mean Chabela would not earn the five dollars, so she moved aside.

Chabela has a green card which allows her to cross the border without difficulty as long as she is not working in the U.S. She crosses at odd hours to minimize attention to herself. When she is asked where she is going, she says shopping, or to the used clothes warehouse. During the period of amnesty, her uncle tried to get her to immigrate legally. This meant she had to give up her green card and not go into Mexico for a full year. She could not bear to do this! This would mean not seeing her daughter, nor her mother, and there was no assurance that her case would be accepted. It could mean risking everything and losing the green card. So she decided not to apply. She says her uncle was very upset with her and now they are not on good speaking terms.

Chabela had one bad experience with an employer. She was working as a housekeeper for a doctor at both his home and his office. She said the doctor's wife accused her of taking a blouse and using it to go to a party. Chabela said she would never do that, and explained that the blouse was at the cleaners. She showed the lady the receipt. It was no use because the lady was abusive and she could not go back to that type of environment. Chabela felt that the lady, formerly from Piedras Negras, Mexico, should be able to understand her, but it was quite the opposite. Her experience has been that the Mexican people treat their own people worse, and they do not try to help each other. She feels the need or the fear of losing what they have makes people insensitive and even cruel.

Other than this experience, Chabela says everyone has been good to her. As with other domestics, Chabela expects little from life. She wants the opportunity to work and to have the health to do so. She hopes for a more promising future for her daughter.

Rosa. Rosa is a 48-year-old, single female. She has worked for a frail, 86-year-old disabled Mexican-American citizen for the past two years. Her employer is totally dependent on others for her livelihood. She is bedridden and requires constant supervision. She has no relatives and only one friend, a neighbor, who hired her. While the elderly lady was able to speak, she had told her friend she never wanted to go to a nursing home. The friend feels obliged to grant her wish and found Rosa to look after her. They live in a dilapidated three-room flame house with no heating or cooling. Rosa earns $40 per week plus room and board. Her previous job was with a doctor's family. She was dismissed from this job because the doctor's wife had a baby, and Rosa was told they did not want the baby to get used to her. Rosa knows that jobs come and go from day to day. She has no notion of job security. When her employer dies, she says she will find someone else who needs her services.

To summarize, foreign domestics live in a marginalized world which sustains them and permits them to experience the inequalities of an economically and socially stratified existence. They may share the same gender, race and ethnicity as their employers, but they are worlds apart. Their life chances are strictly limited by man-made laws. These women work separated from their immediate families, including their children, with no fringe benefits, no legal protections, no job security, no health insurance, no retirement plan in either nation, and no foreseeable better future.

REFERENCES

Davis, M. F. 1993. "Domestic Workers: Out of the Shadows." *Human Rights* 20:14–15.

Fine, D. R. 1988. "Women Caregivers and Home Health Workers: Prejudice and Inequity in Home Health Care." *Sociology of Health Care* 7:105–17.

Hondagneu-Sotelo, P. 1996. "Family and Community in the Migration of Mexican Undocumented Immigrant Women." Pp. 173–85 in *Ethnic Women: A Multiple Status Reality* edited by Vasilikie Demos and Marcia Texler Segal. New York: General Hall, Inc.

Martin, P. L. 1993. *Trade and Migration: NAFTA and Agriculture.* Washington, DC: Institute for International Economics.

Passel, J. S. 1986. "Undocumented Immigration." *Annals of the American Academy of Political and Social Science* 487:181–200.

Romero, M. 1994. "Transcending and Reproducing Race, Class and Gender Hierarchies in the Everyday Interactions between Chicana Private Household Workers and Employers." Pp. 135–44 in *Ethnic Women: A Multiple Status Reality,* edited by V. Demos and M. Texler Segal: New York: General Hall, Inc.

25

Koreans in Small Entrepreneurial Businesses

ILSOO KIM

In contrast to Mexican Americans, Koreans represent a new immigrant group that has achieved significant success and is heading toward upward mobility in American society. Kim's article shows us some of the contrasts separating the conditions of this group from Garza's population, not the least of which is their immigration to America in complete family groups, enabling them to marshall all their attention and resources on their members and life in the new country. Like the Mexican Americans, the Koreans are willing to self-sacrifice and defer gratification so that their children can enjoy a better life, but unlike them, they are usually legal immigrants who do not have to worry about being arrested and deported, enabling them to pool the resources of their extended kinship networks into small business enterprises where they all work. Highly capitalistic and raised in middle-class backgrounds, the Koreans save their money and reinvest it, operating small shops and participating in international trade between America and Asia. Unlike earlier groups of Asian immigrants, these Koreans move beyond their strictly ethnic group affiliations to strongly involve themselves with community institutions in their new urban neighborhoods. Compare and contrast the experiences of Mexican Americans and Korean Americans. Why do you think these two groups have experienced such different histories in American society? Based on these experiences, what do you think is the future of these groups in the United States?

A high proportion of Korean immigrants across the nation have chosen small business as the means for pursuing the American dream in postindustrial America. As in the case of Jewish immigrants at the turn of the century, post-1965 Korean immigrants have developed small businesses as an economic beachhead for their own and their children's further advancement in American society. In fact, Korean immigrants have earned a new racial epithet—"Kew" or "Korean Jew." In the New York metropolitan area alone, Korean immigrants, as of 1985, ran some 9,000 small business enterprises, certainly contributing to the recent vitality of the New York economy.

Given the structural changes in the American, and especially the New York, economy in which small businesses have declined in number (Freedman 1983:103), why have Korean immigrants "inundated" small businesses and reactivated a traditional immigrant path to the American dream? The massive Korean entry into small businesses may be welcomed as a revival of old patterns by those who lament the loss of "rugged individualism" and entrepreneurship in the United States and the dominant trend toward business concentration, centralization, and bureaucratization. But Korean involvement in small business is not a duplication of patterns found among "old" European immigrants earlier in the century. Rather, it must be understood on its own terms—a product of Korean immigrants' homeland-derived socioeconomic and cultural characteristics or resources as well as broader socioeconomic conditions in their new land. . . .

This essay focuses on three modern, structural factors that have supported or been conducive to the proliferation of Korean small businesses in New York. The first factor is the utilization of modern "ethnic class resources" which Koreans brought with them from their home country. These ethnic class resources include advanced education, economic motivation or "success ideology," and money. The second factor is the economic opportunities available in New York where limited employment possibilities in the mainstream economy coincided with opportunities that opened up in New York's economic structure due to demographic, residential, and ethnic changes in the city. Finally, the third factor is the role of such "modern" ethnic institutions as the mass media, churches, and businessmen's associations in the New York Korean community. . . .

THE BACKGROUND OF NEW KOREAN IMMIGRANTS

Migration and Settlement Patterns

The Korean community in the New York metropolitan area is the result of the 1965 Immigration and Nationality Act which ended the severe numerical limits on immigration from Asian countries. According to the 1980 census, 94 percent of the Korean-born population in the New York metropolitan area arrived since 1965.

The reason most Koreans give for emigrating to America is basically "to find a better life." Behind this rather simple answer are a number of complex push factors, including Korean population pressure and political instability. The political uncertainty in South Korea, caused mainly by the military tension between South and North Korea since the Korean War (1950–53), led many middle- and upper-middle-class Koreans to leave for the United States in fear of another Korean war, especially in the 1970s (see Illsoo Kim 1984). Population pressures have also driven South Koreans from their homeland. In a nation roughly the size of the state of Maine, the number of people has grown from 25 million in 1960 to 40 million in 1984. The population density of South Korea is the third highest in the world. This population explosion, along with rapid industrialization and urbanization, has led to typical Third World urban problems—severe crowding, pollution, and intense economic competition for limited resources and opportunities.

South Korea has also witnessed a revolution of rising expectations, a result of the country's rapid economic development through increasing international trade as well as the population's encounters with Western, especially American, mass culture. Koreans are now fascinated by American life-styles, which the vast majority cannot hope to obtain in South Korea. The disparity between their desires and their limited resources and opportunities has tended to make urban, especially college-educated, Koreans restless and dissatisfied and to intensify their conviction that South Korea is not a good place to live. As I will discuss later, the majority of Korean immigrants are drawn from this urban middle class in South Korea. . . .

Korean immigrants have predominantly settled in large metropolitan areas in the United States. The largest Korean community is in the Los Angeles metropolitan area, followed by New York, and then Chicago. Within the New York metropolitan area, Koreans have not formed one single territorial community, although there are heavy concentrations in the white lower-middle-class sections of Flushing, Jackson Heights, Corona, and Elmhurst in Queens. Even in these neighborhoods, Korean immigrants have not established single-block residential enclaves and they live among members of the second or third generation of old European immigrants as well as such new immigrants as Cubans and other Asians (see Illsoo Kim 1981:181–86 for characteristics of the Korean nonterritorial community). Once Koreans make it in New York City—as doctors, engineers, or successful businessmen—they nearly always move to the suburbs.

Socioeconomic Characteristics

The recent Korean immigration is selective and includes a high proportion of well-educated urban middle-class people. A majority of Korean immigrants are drawn from the upper middle or middle classes in the major cities of South Korea such as Seoul, Pusan, and Taegue (Illsoo Kim 1981:38) and thus had experience of living in large modern urban centers before moving to the United States.

Despite the fact that so many Korean immigrants come to New York with high levels of education, professional experience, and an urban middle-class background, most are not able to obtain well-paid professional, white-collar work in the mainstream American occupational structure. Such work requires proficiency in English and a long period of training in large-scale American organizations—insuperable barriers to most Korean immigrants. In urban America, especially in New York, the center of economic activities has shifted to white-collar, service industries; lacking professional service skills, Koreans are handicapped.

Under these circumstances, Korean immigrants, who generally had little propensity for commercial activity in South Korea, often turn to small business in New York. In so doing, they utilize the very ethnic class resources they bring with them—money, advanced education, high economic motivation, a work ethic, and professional or sometimes business skills. I refer to these resources as "ethnic class resources" since they are mainly derived from the social class circumstances of the immigrants in South Korea, a homogeneous ethnic state. Many Korean immigrants in New York come with money they can use to set up small businesses. Their educational background means they have skills relevant to running businesses. And the fact that large numbers held high-level and prestigious jobs in South Korea gives them the confidence and motivation to work hard to succeed.

The Immigrant Family

Koreans have generally come to New York with members of their nuclear family, their basic social unit, although frequently a family is temporarily separated so that a pioneer member can establish an economic base or because of a bureaucratic delay under United States immigration laws. The immigration of families has been made possible by the humane nature of the 1965 U.S. immigration law, which emphasized the reunion of immediate relatives.

The continuation of two- and often three-generation families in New York is another key factor in the proliferation of small businesses among Korean immigrants. First, family members who have come to New York are a major source of capital for small businesses. Some families bring considerable amounts of money from South Korea; others rely heavily on savings accumulated and pooled in New York for their first businesses (compare Kim and Hurh 1985:101–102 on Korean businesses in the Chicago area). The South Korean government allows each emigrant family to take a maximum of $100,000 to the United States in the form of settlement money. Family members planning to emigrate frequently gather a significant sum by combining their savings or by selling personal property. This settlement money represents one of the few ways a Korean family can enter the United States prepared to buy a small business or to pay initial living expenses. Second, many immigrants can rely on family labor in their enterprises in New York. Largely owing to the cultural legacy of Confucian familism, Korean immigrants subscribe to a family-centered success ethic that leads family members to be willing to devote themselves to the family business. That family members are willing to work long hours has been especially important in the greengrocer business, enabling Korean enterprises in New York to compete with supermarket chains (Illsoo Kim 1981:115–116). Here is one family story:

Mr. Yun, his wife, and their two teenage children entered the United States in 1974 after Mr. Yun's brother, a naturalized American citizen, had petitioned the U.S. Immigration and Naturalization Service on their behalf. Upon arrival, Mrs. Yun immediately sent for her parents on a tourist visa, and consequently a three-generation family was established at Mr. Yun's residence. Mr. Yun, a college graduate with a major in business management, had worked for the U.S. Army in South Korea. He had a good command of colloquial English and, when he arrived, was able to get a low-level clerical job in a New York City government agency. His wife and mother-in-law did piecework for New York garment factories. In two years they had saved $20,000, which they used to buy a fruit and vegetable store from another Korean who, in turn, had purchased it from an old Italian.

Thereafter, Mr. Yun arose at 4 A.M. every morning and drove to the Hunt's Point wholesale market in the Bronx to purchase the day's goods. His wife worked from 7 A.M. to 7 P.M. as the store's cashier. Mrs. Yun's mother washed, clipped, and sorted vegetables; her father, who was too old to engage in physical labor, regularly stationed himself on a chair at the store's entrance to help deter shoplifters. The teenagers, too, were part-time but regular workers, helping in the store before and after school. In Mr. Yun's words, "We worked and worked like hell." But he and and his family were successful. Mr. Yun found a Korean partner and in 1980 they bought a gas station on Route 4 in New Jersey. Mr. Yun continued to run the greengrocery business while leaving the management of the gas station to his partner, an auto mechanic.

SMALL BUSINESS AS AN ECONOMIC BASIS OF THE KOREAN COMMUNITY

The extent of Korean immigrants involvement in small business in New York is striking. It is true, of course, that Koreans are found in other occupational spheres in New York—many as medical professionals (doctors, nurses, pharmacists, and technicians) and as engineers, mechanics, and operatives (see Illsoo Kim 1981:40). Nonetheless, small businesses can be said to be the economic foundation of the Korean community in New York and in the United States in general.

Most Korean businesses, especially labor-intensive retail shops such as greengrocery businesses, fish stores, and discount stores want to hire, and actively recruit, Korean immigrants or *kyopo* (fellow countrymen). In the New York Korean community small businesses are the primary source of employment for newcomers or greenhorns who lack proficiency in English and skills that are marketable in the wider economy. These newcomers have no choice but to take jobs with Korean employers, working long hours without such benefits as medical insurance and overtime pay. The availability of this kind of labor is beneficial to business owners who, at the same time, can feel they are helping their compatriots. One Korean business leader, for example, boasted of the small business contribution to the community: "In our *kyopo* (fellow countrymen) society there is no unemployment problem thanks to the jobs created by *kyopo* businessmen. I have never seen unemployed Koreans." Indeed, it is also my observation that unemployed Korean immigrants are few in number.

The very proliferation and success of Korean small businesses has, in itself, had an independent effect in stimulating further business activity in the Korean community. Setting up a small business has become, in a sense, a "cultural fashion" among Koreans, a point that has not received sufficient attention in the literature on immigrant enterprises. A Korean aphorism has it that "running a *jangsa* (commercial business) is the fastest way to get ahead in America," and this saying is widespread among Korean immigrants, including those who have just arrived. In Korean gatherings such as church meetings, alumni meetings and picnics, Koreans devote much of their time to talk about *jangsa*. At one church meeting, a Korean immigrant, who had just opened a fine jewelry store in midtown Manhattan, talked to church members about his business: "Making money is a mysterious magic to me. I never expected such a good profit." Many Korean immigrants have been susceptible to this kind of glorification of small business and, impressed by the success of others, have entered small business. In addition, the ethnic media, especially Korean daily newspapers, frequently present stories about Korean Horatio Algers who have "made it" through small business and this, too, encourages Korean entry into small business. . . .

COMMUNITY INSTITUTIONS AND BUSINESS ENTERPRISES

Korean community institutions play an important role in the proliferation and success of Korean small businesses in New York City, but these differ from the traditional ethnic associations that several works on immigrant enterprises in the United States have emphasized. . . .

In the case of Korean immigrants in New York, it is modern ethnic institutions, many derived from South Korea but all influenced by external forces in New York, that are crucial. These institutions serve many important functions for Korean enterpreneurs and stimulate Korean business activity in New York. The modern, ethnic institutions that support Korean enterprises in the New York metropolitan area include churches, the mass media, and business associations. The analysis below considers the extent to which each contributes to Korean business development.

The Protestant Church as a Community

Protestant churches have flourished among Koreans in New York partly because so many immigrants are drawn from the Christian, especially Protestant, population in South Korea. In addition, the churches have become important community centers for immigrants in the absence of discrete, residential enclaves among them. Indeed, at least for members, church communities have become the substitute for ethnic neighborhoods. . . .

Korean churches are much more than sites for religious services. Because they serve multiple, secular functions they are central places for community activities and they have, in fact, opened up membership to all segments of the Korean population. A Korean engineer living in a "white" New Jersey suburb said: "On Sunday I do not want my children watching TV all day long. At least one day a week I want them to intermingle with other Koreans and learn something about Korea. This is the reason why my family and I attend [Protestant] church even though I am a Buddhist. My offerings are nothing but the payment for the services my family has received from the church." Ministers in the churches perform numerous secular roles, and they are mainly judged by their congregations according to how well they do so. The ministers' extrachurch activities include: "matchmaking, presiding over marriage ceremonies, visiting hospitalized members, assisting moving families, making congratulatory visits to families having a new baby, making airport pick-ups of newly arrived family members, interpreting for "no-English" members, administering job referral and housing services, and performing other similar personal services" (Illsoo Kim 1981:200).

In this context, most Korean Protestants attend churches not only for religious salvation but also for secular—economic and social—reasons. They want to make and meet Korean friends; they are looking for jobs or job information; they want to obtain business information, make business contacts or conduct business negotiations; or they seek private loans or want to organize or participate in *gae,* a Korean rotating credit association.

Ethnic Media and the Business Community

The Korean ethnic media also play a decisive role in maintaining the Korean community in the New York metropolitan area. The Korean ethnic media go beyond delivering news. By informing geographically dispersed Korean immigrants of community events and meetings as well as of commercial sales and news, the media are a powerful means of integrating and sustaining the Korean community. What is pertinent here is that the ethnic media have been crucial in the rapid expansion of Korean enterprises. . . .

The contribution of the newspapers to Korean enterprises is twofold. First, they are an influential agency for socializing Korean immigrants into small business capitalism. They provide immigrants with all kinds of commercial information on tax guidelines, accounting, commercial issues, and prospects for new businesses. They carry articles or essays on successful *kyopo*(fellow countrymen) in business, who serve as role models for newcomers to emulate. They also quickly cover events affecting Korean small businessmen. For example, they have alerted Korean businessmen to immigration officials' crackdowns on Korean illegal aliens employed in Korean businesses and they have promptly published news of frequent conflicts between Korean businessmen and blacks. Second, the community media, especially the daily Korean language newspapers, facilitate the expansion of the Korean subeconomy by carrying advertising for ethnic business products, ethnic services, commodity sales, housing sales or rentals, jobs, and so forth.

Businessmen's Associations and Ethnic Solidarity

Korean businessmen's associations, as their name suggests, are clearly involved in Korean business development. They can be classified into two categories: (1) associations based on business type and (2) associations based on geographical area. Korean businessmen have formed associations of the first type in business lines in which they are active. There are, for example, associations of greengrocers, fish retailers, dry cleaners, garment retailers, garment subcontractors, and gas station operators. As for the second category, Korean businessmen have also established "prosperity associations" in major commercial areas of Korean business concentration. Korean prosperity associations can be found in the following areas of New York City: the South Bronx; Flushing, Sunnyside, and Jamaica in Queens; Church Avenue in Brooklyn; the Lower East Side, Central Harlem, and Washington Heights in Manhattan. . . .

Since Korean immigrants are "modernized" ethnics, they rely upon modern ethnic institutions—churches, the media, and businessmen's associations—in their development of small business capitalism. The Korean utilization of these modern institutions departs from the old pattern of Chinese and Japanese solidarity for business based upon kinship and regional associations. It is my view that studies of ethnic enterprise in America have not paid sufficient attention to the impact of "modern" ethnic community institutions on immigrant enterprises—or, for that matter, to the effect of homeland-derived class resources and economic opportunities in particular American urban contexts. These structural factors, so critical in Korean enterprises in New York, are, I would argue, bound to be important in understanding the involvement of other new immigrant groups in small businesses as well.

REFERENCES

Freedman, Marcia. 1983. "The Labor Market for Immigrants in New York City." *New York Affairs* 7:94–110.

Kim, Illsoo. 1981. *New Urban Immigrants: The Korean Community in New York.* Princeton: Princeton University Press.

———. 1984. "Korean Emigration Connections in Urban America: A Structural Analysis of Premigration Factors in South Korea." Paper presented to the Conference on Asia-Pacific Immigration to the United States, East-West Population Institute, Honolulu, Hawaii.

Kim, Kwang Chung and Won Moo Hurh. 1985. "Ethnic Resources Utilization of Korean Immigrant Entrepreneurs in the Chicago Minority Area." *International Migration Review* 19:82–111.

26

The Decline of Jewish Identity

EDWARD S. SHAPIRO

A well-established group, American Jews have lived in relative peace and prosperity in this country for many decades. Although once the subjects of extreme discrimination, restriction, harassment, and mass elimination, American Jews have found a safe haven in America and flourished economically. Shapiro suggests that, ironically, this diminished prejudice and occupational success have weakened the American Jewish community in a way that threatens its very core more than more hostile environments were able to accomplish. American Jews have experienced significant levels of intermarriage, leading to the diffusion of their population. Religious definitions have been relaxed, permitting less observant behaviors to flourish and weakening restrictions on the definition of membership. Being Jewish in contemporary America has become an identity problem, Shapiro suggests, with many unsure as to whether this is a religious or merely ethnic category. In either case, Shapiro says that the propensity toward intermarriage and the decline of religious attendance and observance may lead to the vanishing of the American Jew. Can you think of any other group whose identity may be diminishing in American society like the Jews? If Jews (and other groups) continue to follow this path, what might be the consequences? Can you think of an alternative path that Jews and others like them can take?

The question of identity has been more problematic for Jews than for any other American subgroup. This is due in the first place to the perplexing nature of what it means to be Jewish. The issue of "Who Is a Jew?" has vexed American Jews just as it has Jews in Israel. Thus the most contentious issue within American Jewish religious circles during the past two decades was the decision of the American Reform movement in the early 1980s to cast aside the definition of Jewishness that had delineated the Jewish community for thousands of years. According to this definition, a Jew had to have a Jewish mother or to have been converted according to Jewish law. Matrilineal descent was used to determine Jewish identity since the identity of one's mother, in contrast to one's father, was never in doubt.

The Reform movement, however, broadened the definition of Jewish identity to include patrilineal descent if intermarried parents, for their part, involved their children in the Jewish community through such things as participation in religious services or enrollment in a religious school. This move was taken not because of any theological revelation but because of the rapid increase in intermarriage between Jews and non-Jews beginning in the 1960s. This resulted in hundreds of thousands of families in which there were doubts about the Jewishness of the children since their mothers had never undergone even a minimal conversion. Leaders from the Orthodox and Conservative wings of American Judaism strongly protested this decision of the Reform movement. They claimed that it would result in a schism within American Judaism since Reform Jews could no longer be automatically considered as appropriate marriage partners for Jews who accepted the traditional definition of who is a Jew and who wished to avoid intermarriage. But the problem of Jewish identity goes beyond this intramural conflict over the religious definition of who is a Jew. It encompasses the broader and more important question of whether Jewishness is a matter of religion, history, culture, or ethnicity.

The identity of other groups is not so muddled. There is no confusion, for example, over the fact that the ethnicity of Irish-Catholics is Irish while their religion is Roman Catholicism. A lapsed Irish-Catholic remains Irish although he or she is no longer a Catholic in good standing. But such clarity of religious and ethnic identity does not exist among Jews. A Jewish atheist or agnostic, such as an Albert Einstein or a Sidney Hook, remains a Jew in good standing. In fact, many of the fiercest critics of Judaism have been Jews, and modern-day Jewish movements such as Jewish socialism and Labor Zionism have opposed traditional Judaism. Jews also do not constitute a language group. Few Jews speak Hebrew, Yiddish, or Ladino (Spanish-Jewish). Nor are Jews defined by being the victims of prejudice and discrimination. This might have been true in Europe and the Arab countries, but it certainly is not true in the United States, where the income, occupational mobility, and social status of Jews is far higher than that of the general population.

This ambiguity regarding the nature of Jewishness is particularly confusing to Americans, who tend to see Jews as comprising a religious group comparable to that of Protestants and Catholics. Books such as Will Herberg's *Protestant-Catholic-Jew* (1955) reinforced this disposition to see religion as the essence of Jewishness. One of the reasons that anti-Semitism has never been as strong in America as elsewhere is that Americans place a high value on religion and because they equate Jewishness with Judaism. This conflating of Jewishness with Judaism is also seen in academia, where courses in Jewish Studies, even when they are concerned with sociology and history, are often located in departments of religion.

Complicating the definition of American Jewish identity is the relationship of Jews to America. Traditional Judaism emphasized the obligation of Jews to be good citizens and to defer to those in power, even if they were anti-Semitic. Thus in the early years of Nazi Germany, Orthodox rabbis in Germany told their followers to respect the edicts of the political authorities. Traditional Judaism, however, also taught that Jews were in exile and would eventually return to the

Promised Land. But the nationalistic and religious impulses encouraging Jews in the late nineteenth and early twentieth centuries to settle in Palestine were not as powerful as the social and economic opportunities of America. For every Jew who left Europe for Palestine in these years, forty emigrated to the United States. Here was a land in which the government did not encourage or tacitly accept anti-Semitism, in which the property and lives of Jews were protected by the local and national governments, and in which there were no official barriers to the social and economic advancement of Jews. As George Washington noted in his famous letter of 1790 to the synagogue of Newport, Rhode Island, in America Jews as well as Christians will "possess alike liberty of conscience and immunities of citizenship."

If America was not the Promised Land, this "*novus ordo seclorum*" was certainly the land of promise for Jews, and they were fiercely loyal to their new country. "This synagogue is our temple, this city our Jerusalem, this happy land our Palestine," Rabbi Gustavus Poznanski told the congregants of Beth Elohim, the Charleston, South Carolina, synagogue, prior to the Civil War. Mary Antin, a Jewish immigrant raised in Boston, also saw America as the new Israel, with Boston being the New Jerusalem. The title of her 1911 autobiography is *The Promised Land*. Another Jew, Irving Berlin, wrote "God Bless America," while a third, Emma Lazarus, wrote "The New Colossus," the sonnet placed at the base of the Statue of Liberty celebrating America as the refuge for the "huddled masses yearning to be free." Lazarus, who was involved in efforts to ameliorate the conditions of eastern European immigrants in New York City, was undoubtedly thinking of her co-religionists when she wrote her poem.

Eager to become part of America, American Jews were skeptical of ideologies and movements that impeded their movement into the American mainstream. Zionism, for example, was unpopular among America's Jews until the 1930s, when it became obvious that a Jewish homeland was necessary for Europe's beleaguered Jews. But American Jews had no intention themselves of migrating to Palestine or, later, to Israel. They feared that the Zionism movement would raise doubts among Americans as to their political loyalties. Rather than rejecting Zionism, American Jews transformed it to conform to American realities. For them, Zionism was not a nationalistic movement encompassing all Jews but rather a philanthropy to succor *other* Jews. This gave rise to the quip that American Zionism was a program in which one group of Jews gave money to another group of Jews to bring a third group of Jews to the Middle East. For Israelis, most notably David Ben-Gurion, the first prime minister of Israel, American Zionism was not truly Zionist since it downplayed the fundamental Zionist principles of the negation of the Diaspora and the ingathering of all exiles to Israel. With the establishment of a Jewish state, American Zionism seemed to be an anachronism to the Israelis. How could one claim to be a Zionist and yet not settle in the Jewish state? The Israeli statesman Abba Eban joked that American Zionism demonstrated the truth of one religious principle: that there could be life after death.

Traditional Judaism suffered the same fate as Zionism in America. The dietary and other restrictions in Judaism that promoted a sharp separation between Jews and Gentiles and discouraged the movement of Jews into the American economic and cultural mainstream fell by the wayside. New religious ideologies, such as Reform Judaism, Conservative Judaism, and Reconstructionism, partially filled the vacuum created by the diminished appeal of Orthodoxy. As one wag put it, Jews gave up Orthodoxy at the drop of a hat. This refusal to cover one's hair or to refrain from eating forbidden foods or to observe the Sabbath was symptomatic of a deeper problem: the fact that only a small minority of American Jews believed that they were a chosen people, elected by God to serve a distinctive purpose. The Bible talks about the Jews being a "peculiar people," and for thousands of years there had been no doubt among Jews as to the source of their distinctiveness. Every morning religious Jews had blessed God for having "chosen us among all peoples and given us thy Torah" and for not having "made me a heathen."

Jews were anxious to be thought of as no different than other Americans. In June 1952, *Look* magazine published Rabbi Morris Kertzer's article "What Is a Jew?" The thrust of Kertzer's piece was that Jews and Christians "share the same rich heritage of the Old Testament. They both believe in the fatherhood of one God, in the sanctity of the Ten Commandments, the wisdom of the prophets and the brotherhood of man." If Kertzer was correct that Jews were like everyone else, then what reason was there for them to maintain their distinctiveness? Why should they choose to remain Jewish if they were not different from their Christian neighbors? Had not the type of thinking expressed by Kertzer deprived Jews and Judaism of the raison d'être for any peculiarity?

In 1960, the sociologist Erich Rosenthal used the phrase "acculturation without assimilation" to describe the process of social adaptation of Jews in America. Jews, he argued, had adopted the values and lifestyles of the general society, but they separated themselves from the rest of America in choosing marriage partners, friends, and places to live. In America even this limited sense of separation could not long be maintained, not when many Jews were attending colleges in which they were a minority and were rapidly moving up the social and economic ladder. Life was simply too attractive and open for Jews to isolate themselves. Yeshiva University could not compete with Columbia or Harvard, nor could the dense Jewish neighborhoods of New York City, Philadelphia, and Chicago compete with the beckoning suburbs. In the case of Kerri Strug, who captivated the country when she won the gold medal for the American women's gymnastic team at the Atlanta Olympics despite having an injured leg, Jewish identity lost out to the attraction of the uneven bars and the balance beam. Jewish newspapers informed their readers that Strug's parents were active in a Tucson synagogue but that their daughter had been too busy with her athletic development to be involved in anything Jewish.

In America, Jews ceased to be a chosen people and instead became a choosing people, and Judaism became a religious persuasion which could be accepted or denied. In America, the sociologist Samuel C. Heilman wrote, being Jewish ceased being "simply a matter of birth, or, more precisely, a matter of irrevocable destiny." Much as patrons at a Chinese restaurant will choose items from column A and column B, so Jews became adept in selecting those aspects of Judaism and Jewishness that harmonized with their identity as modern Americans. The most popular of these did not interfere with acculturation, did not make Jews conspicuous, and were attuned to democratic values. In making such selections, American Jews were exhibiting that sense of individualism that commentators since the time of Alexis de Tocqueville have seen as quintessentially American. In America, a Jew could be any kind of Jew he or she wanted, or even cease being a Jew at all. In America, the historian Robert Seltzer wrote, Jews "have been remarkably free to decide what of their heritage to conserve, reform, and reconstruct." He could have added they were free as well to discard elements of this heritage.

David Gelernter, the polymath Yale professor, protested against this tendency to recast Judaism in order to conform with the latest sociological or intellectual fad. "Like most American Jews," he wrote in 1996,

> I find myself able to observe only a tiny fraction of the Torah's commandments. Unlike some, I believe that the commandments *are* binding. When I fail to perform a religious obligation, I do not want a soothing Reform or Conservative authority to tell me I am in luck—that particular obligation has been dropped from the new edition and I am free to ignore it. I am not free to ignore it and commit a sin when I fail to do it. I acknowledge my failings and recall that God is merciful. . . . This infantile insistence that religious ritual conform to you rather than the other way around is the essence of modern American culture, and is strangling Judaism.

Gelernter's view of American Jewry's future is bleak. Being Jewish, he predicted, will come to mean what " 'being Scottish in America' means: nothing. Certain family names will suggest Jewish or Scottish origins." That is all.

Reinforcing the centrifugal nature of American Jewish identity was the absence of any official rabbinate or politically recognized communal officials with the power to determine who is and is not Jewish and what being Jewish entails. The result was a Jewish community of incredible religious and cultural diversity. America has been the birthplace of new religious ideologies such as Conservative Judaism, Reconstructionist Judaism, the Havurot movement, and Jewish feminism. Though "We Are One," the motto of the United Jewish Appeal in America, might have been an effective fund-raising slogan, it certainly was not an accurate description of the reality of American Jewry. This Jewishness without barriers has had its bizarre side effects as well. Thus there is even a group of "Messianic Jews" or "Jews for Jesus" who claim that they are good Jews despite their belief in the divinity of Jesus.

The enigmatic nature of American Jewish identity is responsible for the curious fact that the three major interpretations of what it means to be an American were provided by Jews. Emma Lazarus's "The New Colossus" argued for integrating the "wretched refuse yearning to be free," and her own assimilated life was a model of what she hoped the immigrants from Europe, and particularly the Jewish immigrants, would conform to. Two decades after the Statue of Liberty was dedicated in 1886, Israel Zangwill published his play *The Melting Pot*. This tale of an intermarriage in New York City between David, a Russian Jew, and Vera, the daughter of a Christian Russian responsible for a pogrom in David's hometown, reflected Zangwill's belief in the beneficence of the ethnic amalgamation that he believed was taking place in America. The glory of America, he emphasized, lay in the ability of Americans, native-born and immigrant alike, to put aside ancient rivalries and to create a new nationality combining the best traits of the various ethnic groups peopling America. Zangwill did not regret the fact that in America the Jew would disappear, as would the Italian, the Irishman, and the Yankee. Zangwill himself had married the daughter of a Protestant clergyman and did not rear his children as Jews.

The most important answer to the melting pot idea was Horace Kallen's 1915 essay "Democracy versus the Melting Pot." Kallen, the son of an Orthodox Jew who had settled in Boston, argued that the most accurate metaphor for the process of Americanization was not a melting pot but an orchestra. Just as each instrument in an orchestra made a distinctive contribution to the symphony, so each ethnic group made a distinctive contribution to American life. And just as it would be foolish to melt down the instruments of the orchestra, so it would be equally foolish to melt down America's ethnic groups. Out of this ethnic diversity, Kallen predicted, there was emerging a new "symphony of civilization" in which "each nationality would have its emotional and involuntary life, its own peculiar dialect or speech, its own individual and inevitable esthetic and intellectual forms." And just as a democratic government was obligated to safeguard the rights of the citizens to join whatever religious and social groups they chose, so a democratic government should encourage the people to preserve their ethnic identities, which, Kallen mistakenly believed, were inalienable: "There are human capacities which it is the function of the state to liberate and to protect in growth," Kallen said, "and the failure of the state as a government to accomplish this automatically makes for its abolition."

The pluralism expressed by Horace Kallen has been American Jewry's greatest strength and its greatest potential weakness. Elaine Marks, a professor of European literature at the University of Wisconsin, recently showed just how far a Jewishness without boundaries can be stretched: "I am Jewish precisely because I am not a believer," she said paradoxically, "because I associate from early childhood the courage not to believe with being Jewish." For Marks, choosing to deny Judaism is thus a quintessential Jewish act, and the Jew who rejects Judaism is transformed into the most committed Jew.

With Jewish identity increasingly a matter of prescription rather than ascription, what guarantee is there that a sufficient number of Jews will ever choose the same things? What will be the source of Jewish communal affiliation when many Jews have come to believe that Judaism sanctions whatever they believe or wish to do? What is the lowest common denominator of Jewish belief and practice that can act as the cement of Jewish identity? Certainly traditional Judaism no longer fills that role. No more than one-quarter of American Jews observe the dietary laws, less than 10 percent keep the Sabbath as a day of rest, and over half do not light Sabbath candles (90 percent even think that a Jew could be religious without being observant). These same persons claim that being Jewish was very important to them and that they consider themselves to be "very good" Jews. For them, being Jewish has little to do with practicing Judaism. In their own minds, they are very good Jews because their version of Jewishness demands nothing of them. It is a Jewishness without content.

Neither Israel nor the memory of the Holocaust can be the core of an American Jewish identity. In an age of jet travel, when it is easy and relatively cheap to travel to Israel, less than 30 percent of American Jews have even visited the Jewish state. As time goes on, the heroic memories of Israel during 1948, 1967, and 1973 will resonate less clearly for new generations of American Jews. And the same thing is true of memories of the Holocaust. Furthermore, the memory of the Holocaust is certainly not a sound basis for inculcating Jewish identity. Not many people will be attracted to a Judaism that holds out victimhood status.

For decades, the lowest common denominator of Jewish identity in America was contributing to Jewish causes, the most important of which was the United Jewish Appeal (UJA). The UJA's major selling point was the aid it provided to the struggling state of Israel. But the Israel of 1997 is not the undeveloped and beleaguered Israel of the 1950s and 1960s. The Jewish state has grown up. With a gross national product of more than $80 billion dollars and with a per capita income approximately that of England and Italy, Israel is not desperate for American dollars. With the financial support for Israel having less emotional appeal, what will play its role in recharging the ethnic and religious identities of American Jews?

A survey among Los Angeles Jews in the 1980s revealed that the most important element in their Jewish identity was not the ritual obligations of Judaism or even support for the state of Israel. It was, instead, a commitment to social equality. This belief in liberalism as the essence of Jewishness was the theme of Leonard Fein's 1988 book *Where Are We? The Inner Life of America's Jews.* Only a commitment to economic and social justice, Fein said, "can serve as our preeminent motive, the path through which our past is vindicated, our present warranted, and our future affirmed."

The question of Jewish identity is particularly important today when it can no longer be assumed that Jewishness, however it might be defined, is being automatically passed on from one generation to the next. In his 1959 apologia, *This Is My God,* the novelist Herman Wouk, an Orthodox Jew, described a mythical Mr. Abramson, a Jewish amnesiac, "pleasantly vanishing down a broad highway at the wheel of a high-power station wagon, with the golf clubs piled in the back." When his amnesia clears "he will be Mr. Adamson, and his wife and children will join him, and all will be well. But the Jewish question will be over in the United States."

For Jewish survivalists, the most troubling aspect of American life was the rapid increase in intermarriage beginning in the 1960s. The *1990 National Jewish Population Survey,* an important demographic study of contemporary American Jewry, reported that 52 percent of Jews were then choosing non-Jewish marriage partners. But could anything be done to reverse this development, or would Jews have to learn to live with it? In other words, was intermarriage a problem for which there was a solution, or was it a condition that could not be changed? For Conservative Rabbi Robert Gordis it was the latter. In *Judaism in a Christian World* (1966), Gordis argued that "intermarriage is part of the price that modern Jewry must pay for freedom and equality in an open society." This was scant comfort for Jewish survivalists.

Such statistics seemed to confirm the apprehensions of pessimists such as Hayim Greenberg regarding the future of American Jewry. Greenberg, a Labor

Zionist, had feared as early as 1950 that American Jews were in "grave danger of becoming merely an ethnic group in the conventional sense of the term . . . only a group with a long and heroic history which, when cultivated, can arouse much justified pride . . . but without the consciousness of a specific drama and tension in its life." During the optimistic postwar years, Jews viewed Greenberg's fears as unduly alarmist. By the 1980s and 1990s, however, pessimism was in style. Thus Samuel Heilman ended his 1995 volume, *Portrait of American Jews: The Last Half of the 20th Century,* by speculating on whether Judaism and Jewishness had any future in America: "If I am to be certain that my children and their children will continue to be actively Jewish," he wrote, "then the boat that brought my family here to America in 1950 may still have another trip to make."

Certainly there is much to justify the pessimism of Gelernter, Greenberg, and Heilman. But there is also much that belies their gloom as well. Despite Greenberg's fears, Jews have not become a "conventional" ethnic group. During the four and a half decades since Greenberg uttered his warning, American Jews have experienced an unexpected flowering of institutional and intellectual life. During these years, over five hundred all-day Jewish schools were established, American Jewry produced a home-grown intelligentsia that now staffs the country's many rabbinical seminaries and university programs in Jewish Studies, and Jewish fund-raising expanded into an operation collecting well over a billion dollars a year. If most of the nation's Jews can be described as Jewish Americans, there still remains a sizable minority of American Jews. The key question is whether, to quote Heilman, "these few can continue to exert such influence and define the character of American Judaism, and whether they can continue to be actively Jewish while the majority drifts away toward a peripheral involvement with Judaism."

In *The Ambivalent American Jew* (1973), the sociologist Charles Liebman noted that American Jews were "torn between two sets of values—those of integration and acceptance into American society and those of Jewish group survival." Jerold A. Auerbach, a historian at Wellesley College, agreed with Liebman. In *Rabbis and Lawyers: The Journey from Torah to Constitution* (1990), Auerbach argued that American Jews were heirs to two disparate and, at times, contradictory traditions—Jewish and American—and that committed American Jews were fated to live in two competing and discordant worlds. "The synthesis of Judaism and Americanism," he said, was "a historical fiction." The symposium in the August 1996 issue of *Commentary* magazine on "What Do American Jews Believe?" indicated that a significant number of America's leading Jewish religious thinkers agree.

If the buzzwords of the 1930s and 1940s for American Jews were "survival" and "anti-Semitism," the buzzwords of the 1980s and 1990s have been "continuity" and "identity." While Jews do not fear for their physical safety, they are concerned about the viability of a Jewish population that is experiencing major demographic hemorrhaging due to religious apathy, a low birth rate, a high rate of exogamy, and the lure of a secular culture emphasizing individual autonomy and personal gratification rather than religious obligations and communal commitment. History should instill caution on the ability to predict the future. Who could have predicted in 1988 the demise of the Soviet Union, a peace treaty between Israel and the Palestine Liberation Organization, and the election of an obscure governor of Arkansas as president? In May, 1964, *Look* magazine published an article entitled "The Vanishing American Jew." Well, the American Jew did not vanish, but *Look* soon did. But whether the editors of *Look* will have the last laugh remains to be seen. The experience of Jews, more than that of America's other ethnic groups, "is the supreme test of how far acculturation can go without eroding the sense of distinctiveness," Stephen J. Whitfield wrote recently. "So far American Jewry as a whole has not flunked this test of an open society. Sometimes they are in the dark, however."

27

Girls' and Boys' Popularity

PATRICIA A. ADLER
AND PETER ADLER

We have long known that boys and girls are socialized into different gendered cultures that intermediate between them as individuals and the larger society. Although gender roles have evolved significantly over the last generation, children may represent a more conservative age group for the display of gender, as they focus more fixedly on learning and conforming to gender roles before they develop the sense of confidence to let themselves develop as individuals. The core values of preadolescent boys' and girls' school culture are those they have extracted from the larger adult culture, and serve as the factors promoting popularity, that omnipresent concern of our youth. In this selection taken from our study of preadolescents, we trace the factors promoting popularity in boys' and girls' worlds, and then analyze the underlying images and trends in the larger society that these forces represent. Do the factors we identify for girls and boys make sense to you? Are they the same factors that made you and/or your classmates popular? How does popularity operate more generally in social life? What gender differences do you see in adults?

Considerable effort has been invested in the past two decades toward understanding the nature of gender differences in society. Critical to this effort is knowledge about where gender differences begin, where they are particularly supported, and how they become entrenched.

Elementary schools are powerful sites for the construction of culturally patterned gender relations. In what has been called the "second curriculum" (Best 1983) or the "unofficial school" (Kessler et al. 1985), children create their own norms, values, and styles within the school setting that constitute their peer culture, what Glassner (1976) called "kid society." It is within this peer culture that they do their "identity work" (Wexler 1988), learning and evaluating roles and values for their future adult behavior, of which their "gender regimes" (Kessler et al. 1985) are an important component. . . .

In educational institutions, children develop a stratified social order that is determined by their interactions with peers, parents, and others (Passuth 1987). According to Corsaro (1979), children's knowledge of social position is influenced by their conception of status, which may be defined as popularity, prestige, or "social honor" (Weber 1946). This article focuses primarily on the concept of popularity, which can be defined operationally as the children who are liked by the greatest number of their peers, who are the most influential in setting group opinions, and who have the greatest impact on determining the boundaries of membership in the most exclusive social group. In the school environment, boys and girls have divergent attitudes and behavioral patterns in their gender-role expectations and the methods they use to attain status, or popularity, among peers.

BOYS' POPULARITY FACTORS

Boys' popularity, or rank in the status hierarchy, was influenced by several factors. Although the boys' popularity ordering was not as clearly defined as was the girls', there was a rationale underlying the stratification in their daily interactions and group relations.

Athletic Ability

The major factor that affected the boys' popularity was athletic ability (cf. Coleman 1961, Eder and Parker 1987, Eitzen 1975, Fine 1987, Schofield 1981). Athletic ability was so critical that those who were proficient in sports attained both peer recognition and upward social mobility. In both schools we observed, the best athlete was also the most popular boy in the grade. Two third- and fourth-grade boys considered the question of what makes kids popular:

Nick: Craig is sort of mean, but he's really good at sports, so he's popular.

Ben: Everybody wants to be friends with Gabe, even though he makes fun of most of them all the time. But they still all want to pick him on their team and have him be friends with them because he's a good athlete, even though he brags a lot about it. He's popular.

In the upper grades, the most popular boys all had a keen interest in sports even if they were not adept in athletics. Those with moderate ability and interest in athletic endeavors fell primarily into lower status groups. Those who were least proficient athletically were potential pariahs. . . .

Coolness

Athletics was a major determinant of the boys' social hierarchy, but being good in sports was not the sole variable that affected their popularity. For boys, being "cool" separated a great deal of peer status. As Lyman and Scott (1989, p. 93) noted, "a display of coolness is often a prerequisite to entrance into or maintenance of membership in certain social circles." Cool was a social construction whose definition was in constant flux. Being cool involved individuals' self-presentational skills, their accessibility to expressive equipment, and their impression-management techniques (Fine 1981). Various social forces were involved in the continual negotiation of cool and how the students came to agree on its meaning. As a sixth-grade teacher commented:

> The popular group is what society might term "cool." You know they're skaters, they skateboard, they wear more cool clothes, you know the "in" things you'd see in ads right now in magazines. If you look at our media and advertising right now on TV, like the Levi commercials, they're kinda loose, they skate and they're doing those things. The identity they created for themselves, I think, has a lot to do with the messages the kids are getting from the media and advertising as to what's cool and what's not cool.

There was a shared agreement among the boys as to what type of expressive equipment, such as clothing, was socially defined as cool. Although this type of apparel was worn mostly by the popular boys, boys in the other groups also tried to emulate this style. . . .

Toughness

In the schools we studied, the popular boys, especially in the upper grades, were defiant of adult authority, challenged existing rules, and received more disciplinary actions than did boys in the other groups. They attained a great deal of peer status from this type of acting out. This defiance is related to what Miller (1958) referred to as the "focal concerns" of lower-class culture, specifically "trouble" and "toughness." Trouble involves rule-breaking behavior, and, as Miller (1958, p. 176) noted, "in certain situations, 'getting into trouble' is overtly recognized as prestige conferring." Boys who exhibited an air of nonchalance in the face of teacher authority or disciplinary measures enhanced their status among their peers. Two fourth-grade boys described how members of the popular group in their grade acted:

> **Mark:** They're always getting into trouble by talking back to the teacher.
>
> **Tom:** Yeah, they always have to show off to each other that they aren't afraid to say anything they want to the teacher, that they aren't teachers' pets. Whatever they're doing, they make it look like it's better than what the teacher is doing, 'cause they think what she's doing is stupid.
>
> **Mark:** And one day Josh and Allen got in trouble in music 'cause they told the teacher the Disney movie she wanted to show sucked. They got pink [disciplinary] slips.
>
> **Tom:** Yeah, and that's the third pink slip Josh's got already this year, and it's only Thanksgiving.

Toughness involved displays of physical prowess, athletic skill, and belligerency, especially in repartee with peers and adults. In the status hierarchy, boys who exhibited "macho" behavioral patterns gained recognition from their peers for being tough. Often, boys in the high-status crowd were the "class clowns" or "troublemakers" in the school, thereby becoming the center of attention.

In contrast, boys who demonstrated "effeminate" behavior were referred to by pejorative terms, such as "fag," "sissy," and "homo," and consequently lost status (cf. Thorne and Luria 1986). One boy was constantly derided behind his back because he got flustered easily, had a "spaz" (lost his temper, slammed things down on his desk, stomped around the classroom), and then would start to cry. Two fifth-grade boys described a classmate they considered the prototypical "fag":

Travis: Wren is such a nerd. He's short and his ears stick out.

Nikko: And when he sits in his chair, he crosses one leg over the other and curls the toe around under his calf, so it's double-crossed, like this [*shows*]. It looks so faggy with his "girly" shoes. And he always sits up erect with perfect posture, like this [*shows*].

Travis: And he's always raising his hand to get the teacher to call on him.

Nikko: Yeah, Wren is the kind of kid, when the teacher has to go out for a minute, she says, "I'm leaving Wren in charge while I'm gone."

Savoir-Faire

Savoir-faire refers to children's sophistication in social and interpersonal skills. These behaviors included such interpersonal communication skills as being able to initiate sequences of play and other joint lines of action, affirmation of friendships, role-taking and role-playing abilities, social knowledge and cognition, providing constructive criticism and support to one's peers, and expressing feelings in a positive manner. Boys used their social skills to establish friendships with peers and adults both within and outside the school, thereby enhancing their popularity.

Many of the behaviors composing savior-faire depended on children's maturity, adroitness, and awareness of what was going on in the social world around them. Boys who had a higher degree of social awareness knew how to use their social skills more effectively. This use of social skills manifested itself in a greater degree of sophistication in communicating with peers and adults. . . .

Group leaders with savoir-faire often defined and enforced the boundaries of an exclusive social group. Although nearly everyone liked them and wanted to be in their group, they included only the children they wanted. They communicated to other peers, especially unpopular boys, that they were not really their friends or that play sessions were temporary. This exclusion maintained social boundaries by keeping others on the periphery and at a marginal status.

In contrast, those with extremely poor savoir-faire had difficult social lives and low popularity [cf. Asher and Renshaw 1981]. Their interpersonal skills were awkward or poor, and they rarely engaged in highly valued interaction with their peers. Some of them were either withdrawn or aggressively antisocial. Others exhibited dysfunctional behavior and were referred to as being "bossy" or mean. These boys did not receive a great deal of peer recognition, yet often wanted to be accepted into the more prestigious groups. . . .

Cross-Gender Relations

Although cross-gender friendships were common in the preschool years, play and games became mostly sex segregated in elementary school, and there was a general lack of cross-sex interaction in the classroom (cf. Hallinan 1979). . . .

Sometime during the fourth or fifth grade, both boys and girls began to renegotiate the social definition of intergender interactions because of pubertal changes and the emulation of older children's behavior[1] (cf. Thorne 1986). . . .

By the sixth grade, the boys began to display a stronger interest in girls, and several of the more popular boys initiated cross-gender relations. As one teacher remarked: "The big thing I think is that they are with the girls. They've got some relationships going with the girls in the class, whereas the less popular group does not have that at all."

As Fine (1987) pointed out, sexual interest is a sign of maturity in preadolescent boys, yet it is difficult for inexperienced boys who are not fully cognizant of the norms involved. For safety, boys often went through intermediaries (cf. Eder and Sanford 1986) in approaching girls to find out if their interests were reciprocated. They rarely made such dangerous forays face to face. Rather, they gathered with a friend after school to telephone girls for each other or passed notes or messages from friends to the girls in question. When the friends confirmed that the interest was mutual, the interested boy would then ask the girl to "go" with him. One sixth-grade boy described the Saturday he spent with a friend:

> We were over at Bob's house and we started calling girls we liked on the phone, one at a time. We'd each call the girl the other one liked and ask if she wanted to go with the other one. Then we'd hang up. If she didn't say yes, we'd call her back and ask why. Usually they wouldn't say too much. So sometimes we'd call her best friend to see if she could tell us anything. Then they would call each other and call us back. If we got the feeling after a few calls that she really was serious about no, then we might go on to our next choice, if we had one. . . .

A boy who was successful in getting a girl to go with him developed the reputation of being a "ladies' man" and gained status among his peers.

Academic Performance

The impact of academic performance on boys' popularity was negative for cases of extreme deviation from the norm, but changed over the course of their elementary years for the majority of boys from a positive influence to a potentially degrading stigma.

At all ages, boys who were skewed toward either end of the academic continuum suffered socially. Thus, those who struggled scholastically, who had low self-confidence in accomplishing educational tasks, or who had to be placed in remedial classrooms lost peer recognition. For example, one third-grade boy who went to an afterschool tutoring institute shielded this information from his peers, for fear of ridicule. Boys with serious academic problems were liable to the pejorative label "dummies." At the other end of the continuum, boys who were exceedingly smart but lacked other status-enhancing traits, such as coolness, toughness, or athletic ability, were often stigmatized as "brainy" or "nerdy." The following discussion by two fifth-grade boys highlighted the negative status that could accrue to boys with excessive academic inclinations and performance:

Mark: One of the reasons they're so mean to Seth is because he's got glasses and he's really smart. They think he's a brainy-brain and a nerd.

Seth: You're smart, too, Mark.

Mark: Yeah, but I don't wear glasses, and I play football.

Seth: So you're not a nerd. [What makes Seth a nerd?]

Mark: Glasses, and he's a brainy-brain. He's really not a nerd, but everybody always makes fun of him 'cause he wears glasses.

In the early elementary years, academic performance in between these extremes was positively correlated with social status. Younger boys took pride in their work, loved school, and loved their teachers. Many teachers routinely hugged their students at day's end as they sent them out the door. Yet sometime during the middle elementary years, by around third grade, boys began to change their collective attitudes about academics. This change in attitude coincided with a change in their orientation, away from surrounding adults and toward the peer group.

The boys' shift in attitude involved the introduction of a potential stigma associated with doing too well in school. The macho attitudes embodied in the coolness and toughness orientations led them to lean more toward group identities as renegades or rowdies and affected their exertion in academics, creating a ceiling level of effort beyond which it was potentially dangerous to reach. Boys who persisted in their pursuit of academics while lacking other social skills were subject to ridicule as "cultural dopes" (Garfinkel 1967). Those who had high scholastic aptitude, even with other culturally redeeming traits, became reluctant to work up to their full potential for fear of exhibiting low-value behavior. By diminishing their effort in academics, they avoided the disdain of other boys. One fifth-grade boy explained why he put little more than the minimal work into his assignments:

> I can't do more than this. If I do, then they'll [his friends] make fun of me and call me a nerd. Jack is always late with his homework, and Chuck usually doesn't even do it at all [two popular boys]. I can't be the only one. . . .

GIRLS' POPULARITY FACTORS

The major distinction between the boys' and girls' status hierarchies lay in the factors that conferred popularity. Although some factors were similar, the girls used them in a different manner to organize their social environment. Consequently, the factors had different effects on the girls' and boys' status hierarchies.

Family Background

Similar to the middle-school girls studied by Eder (1985), the elementary school girls' family background was a powerful force that affected their attainment of popularity in multiple ways. Their parents' socioeconomic status (SES) and degree of permissiveness were two of the most influential factors.

SES Maccoby (1980) suggested that among the most powerful and least understood influences on a child are the parents' income, education, and occupation (SES). In general, many popular girls came from upper-class and upper-middle-class families and were able to afford expensive clothing that was socially defined as "stylish" and "fashionable." These "rich" girls had a broader range of material possessions, such as expensive computers or games, a television in their room, and a designer phone with a separate line (some girls even had a custom acronym for the number). They also participated in select extracurricular activities, such as horseback riding and skiing and vacationed with their families at elite locations. Some girls' families owned second houses in resort areas to which they could invite their friends for the weekend. Their SES gave these girls greater access to highly regarded symbols of prestige. Although less privileged girls often referred to them as "spoiled," they secretly envied these girls' life-styles and possessions. As two fourth-grade girls in the unpopular group stated:

Alissa: If your Mom has a good job, you're popular, but if your Mom has a bad job, then you're unpopular.

Betty: And, if, like, you're on welfare, then you're unpopular because it shows that you don't have a lot of money.

Alissa: They think money means that you're great—you can go to Sophia's [a neighborhood "little store" where popular people hang out] and get whatever you want and stuff like that. You can buy things for people.

Betty: I have a TV, but if you don't have cable [TV] then you're unpopular because everybody that's popular has cable. . . .

Laissez-faire Laissez-faire refers to the degree to which parents closely supervised their children or were permissive, allowing them to engage in a wide range of activities. Girls whose parents let them stay up late on sleepover dates, go out with their friends to all kinds of social activities, and gave them a lot of freedom while playing in the house were more likely to be popular. Girls who had to stay home (especially on weekend nights) and "get their sleep," were not allowed to go to boy-girl parties, had strict curfews, or whose parents called ahead to parties to ensure that they would be adult "supervised" were more likely to be left out of the wildest capers and the most exclusive social crowd. . . .

Physical Appearance

Another powerful determinant of girls' location in the stratification system was their physical attractiveness. Others (Coleman 1961, Eder and Parker 1987, Eder and

Sanford 1986, Schofield 1981) have noted that appearance and grooming behavior are not only a major topic of girls' conversation, but a source of popularity. . . .

In the upper grades, makeup was used as a status symbol, but as Eder and Sanford (1986) observed, too much makeup was highly criticized by other members of the group, thereby inhibiting social mobility. Finally, girls who were deemed pretty by society's socially constructed standards were attractive to boys and had a much greater probability of being popular. . . .

The perception that popularity was determined by physical traits was fully evidenced by these kindergartners. These aspects of appearance, such as clothing, hairstyles, and attractiveness to boys, were even more salient, with the girls in the upper grades. As an excerpt from one of our field notes indicated:

> I walked into the fifth-grade coat closet and saw Diane applying hairspray and mousse to Paula's and Mary's hair. Someone passed by and said. "Oh, Mary I like your hair," and she responded, "I didn't do it; Diane did it." It seemed that Diane, who was the most popular girl in the class, was socializing them to use the proper beauty supplies that were socially accepted by the popular clique. I asked what made girls unpopular, and Diane said, "They're not rich and not pretty enough. Some people don't use the same kind of mousse or wear the same style of clothing."

As girls learn these norms of appearance and associate them with social status, they form the values that will guide their future attitudes and behavior, especially in cross-gender relationships (cf. Eder and Sanford 1986). This finding correlates with other research (Hatfield and Sprecher 1986) that suggested that physical appearance is closely related to attaining a mate, that people who perceive themselves as being unattractive have difficulty establishing relationships with others, and that there is a correlation between opportunities for occupational success and physical attractiveness.

Social Development

Social factors were also salient to girls' popularity. Like the boys, the most precocious girls achieved dominant positions, but they were also more sensitive to issues of inclusion and exclusion. Precocity and exclusivity were thus crucial influences on girls' formation of friendships and their location on the popularity hierarchy.

Precocity Precocity refers to girls' early attainment of adult social characteristics, such as the ability to express themselves verbally, to understand the dynamics of intra- and intergroup relationships, to convince others of their point, and to manipulate others into doing what they wanted, as well as interest in more mature social concerns (such as makeup and boys). As with the boys, these social skills were only partly developmental; some girls just seemed more precocious from their arrival in kindergarten. One teacher discussed differences in girls' social development and its effects on their interactions:

> Communication skills, I can see a definite difference. There is not that kind of sophistication in the social skills of the girls in the unpopular group. The popular kids are taking on junior high school characteristics pretty fast just in terms of the kinds of rivalries they have. They are very active after school, gymnastics especially. Their conflicts aren't over play as much as jealousy. Like who asked who over to their house and who is friends. There is some kind of a deep-running, oh, nastiness, as opposed to what I said before. The popular group—they seem to be maturing, I wouldn't call them mature, but their behavior is sophisticated. The unpopular girls seem to be pretty simple in their ways of communicating and their interests.

The most precocious girls showed an interest in boys from the earliest elementary years.[2] They talked about boys and tried to get boys to pay attention to them. This group of girls was usually the popular crowd, with the clothes and appearance that boys (if they were interested in girls) would like. These girls told secrets and giggled about boys and passed boys notes in class and in the halls that embarrassed but excited the boys. They also called boys on the phone, giggling at them, asking them mundane or silly questions, pretending they were the teacher, singing radio jingles to them, or blurting out "sexy" remarks. . . .

By around the fourth to sixth grade, it became more socially acceptable for girls to engage in cross-gender interactions without being rebuked by their peers. The more precocious girls began to experiment further by flirting with boys; calling them on the telephone; "going" with them; going to parties; and, ultimately, dating. Although some girls were adventurous enough to ask a boy out, most followed traditional patterns and waited for boys to commit themselves first. One fourth-grade girl described what it meant to "go" with a boy: "You talk. You hold hands at school. You pass notes in class. You go out with them, and go to movies, and go swimming. . . . We usually double date."

In the upper grades, if a girl went with a popular boy, she was able to achieve a share of his prestige and social status. Several girls dreamed of this possibility and even spoke with longing or anticipation to their friends about it.[3] When popular girls went with popular boys, it reinforced and strengthened the status of both. This was the most common practice, as a fifth-grade girl noted: "It seems that most of the popular girls go out with the popular boys: I don't know why." One fourth-grade girl referred to such a union as a "Wowee" (a highly prestigious couple), because people would be saying "Wow!" at the magnitude of their stardom. Yet, to go out with a lower-status boy would diminish a girl's prestige. Several fourth-grade girls responded to the question, "What if a popular girl went with an unpopular boy?" this way:

> **Alissa:** *Down!* The girl would move down, way down.
>
> **Betty:** They would not do it. No girl would go out with an unpopular boy.
>
> **Lisa:** If it did happen, the girl would move down, and no one would play with her either.

A high-status girl would thus be performing a form of social suicide if she interacted with a low-status boy in any type of relationship. Although the girls acknowledged that they were sensitive to this issue, they were doubtful whether a popular boy's rank in the social hierarchy would be affected by going with a girl from a lower stratum. They thought that boys would not place as much weight on such issues.

Exclusivity Exclusivity refers to individuals' desire, need, and ability to form elite social groups using such negative tactics as gossiping, the proliferation of rumors, bossiness, and meanness. One or two elite groups of girls at each grade level jointly participated in exclusionary playground games and extracurricular activities, which created clearly defined social boundaries because these girls granted limited access to their friendship circles. In one fourth-grade class, a clique of girls had such a strong group identity that they gave themselves a name and a secret language. As they stated:

Anne: We do fun things together, the "Swisters" here, 'um we go roller skating a lot, we walk home together and have birthday parties together.

Carrie: We've got a secret alphabet.

Anne: Like an A is a different letter.

Debbie: We have a symbol and stuff.

Anne: But we don't sit there and go like mad and walk around and go, "We're the Swisters and you're not and you can't be in and anything."

Carrie: We don't try to act cool; we just stick together, and we don't sit there and brag about it.

This group of girls restricted entrée to their play and friendship activities, although they did not want to be perceived as pretentious and condescending. Many girls in the less popular groups did not like the girls in the highest-status crowd, even though they acknowledged that these girls were popular (cf. Eder 1985). . . .

Academic Performance

In contrast to the boys, the girls never seemed to develop the machismo culture that forced them to disdain and disengage from academics. Although not all popular girls were smart or academic achievers, they did not suffer any stigma from performing well scholastically. Throughout elementary school, most girls continued to try to attain the favor of their teachers and to do well on their assignments. They gained status from their classmates for getting good grades and performing difficult assignments. The extent to which a school's policies favored clumping students of like abilities in homogeneous learning groups or classes affected the influence of academic stratification on girls' cliques. Homogeneous academic groupings were less common during the early elementary years, but increased in frequency as students approached sixth grade and their performance curve spread out wider. By fifth or sixth grade, then, girls were more likely to become friends with others of similar scholastic levels. Depending on the size of the school, within each grade there might be both a clique of academically inclined popular girls and a clique composed of popular girls who did not perform as well and who bestowed lower salience on schoolwork.

DISCUSSION

One of the major contributions of this work lies in its illustration of the role of popularity in gender socialization. Gaining and maintaining popularity has enormous significance on children's lives (cf. Eder 1985), influencing their ability to make friends, to be included in fun activities, and to develop a positive sense of self-esteem. In discerning, adapting to, and creatively forging these features of popularity, children actively socialize themselves to the gender roles embodied in their peer culture.

NOTES

1. Children with older siblings were often more precocious than were others, overcoming their reluctance to approach girls and initiating rites of flirtation and dating.

2. In the second grade, a group of popular girls, centered on an extremely precocious ringleader, regularly called boys. They asked the boys silly questions, giggled, and left long messages on their telephone answering machines. At one school outing, the dominant girl bribed a boy she liked with money and candy to kiss her, but when he balked at the task (after having eaten the candy and spent the money), she had to pretend to her friends that he had done so, to avoid losing face.

3. One girl even lied to her friends about it, pretending to them that she was going with a popular boy. When they found out that she had fabricated the story, they dropped her, and she lost both her status and her friends.

REFERENCES

Asher, Steven R. and Peter D. Renshaw. 1981. "Children without Friends: Social Knowledge and Social Skill Training." Pp. 273–96 in *The Development of Children's Friendships,* edited by S. R. Asher and J. M. Gottman. New York: Cambridge University Press.

Best, Raphaela. 1983. *We've All Got Scars.* Bloomington: Indiana University Press.

Coleman, James. 1961. *The Adolescent Society.* Glencoe: Free Press.

Corsaro, William. 1979. "Young Children's Conceptions of Status and Role." *Sociology of Education* 52: 46–59

Eder, Donna. 1985. "The Cycle of Popularity: Interpersonal Relations Among Female Adolescents." *Sociology of Education* 58: 154–65.

Eder, Donna and Stephen Parker. 1987. "The Cultural Production and Reproduction of Gender: The Effect of Extracurricular Activities on Peer-Group Culture." *Sociology of Education* 60: 200–213.

Eder, Donna and Stephanie Sanford. 1986. "The Development and Maintenance of Interactional Norms Among Early Adolescents." Pp. 283–300 in *Sociological Studies of Child Development,* Vol. I, edited by P. A. Adler and P. Adler, Greenwich, CT: JAI Press.

Eitzen, D. Stanley. 1975. "Athletics in the Status System of Male Adolescents: A

Replication of Coleman's *The Adolescent Society.*" *Adolescence* 10: 267–76.

Fine, Gary Alan. 1987. *With the Boys.* Chicago: University of Chicago Press.

———. 1981. "Friends, Impression Management, and Preadolescent Behavior." Pp. 29–52 in *The Development of Children's Friendships,* edited by S. Asher and J. Gottman. New York: Cambridge University Press.

Garfinkel, Harold. 1967. *Studies in Ethnomethodology.* Englewood Cliffs, NJ: Prentice-Hall.

Glassner, Barry. 1976. "Kid Society." *Urban Education* 11: 5–22.

Hallinan, Maureen. 1979. "Structural Effects on Children's Friendships and Cliques." *Social Psychology Quarterly* 42: 43–54.

Hatfield, Elaine and Susan Sprecher. 1986. *Mirror, Mirror . . . The Importance of Looks in Everyday Life.* Albany: State University of New York Press.

Kessler, S., D. J. Ashenden, R. W. Connell, and G. W. Dowsett. 1985. "Gender Relations in Secondary Schooling." *Sociology of Education* 58: 34–48.

Lyman, Stanford and Marvin Scott. 1989. *A Sociology of the Absurd,* 2nd ed. Dix Hills, NY: General Hall.

Maccoby, Eleanor. 1980. *Social Development.* New York: Harcourt, Brace and Jovanovich.

Miller, Walter. 1958. "Lower Class Culture and Gang Delinquency." *Journal of Social Issues* 14:5–19.

Passuth, Patricia. 1987. "Age Hierarchies with Children's Groups." Pp. 185–203 in *Sociological Studies of Child Development* (V.2), edited by P. A. Adler and P. Adler. Greenwich, CT: JAI.

Schofield, Janet W. 1981. *Black and White in School.* New York: Praeger.

Thorne, Barrie. 1986. "Girls and Boys Together, But Mostly Apart: Gender Arrangements in Elementary Schools." Pp. 167–84 in *Relationships and Development,* edited W. Hartup and Z. Rubin. Hillsdale, NJ: Lawrence Erlbaum.

Thorne, Barrie and Zella Luria. 1986. "Sexuality and Gender in Children's Daily Worlds." *Social Problems* 33: 176–90.

Weber, Max. 1946. "Class, Status, and Party." Pp. 180–95 in *From Max Weber,* edited by H. Gerth and C. W. Mills. New York: Oxford University Press.

Wexler, Philip. 1988. "Symbolic Economy of Identity and Denial of Labor: Studies in High School Number 1." Pp. 302–15 in *Class, Race and Gender in American Education,* edited by L. Weis. Albany: SUNY Press.

28

Men and Women in Conversation

DEBORAH TANNEN

Popular explanations of gender differences have recently informed Americans that "women are from Venus and men are from Mars," an insight into gendered conversations that helps explain the communication gap between them. Tannen, the foremost scholar of gendered communication, offers a powerful analysis of the different ways men and women talk to each other and what they want to hear. These patterns are heavily influenced by their differential gender socialization, leading men toward instrumental and women toward affective patterns. Participants and observers of everyday talk will find great insight in this fascinating and sensitive selection that explains both why men and women talk at cross-purposes against each other and what they can ultimately do about it. Listen to conversation in the cafeteria or dorm. Do you see the patterns that Tannen describes? Why do you suppose, even in this era of great gender equality, that these subtle, minute differences still exist? What do you think this tells us about the future of gender relations?

Conversation is a ritual. We say things that seem obviously the thing to say, without thinking of the literal meaning of our words, any more than we expect the question "How are you?" to call forth a detailed account of aches and pains.

Unfortunately, women and men often have different ideas about what's appropriate, different ways of speaking. Many of the conversational rituals common among women are designed to take the other person's feelings into account, while many of the conversational rituals common among men are designed to maintain the one-up position, or at least avoid appearing one-down. As a result, when men and women interact—especially at work—it's often women who are at the disadvantage. Because women are not trying to avoid the one-down position, that is unfortunately where they may end up.

Here, the biggest areas of miscommunication.

1. APOLOGIES

Women are often told they apologize too much. The reason they're told to stop doing it is that, to many men, apologizing seems synonymous with putting oneself down. But there are many times when "I'm sorry" isn't self-deprecating, or even an apology; it's an automatic way of keeping both speakers on an equal footing. For example, a well-known columnist once interviewed me and gave me her phone number in case I needed to call her back. I misplaced the number and had to go through the newspaper's main switchboard. When our conversation was winding down and we'd both made ending-type remarks, I added "Oh, I almost forgot—I lost your direct number, can I get it again?" "Oh, I'm sorry," she came back instantly, even though she had done nothing wrong and *I* was the one who'd lost the number. But I understood she wasn't really apologizing; she was just automatically reassuring me she had no intention of denying me her number.

Even when "I'm sorry" *is* an apology, women often assume it will be the first step in a two-step ritual: I say "I'm sorry" and take half the blame, then you take the other half. At work, it might go something like this:

A: When you typed this letter, you missed this phrase I inserted.

B: Oh, I'm sorry. I'll fix it.

A: Well, I wrote it so small it was easy to miss.

When both parties share blame, it's a mutual face-saving device. But if one person, usually the woman, utters frequent apologies and the other doesn't, she ends up looking as if she's taking the blame for mishaps that aren't her fault. When she's only partially to blame, she looks entirely in the wrong.

I recently sat in on a meeting at an insurance company where that sole woman, Helen, said "I'm sorry" or "I apologize" repeatedly. At one point she said, "I'm thinking out loud. I apologize." Yet the meeting was intended to be an informal brain-storming session, and *everyone* was thinking out loud.

The reason Helen's apologies stood out was that she was the only person in the room making so many. And the reason I was concerned was that Helen felt the annual bonus she had received was unfair. When I interviewed her colleagues, they said that Helen was one of the best and most productive workers—yet she got one of the smallest bonuses. Although the problem might have been outright sexism, I suspect her speech style, which differs from that of her male colleagues, masks her competence.

Unfortunately, not apologizing can have its price too. Since so many women use ritual apologies, those who don't may be seen as hard-edged. What's important is to be aware of how often you say you're sorry (and why), and to monitor your speech based on the reaction you get.

2. CRITICISM

A woman who cowrote a report with a male colleague was hurt when she read a rough draft to him and he leapt into a critical response—"Oh, that's too dry! You have to make it snappier!" She herself would have been more likely to say, "That's a really good start. Of course, you'll want to make it a little snappier when you revise."

Whether criticism is given straight or softened is a matter of convention. In general, women use more softeners. I noticed this difference when talking to an editor about an essay I'd written. While going over changes she wanted to make, she said, "There's one more thing. I know you may not agree with me. The reason I noticed the problem is that your other points are so lucid and elegant." She went on hedging for several more sentences until I put her out of her misery: "Do you want to cut that part?" I asked—and of course she did. But I appreciated her tentativeness. In contrast, another editor (a man) I once called summarily rejected my idea for an article by barking. "Call me when you have something new to say."

Those who are used to ways of talking that soften the impact of criticism may find it hard to deal with the right-between-the-eyes style. It has its own logic, however, and neither style is intrinsically better. People who prefer criticism given straight are operating on an assumption that feelings aren't involved. "Here's the dope. I know you're good; you can take it."

3. THANK-YOUS

A woman manager I know starts meetings by thanking everyone for coming, even though it's clearly their job to do so. Her "thank-you" is simply a ritual.

A novelist received a fax from an assistant in her publisher's office; it contained suggested catalogue copy for her book. She immediately faxed him her suggested changes and said, "Thanks for running this by me," even though her contract gave her the right to approve all copy. When she thanked the assistant, she fully expected him to reciprocate: "Thanks for giving me such a quick response." Instead, he said, "You're welcome." Suddenly, rather than an equal exchange of pleasantries, she found herself positioned as the recipient of a favor. This made her feel like responding. "Thanks for nothing!"

Many women use "thanks" as an automatic conversation starter and closer; there's nothing literally to thank you for. Like many rituals typical of women's conversation, it depends on the goodwill of the other to restore the balance. When the other speaker doesn't reciprocate, a woman may feel like someone on a seesaw whose partner abandoned his end. Instead of balancing in the air, she has plopped to the ground, wondering how she got there.

4. FIGHTING

Many men expect the discussion of ideas to be a ritual fight—explored through verbal opposition. They state their ideas in the strongest possible terms, thinking that if there are weaknesses someone will point them out, and by trying to argue against those objections, they will see how well their ideas hold up.

Those who expect their own ideas to be challenged will respond to another's ideas by trying to poke holes and find weak links—as a way of *helping*. The logic is that when you are challenged you will rise to the occasion: Adrenaline makes your mind sharper, you get ideas and insights you would not have thought of without the spur of battle.

But many women take this approach as a personal attack. Worse, they find it impossible to do their best work in such a contentious environment. If you're not used to ritual fighting, you begin to hear criticism of your ideas as soon as they are formed. Rather than making you think more clearly, it makes you doubt what you know. When you state your ideas, you hedge in order to fend off potential attacks. Ironically, this is more likely to *invite* attack because it makes you look weak.

Although you may never enjoy verbal sparring, some women find it helpful to learn how to do it. An engineer who was the only woman among four men in a small company found that as soon as she learned to argue, she was accepted and taken seriously. A doctor attending a hospital staff meeting made a similar discovery. She was becoming more and more angry with a male colleague who'd loudly disagreed with a point she'd made. Her better judgment told her to hold her tongue, to avoid making an enemy of this powerful senior colleague. But finally she couldn't hold it any longer, and she rose to her feet and delivered an impassioned attack on his position. She sat down in a panic, certain she had permanently damaged her relationship with him. To her amazement, he came up to her afterward and said, "That was a great rebuttal. I'm really impressed. Let's go out for a beer after work and hash out our approaches to this problem."

5. PRAISE

A manager I'll call Lester had been on his new job six months when he heard that the women reporting to him were deeply dissatisfied. When he talked to them about it, their feelings erupted; two said they were on the verge of quitting because he didn't appreciate their work, and they didn't want to wait to be fired. Lester was dumbfounded: He believed they were doing a find job. Surely, he thought, he had said nothing to give them the impression he didn't like their work. And indeed he hadn't. That was the problem. He had said *nothing*—and the women assumed he was following the adage "If you can't say something nice, don't say anything." He thought he was showing confidence in them by leaving them alone.

Men and women have different habits in regard to giving praise. For example, Deidre and her colleague William both gave presentations at a conference. Afterward, Deidre told William, "That was a great talk." He thanked her. Then she asked, "What did you think of mine?" and he gave her a lengthy and detailed critique. She found it uncomfortable to listen to his comments. But she assured herself that he meant well, and that his honesty was a signal that she, too, should be honest when he asked for a critique of his performance. As a matter of fact, she had noticed quite a few ways in which he could have improved his presentation. But she never got a chance to tell him because he never asked—and she felt put down. The worst part was that it seemed she had only herself to blame, since she *had* asked what he thought of her talk.

But had she really asked for his critique? The truth is, when she asked for his opinion, she was expecting a compliment, which she felt was more or less required following anyone's talk. When he responded with criticism, she figured, Oh, he's playing 'Let's critique each other'—not a game she'd initiated, but one which she was willing to play. Had she realized he was going to criticize her and not ask her to reciprocate, she would never have asked in the first place.

It would be easy to assume that Deidre was insecure, whether she was fishing for a compliment or soliciting a critique. But she was simply talking automatically, performing one of the many conversational rituals that allow us to get through the day. William may have sincerely misunderstood Deidre's intention—or may have been unable to pass up a chance to one-up her when given the opportunity.

6. COMPLAINTS

"Troubles talk" can be a way to establish rapport with a colleague. You complain about a problem (which shows that you are just folks) and the other person responds with a similar problem (which puts you on equal footing). But while such commiserating is common among women, men are likely to hear it as a request to *solve* the problem.

One woman told me she would frequently initiate what she thought would be pleasant complaint-airing sessions at work. She'd just talk about situations that bothered her just to talk about them, maybe to understand them better. But her male office mate would quickly tell her how she could improve the situation. This left her feeling condescended to and frustrated. She was delighted to see this very impasse in a section in my book *You Just Don't Understand,* and showed it to him. "Oh," he said, "I see the problem. How can we solve it?" Then they both laughed, because it had happened again: He short-circuited the detailed discussion she'd hoped for and cut to the chase of finding a solution.

Sometimes the consequences of complaining are more serious: A man might take a woman's lighthearted griping literally, and she can get a reputation as a chronic malcontent. Further, she may be seen as not up to solving the problems that arise on the job.

7. JOKES

I heard a man call in to a talk show and say, "I've worked for two women and neither one had a sense of humor. You know, when you work with men, there's a lot of joking and teasing." The show's host and the guest (both women) took his comment at face value and assumed the women this man worked for were humorless. The guest said, "Isn't it sad that women don't feel comfortable enough with authority to see the humor?" The host said, "Maybe when more women are in authority roles, they'll be more comfortable with power." But although the women this man worked for *may* have taken themselves too seriously, it's just as likely that they each had a terrific sense of humor, but maybe the humor wasn't the type he was used to. They may have been like the woman who wrote to me: "When I'm with men, my wit or cleverness seems inappropriate (or lost!) so I don't bother. When I'm with my women friends, however, there's no hold on puns or cracks and my humor is fully appreciated."

The types of humor women and men tend to prefer differ. Research has shown that the most common form of humor among men is razzing, teasing, and mock-hostile attacks, while among women it's self-mocking. Women often mistake men's teasing as genuinely hostile. Men often mistake women's mock self-deprecation as truly putting themselves down.

Women have told me they were taken more seriously when they learned to joke the way the guys did. For example, a teacher who went to a national conference with seven other teachers (mostly women) and a group of administrators (mostly men) was annoyed that the administrators always found reasons to leave boring seminars, while the teachers felt they had to stay and take notes. One evening when the group met at a bar in the hotel, the principal asked her how one such seminar had turned out. She retorted, "As soon as you left, it got much better." He laughed out loud at her response. The playful insult appealed to the men—but there was a trade-off. The women seemed to back off from her after this. (Perhaps they were put off by her using joking to align herself with the bosses.)

There is no "right" way to talk. When problems arise, the culprit may be style differences—and *all* styles will at times fail with others who don't share or understand them, just as English won't do you much good if you try to speak to someone who knows only French. If you want to get your message across, it's not a question of being "right"; it's a question of using language that's shared—or at least understood.

29

Sex Differences in Moving Up and Taking Charge

BARBARA RESKIN
AND IRENE PADAVIC

Significant numbers of American women have now entered the labor force, although statistics show that too many are still underemployed and underpaid. Research has shown that women are ghettoized into certain kinds of occupations and given stereotypical roles that they are expected to fill within these. While men gain valuable connections and knowledge from existing "good old boys' networks," these rarely exist for women. And while men usually advance their careers unencumbered, women are often responsible for taking care of children when they are sick, sometimes landing on the "mommy track" to do so, thus scuttling their career advancement. Reskin and Padavic offer a sweeping overview of promotion trends in this field, explaining the glass ceilings that block women's advancement, the glass elevators that facilitate men's, their different kinds of job experiences, and the way these lead to segregation, promotion, and wage differentials. While none of these revelations will come as a great shock, it is important to understand the complex variety of mechanisms that contribute to maintaining the gender gap in the occupational arena. What trends can you see in your own world about changes (or lack thereof) in occupational opportunities of men and women? When you leave college, will gender play an important role in the type of career you select or the jobs that you will get? What are the implications for society if Reskin and Padavic's research is correct? Do you think that the coming decade will witness a further change in these trends?

Until 20 years ago, few employers even considered women for promotions that would take them outside the female clerical or assembly-line ghetto. Although more women have been promoted in recent years, women still have a long way to go. Now they may get some help from the legal system, however. For instance, the Supreme Court ruled in 1990 that Ann Hopkins should be promoted to partner status at the Big Six accounting firm of Price Waterhouse. The decision put other companies on notice that discriminating in promotions could be costly. Then in 1991 a jury awarded $6.3 million to a woman whom Texaco had twice passed over for promotion. On the legislative front, the Civil Rights Act of 1991 now allows employees to sue not merely for lost wages and litigation expenses but also for punitive damages. Similarly, the Glass Ceiling Act of 1991—designed to encourage employers to remove barriers to the progress of women and minorities—reflects increasing public concern with barriers to job mobility. But how effective this legislation will be remains to be seen.

As for access to authority, women have also made some strides. In 1940 many firms explicitly prohibited women from occupying managerial positions (Goldin 1990). By the 1970s, with women flooding the labor market and the number of managerial jobs expanding dramatically, unprecedented numbers of women were entering the ranks of management. Since then, American women have increased their representation in management ranks from 18 percent of all managers in 1970 to 30 percent in 1980 and 40 percent in 1990. These figures indisputably show that thousands of women are gaining access to jobs that usually confer organizational power. Whether women in these positions actually are able to act on the authority typical of managerial positions is a question we address in this chapter.

THE PROMOTION GAP

In the real world of work, just a handful of women reach the top of the corporate hierarchy. In 1990, only 19—or fewer than 0.5 percent—of the 4,012 highest-paid officers and directors in top companies were women (Fierman 1990). Within Fortune 500 companies, women and minorities held fewer than 5 percent of senior management posts (Fierman 1990), indicating snail-like progress from the early 1970s, when 1 to 3 percent of senior managers were women (Segal 1992).

A *glass ceiling* blocks the on-the-job mobility of women of all classes, as well as minorities of both sexes. Indeed, in some organizations the glass ceiling may be quite low; for many women of all races, the problem is the *sticky floor,* which keeps them trapped in low-wage, low-mobility jobs (Berheide 1992). Data from a study conducted in the early 1980s in Illinois found that the average man had 0.83 promotions and the average woman, 0.47 promotions (Spaeth 1989). National data for 1991 show a smaller but still significant promotion gap: 48 percent of men had been promoted by their current employer but only 34 percent of women (Reskin and Kalleberg 1993). Both these studies underestimate the promotion gap by failing to distinguish between small-step promotions (for example, clerk-typist 1 to clerk-typist 2) and larger ones (for example, sales representative to sales manager).

Several studies have indicated that men do not confront blocked opportunities because of their sex; in fact, as one sociologist noted, men tend to "rise to the top like bubbles in wine bottles" (Grimm 1978). Christine Williams (1992) found that employers singled out male workers in traditionally female jobs—nurse, librarian, elementary school teacher, social worker, and the like—for an express ride to the top on a "glass escalator." Some of the men in Williams's study faced pressure to accept promotions, like the male children's librarian who received negative evaluations for not "shooting high enough."

Although American women lag behind men in promotions, compared to women in most other countries, American women are doing relatively well. American women were four times more likely than French women to hold administrative and managerial jobs and six and half times more likely to do so than British women (Crompton and Sanderson 1990:176). In no country, however, were women represented in top-level jobs on a par with their numbers in administrative and managerial positions (Farley 1993). In Denmark, women were 14.5 percent of administrators and managers in 1987 but only between 1 and 5 percent of top management; in Japan, women were 7.5 percent of administrators and managers but only 0.3 of a percent of top management in the private sector (Antal and Izraeli 1993:58; Steinhoff and Kazuko 1988). In all the countries for which information was available, women were vastly underrepresented in top-level jobs (Antal and Izraeli 1993).

Consequences of the Promotion Gap

Does it matter that women are locked out of the higher-level jobs? Yes. First of all, the practice is unfair. Americans of both sexes value promotions as a path to greater pay, authority, autonomy, and job satisfaction (Markham et al. 1987:227). And both sexes are ready to work hard for a promotion. In a recent survey of federal employees, 78 percent of the women and 74 percent of the men agreed that they were willing to devote whatever time was necessary in order to advance in their career (U.S. Merit Systems Protection Board 1992). Among minority women, 86 percent were willing to devote as much time as it takes (minority men's responses were not reported separately). Upward mobility is the heart of the American dream, and its denial to women reflects poorly on our society. A second reason to be concerned about the promotion gap is that it depresses women's wages. At a time when women's wages average only 70 percent of men's, this is a serious consideration. Third, promotion barriers reduce women's opportunities to exercise authority on the job (as we discuss later) and to have autonomy from close supervision. *Autonomy*—the freedom to design aspects of one's work, to decide the pace and hours of work, and to not have others exercising authority over oneself (M. Adler 1993)— enhances job satisfaction. A fourth consequence of women's blocked mobility is that it often leads women to quit in frustration.

Some women try to get around blocked mobility by starting their own business. In 1990 women owned 30 percent of all small businesses, and the Small Business Administration expects the number to rise to 40 percent by the year 2000 (Shellenbarger 1993). Yet women are less successful than men in these ventures; in 1982 the average business run by a woman grossed only 35 percent of what the average man-run business grossed (U.S. Small Business Administration 1988). A partial explanation is that women-run businesses are usually economically marginal. Many women business owners have the legal status of entrepreneur but are *independent contractors:* workers hired on a freelance basis to do work that regular employees otherwise would do in-house (Christensen 1989). Far fewer women own a business that employs others. The median hourly wage for a full-time, self-employed woman was $3.75 in 1987, compared to $8.08 for a full-time female employee (Collins 1993). Often self-employed women provide services that help other employed women cope with domestic work, such as catering, housecleaning, caring for children, and being a "mother's helper."

Explanations for the Promotion Gap

Many factors impede women's mobility. Although women are as committed to their careers as men are, women have less of the education, experience, and training that employers desire. Women also tend to be located in jobs that do not offer the same diversity of experience or the same opportunities for upward mobility as men's jobs. Finally, in making promotion decisions, employers discriminate against women. These are the three basic explanations of why companies still promote more white men than women and minorities.

Human-Capital Inequities and Promotion Human-capital theorists claim that sex differences in promotion rates are due to sex differences in commitment, education, and experience. These differences are presumed to make women less productive than men.

The claim that men are more committed than women to their jobs is based on the idea that women place family responsibilities ahead of career commitment. According to this reasoning, family demands do not allow women to devote as much time to their careers as men, and therefore women are unable to do all the things necessary to get promoted rapidly. Although employers act on this stereotype of women's lesser commitment, it is not founded in reality. . . . Women's career commitment does not differ from men's.

The second human-capital claim is that educational differences account for the promotion gap. Indeed, in 1992, women earned 47 percent of the bachelor's degrees in business administration but only 34 percent of the master's degrees. Thus educational differences do contribute to the sex disparity in promotion rates. However, they do not account for all of the disparity. Women with the same educational credentials as men are not attaining top-management jobs at the same rate. When told that women will slowly make their way to the top as more get advanced degrees, one woman manager countered, "My generation came out of graduate school 15 or 20 years ago. The men are now in line to run major corporations. The women are not. Period" (Fierman 1990).

As for the third human-capital claim, women do receive less training and have less experience both within a firm and within the labor force. It is implausible, however, that women voluntarily acquire less experience. In many settings, employers prevent women from acquiring the essential experiences needed for advancement. For example, military promotion to the rank of commissioned officer is usually reserved for people with combat experience, but Congress and the military have banned women from combat positions. In banking, managers who hope for a top spot need extensive experience in commercial lending. But until recently, most women bank managers were not given the chance to work in commercial lending, so few women could acquire the expertise needed to rise beyond middle management. In the same way, the sex segregation of blue-collar production jobs denies women the experience they need to rise to management positions in manufacturing firms.

Segregation and Promotion The segregation explanation of the promotion gap focuses on differences between men's and women's organizational locations. A key concept in this explanation is the *internal labor market* or a firm's system for filling jobs by promoting experienced employees. Internal labor markets are composed of related jobs (or job families) connected by *job ladders,* which are promotion or transfer paths that connect lower- and higher-level jobs. These ladders may have only two rungs (as in a take-out restaurant that promotes counter workers to delivery persons), or they may span an entire organization. Job ladders also differ in shape. Some are shaped like a ladder—for example, a company's sales division whose job ladder includes one stock clerk, one sales trainee, one sales representative, one assistant sales manager, and one sales manager. When the vice president in this division retires, everyone moves up one step. In contrast, other job ladders are shaped like a pyramid, with many entry-level jobs feeding into smaller and smaller numbers of jobs at progressively higher levels of the organization, so many workers compete for relatively few jobs. The broader the base of the pyramid, the smaller a worker's odds of being promoted.

The basic idea behind the segregation explanation is that women are promoted less often than men partly because access to internal labor markets is gendered. Women workers are more likely than men to be in jobs with short job ladders or to be in dead-end jobs. This sort of sex segregation begins with entry-level jobs. Men, but not women, are more often placed in jobs with long job ladders and chances at top jobs.

Segregation also contributes to the promotion gap by making women's accomplishments invisible. In prestigious Wall Street law firms, for example, senior partners have tended to assign women lawyers to research rather than litigation. Appearing in the courtroom less often than their male peers made women less visible, thus hampering their promotion to partner (Epstein 1993). In corporations, women managers are concentrated in staff positions, such as personnel or public relations, and men are concentrated in the more visible and important line positions, such as sales or production. Staff positions involve little risk and therefore provide few opportunities for workers to show senior management their full capabilities. When top executives are looking for people to promote to senior management, they seldom pick vice presidents of personnel management or public relations. They usually pick vice presidents in product management or sales, who more often than not are male.

Discrimination and Promotion Sex discrimination by employers is a third explanation for the promotion gap. In brief, it recognizes that employers reward men's qualifications more than women's. The case of Ann Hopkins illustrates blatant discrimination. Despite having brought more money into Price Waterhouse than any other contender for promotion had done, Hopkins was denied promotion to partner. According to the court that ruled in her favor, the senior partners based their evaluation on her personality and appearance and ignored her accomplishments. Her male mentor had advised Hopkins that her chances of promotion would improve if she would "walk more femininely, talk more femininely, wear makeup, have [her] hair styled, and wear jewelry" (White 1992:192).

Actually, appearing feminine does not help a woman get a promotion. In fact, the mere fact of being a woman may be an insurmountable barrier. A male personnel officer told a female bank manager that her career had stalled because

> what the chairman and presidents want are people that they are comfortable with, and they are not . . . comfortable with women. It doesn't even get to the . . . level of you as an individual and [your] personality; *it is the skirt that's the problem.* (Reskin and Roos 1990:156)

Top-level management is a male environment, and some men feel uncomfortable with women. The result is discrimination. The Federal Bureau of Investigation (FBI), for example, refused to even hire women as agents until the death of Director J. Edgar Hoover in 1972 (Johnston 1993). The lingering effect of this discrimination is that the highest-ranking woman is the agent in charge of the FBI's smallest field office, in Anchorage, Alaska.

Statistical discrimination comes into play as well. Employers statistically discriminate against women when they stereotype them as disinterested in advancement or as lacking the attributes needed in higher-level jobs. According to Rosabeth Moss Kanter (1977), when managers are uncertain about the qualifications necessary to do a job, they prefer to promote people who have social characteristics like their own. Kanter called this practice *homosocial reproduction.* Presumably managers believe that similar people are likely to make the same decisions they would. Thus they seek to advance others who are the same sex, race, social class, and religion; who belong to the same clubs; who attended the same colleges; and who enjoy the same leisure activities. Homosocial reproduction is especially likely in risky ventures, like launching a new TV series. Denise Bielby and William Bielby (1992) argued that, because most studio and network executives are male, they view male writers and producers as "safer" than women with equally strong qualifications and are more likely to give men rather than women long-term deals and commitments for multiple series. The fear that someone who seems different—a woman, perhaps, or a Hispanic—will conform to executives' negative stereotypes results in a cadre of top managers who look alike and think alike.

SUMMARY

In the past 20 years, women have made progress in closing the promotion and authority gaps, but they still have a long way to go. Will the outlook be better for the college students of today? We don't know. History shows that women's job options do not improve automatically. They improved during the 1970s through the efforts of federal agencies enforcing new laws, advocacy groups, and companies voluntarily establishing Equal Employment Opportunity programs (Reskin and Hartmann 1986:97). Further progress depends on similar efforts in the 1990s and beyond.

REFERENCES

Adler, Marina A. 1993. "Gender Differences in Job Autonomy: The Consequences of Occupational Segregation and Authority Position." *Sociological Quarterly* 34:449–66.

Antal, Ariane B. and Dafna N. Izraeli. 1993. "A Global Comparison of Women in Management: Women Managers in Their Homelands and as Expatriates." Pp. 52–96 in Ellen A. Fagenson (ed.), *Women in Management: Trends, Issues and Challenges in Managerial Diversity.* Newbury Park, CA: Sage.

Berheide, Catherine W. 1992 "Women Still 'Stuck' in Low-Level Jobs." *Women in Public Services: A Bulletin for the Center for Women in Government* 3 (Fall).

Bielby, Denise D. and William T. Bielby. 1992. "Cumulative Versus Continuous Disadvantage in an Unstructured Labor Market." *Work and Occupations* 19:366–87.

Christensen, Kathleen. 1989. "Flexible Staffing and Scheduling in U.S. Corporations." *Research Bulletin No. 240.* New York: The Conference Board.

Collins, Nancy. 1993. "Self-Employment Versus Wage and Salary Jobs: How Do Women Fare?" *Research-in-Brief.* Washington, DC: Institute for Women's Policy Research.

Crompton, Rosemary and Kay Sanderson. 1990. *Gendered Jobs and Social Change.* Boston: Unwyn Hyman.

Epstein, Cynthia F. 1993. *Women in Law.* Urbana: University of Illinois Press.

Farley, Jennie. 1993. "Commentary." Pp. 97–102 in Ellen A. Fagenson (ed.), *Women in Management: Trends, Issues and Challenges in Managerial Diversity.* Newbury Park, CA: Sage.

Fierman, Jaclyn. 1990. "Why Woman Still Don't Hit the Top." *Fortune* (July 30):40,42,46,50,54,58,62.

Goldin, Claudia. 1990. *Understanding the Gender Gap.* New York: Oxford University Press.

Grimm, James W. 1978. "Women in Female-Dominated Professions." Pp. 293–315 in Ann Stromberg and Shirley Harkess (eds.), *Women Working: Theories and Facts in Perspective.* Palo Alto, CA: Mayfield.

Johnston, David. 1993. "FBI Agent to Quit Over Her Treatment in Sexual Harassment Case." *New York Times* (October 11):A7.

———. 1977. *Men and Women of the Corporation.* New York: Basic Books.

Markham, William T., Sharon Harlan, and Edward J. Hackett. 1987. "Promotion Opportunity in Organizations." *Research in Personnel and Human Resource Management* 5:223–87.

Reskin, Barbara F. and Heidi Hartmann. 1986. *Women's Work, Men's Work: Sex Segregation on the Job.* Washington, DC: National Academy Press.

Reskin, Barbara F. and Arne L. Kalleberg. 1993. "Sex Differences in Promotion Experiences in the United States and Norway." R.C. No. 28, Presented at the International Sociological Association meeting, Durham, NC.

Reskin, Barbara F. and Patricia A. Roos. 1990. *Job Queues, Gender Queues: Explaining Women's Inroads Into Male Occupations.* Philadelphia: Temple University Press.

Segal, Amanda T. with Wendy Zellner. 1992. "Corporate Women." *Business Week* (June 8):74–8.

Shellenbarger, Sue. 1993. "Work and Family: Women Start Younger at Own Businesses." *Wall Street Journal* (March 15):B1.

Spaeth, Joe L. 1989. *Determinants of Promotion in Different Types of Organizations.* Unpublished manuscript. Urbana: University of Illinois.

Steinhoff, P. G. and T. Kazuko. 1988. "Woman Managers in Japan." Pp. 103–21 in Nancy J. Adler and Dafna N. Izraeli (eds.), *Women in Management Worldwide.* New York: M. E. Sharpe.

U.S. Merit Systems Protection Board. 1992. *A Question of Equity: Women and the Glass Ceiling in the Federal Government. A Report to the President and Congress by the U.S. Merit Systems Protection Board.* Washington, DC: U.S. Merit Systems Protection Board.

U.S. Small Business Administration. 1988. *Small Business in the American Economy.* Washington, DC: U.S. Government Printing Office.

White, Jane. 1992. *A Few Good Women: Breaking the Barriers to Top Management.* Englewood Cliffs, NJ: Prentice-Hall.

Williams, Christine L. 1992. "The Glass Escalator: Hidden Advantages for Men in the 'Female' Professions'" *Social Problems* 39:253–67.

PART IV

✦

Social Institutions

In this concluding Part, we return to the major institutions of society, examining the core building blocks that make up the integral components of our world. Here, we examine the leading transformations of these core institutions, recent trends and developments influencing their current state, and significant controversial issues occurring within each.

The **Family** represents the foundational *primary group* of society, within which children are born and raised. It also serves as the fundamental economic unit, with adults working to support the family and arranging for the food, clothing, and shelter of its members. The family is a micro *social system,* sanctioning the legitimate conduct of sexuality and pairing people off for reproduction. Within modern society, we generally observe cultural values of monogamy (fidelity among two partners), freedom of partnership choice, bilateral kinship descent (lines of inheritance follow through both men and women), neolocal residence (neither with the groom's nor bride's family, but in a location of their own), moderate gender equality in interpersonal power relationships, and a nuclear rather than extended family structure.

Several important norms guide the character of family formation: *endogamy* (that marital partners should be similar to each other), where we expect some degree of similarity between partners in terms of age, race/ethnicity, class, and religion; *exogamy* (that marital partners should be different from each other in other ways), where we expect people to marry outside their gender and family (the incest taboo); and *homogamy* (that marital partners should be similar to each other in respect to characteristics), where we expect some overlap in personality. These are in a constant state of evolution as the definitions of acceptability shift with new social developments. Our society has seen several changes in existing family forms during the second half of the twentieth century that have shaken it as an institution and raised serious questions about its functioning and future. These include divorce, family violence, remarriage, blended families, single-parent families, gay families, dual-career families, commuter families, childless families, and non-married families. In our first selection, Steven Nock discusses this issue, focusing on gender roles in the family and their consequences. Times have changed, Nock suggests, but not within the family. Barbara Risman searches for families where gender roles have become truly equalized in contemporary society, examining the way these couples arrange their work and home lives. But as we have learned, the advent of women into the workforce has generally not freed women from their child care responsibilities. Sharon Hays discusses the model we hold up in society for the ideal mother and how women strive to meet it. This construct is revealed as racially and economically bounded in William Brown's study of inner-city African American mothers and their gang daughters. Even though these mothers are prevented by economic necessity and the culture of the streets from embodying the ideal middle-class model of motherhood, we see the strength of these mother-daughter ties, the sacrifices these mothers make for their daughters, and the way the daughters view their mothers as important role models.

Religion serves as another building block of society. All religions offer a *dogma,* a set of tenets, or *belief system,* that explains the meaning of life. People accept these convictions on faith and use them as guiding principles. Religious beliefs separate objects and behavior into the realms of the *sacred* and the *profane,* with the former encompassing the divine or supernatural and the latter the earthly or secular. Sometimes these overlap in the area of *civil religion,* where church and state come into union. Countries characterized by civil religion have religious beliefs guiding their policies and often think that their nation is divinely favored, that it is guided by moral standards of behavior, and that it has a mission to perform in the world. We often see this in social institutions that take on a religious-like dimension, such as sport, which Michael Novak discusses in one of our selections, when athletic teams pray before games for divine guidance and victory over their opponents. The belief that one country, religion, or group is supernaturally chosen over others, and has a moral mandate to convert others, has created much animosity and conflict in the world, dating back centuries. This continues to be an important topic for the world to address.

In our lead selection here, Jay Demerath offers us an overview of some contemporary trends in American religious life, addressing issues such as religiosity, old- versus new-line religions, and the real religious practices of the American people. Andrew

Billingsley and Cleopatra Howard Caldwell examine a distinctly American religious institution: the black church, and its long-standing role in the community. On a lighter note, we close with Novak's comparison between two seemingly disparate institutions: religion and sport. Those of you who are torn between the two may find this intriguing.

We turn, next, to the role of **Education,** conceived by many as the great route of upward mobility. America is one of the most highly educated countries in the world, and people from all over come here to study in our colleges and universities. Over the years, we have turned increasingly to the school system to teach our children about morals and values, health and sex education, independent living in society, and a host of other matters that were once handled in the home and church. We also want our schools to teach diversity, conflict resolution, tolerance, and to impart the American culture to our new immigrants. Yet public funding for education remains a controversial issue, as we vie between investing in our next generation or existing ones, and between supporting public schools or letting people transfer their property tax dollars into the private domain. Curricular matters are also a battlefield where sacred and secular views clash, as witnessed by the recent debate over teaching evolutionism versus creationism. The same vehemence can be witnessed in the way our citizens approach forced busing, integration, prayer in schools, and school violence, topics that divide liberals and conservatives across the nation. Thus, we are increasingly asking our schools to do more, but supporting them less.

With so many important issues abounding in this field, we can only address a limited few. Jonathan Kozol poignantly depicts the differential opportunities children from different geographic and economic groups find when they get to their public schools. As a nation, we are torn between our principles of equality of opportunity and the very real costs of making this a reality. David Karp and William Yoels examine what actually goes on inside the types of classes you might be sitting in, by looking at who talks, who listens, and the subtext of educational messages in the college classroom. Finally, Ruth Sidel brings the issue of diversity to the fore in discussing the struggle over access to a college education, highlighting some of the vitally important social factors that lie outside of and frame our classrooms.

Like our educational system, meaningful aspects of **Health Care** delivery function at all levels, from the doctor-patient relationship to the hospitals and medical delivery organizations, to the economics of medicine. We live in an era of modern medicine, where many diseases have been eradicated, science and technology are battling others, and sterilization and sanitation procedures are highly advanced. As a result, people are living longer and enjoying a better quality of life into their later years. The science of *epidemiology* has flourished, with advanced research being conducted into the patterns by which diseases thrive and spread throughout populations. Other issues in the sphere of public health include prevention (wellness) versus treatment, harm reduction versus criminalization (drugs), and medical insurance coverage. Compared to some countries, the United States does not make medical treatment accessible to all its citizens, but the level of health care available is better here than in many others. Like many other dimensions we have examined, however, we see grave inequalities here, with some people getting the most advanced treatment while others suffer deprivation. Another important trend in medicine has been the expansion of *medicalization* into previously nonmedical spheres. We have seen many behaviors redefined as medical issues such as addiction (physical and psychological), childbirth, anxiety, classroom misbehavior, and learning differences. All of these swell the influence of medical professionals and expand the domain of the *sick role* in society. While this may make us a more humane nation, it may also lead to the overmedicalization of our population: doctors may be prescribing more drugs now than ever before.

Debra McPhee addresses the complex issue that our government has been unable to resolve in her overview of access to medical care in America and health insurance. Howard Waitzkin takes us inside some doctor-patient encounters and reflects on how these are influenced by the biggest shift we are experiencing in the structure of medicine: the movement from a fee-for-service system to the Health Maintenance Organization (HMO). Ken Silverstein offers a view of international issues in medicine by looking at the system of profit and its affect on the research, production, and dissemination of drugs among the rich and poor nations.

30

The Problem with Marriage

STEVEN L. NOCK

Nock takes a sweeping view of the institution of marriage in contemporary American society and pinpoints one of its main challenges: changing gender roles. While the workplace has grudgingly accommodated women's emancipated roles, Nock suggests that the family has been less yielding in offering flexibility from tradition. Yet women, to a greater extent than men, are now looking for marriages that differ importantly from those of their grandparents. Nock suggests that the American family's strength is based, in part, on a traditional gender role model, with its norms of heterosexuality, fidelity and monogamy, parenthood, and husband as head, and that such innovations as cohabitation, unmarried childbearing, and female independence weaken it considerably. How marriage adapts to contemporary changes may determine its future appearance and vigor. Do you think the institution of marriage has suffered since the onset of more women in the labor force? Do you think that the way we marry and raise families will look considerably different in the coming years? If so, does this frighten you or excite you? Is there a "problem" with marriage, as Nock suggests, or do you think this is much ado about nothing?

A Call to Civil Society warns that the institutions most critical to democratic society are in decline. "What ails our democracy is not simply the loss of certain organizational forms, but also the loss of certain organizing ideals—moral ideals that authorize our civic creed, but do not derive from it. At the end of this century, our most important challenge is to strengthen the moral habits and ways of living that make democracy possible." I suggest that American institutions have traditionally been organized around gender and that the loss of this organizing principle explains many of the trends discussed in the report. Specifically, the continued centrality of gender in marriage—and its growing irrelevance everywhere else—helps explain many contemporary family problems. The solution is to restore marriage to a privileged status from which both spouses gain regardless of gender.

Reprinted by permission of Transaction Publishers.

The family trends we are now seeing reflect a conflict between the ideals central to marriage and those that define almost all other institutions. Growing numbers of Americans reject the idea that adults should be treated differently based on their gender. But it is difficult to create a new model of marriage based on such a premise. For many people, assumptions about gender equality conflict with the reality of their marriages. It may hardly matter if one is male or female in college, on the job, at church, in the voting booth, or almost anywhere else in public. But it surely matters in marriage. The family, in short, is still organized around gender while virtually nothing else is. Alternatively, marriage has not been redefined to accommodate the changes in male-female relations that have occurred elsewhere. This, I believe, is the driving force behind many of the problematic trends identified in *A Call to Civil Society.*

Stable marriages are forged of extensive dependencies. Yet trends toward gender equality and independence have made the traditional basis of economic dependency in marriage increasingly problematic. The challenge is to reinvent marriage as an institution based on dependency that is not automatically related to gender. Both partners, that is, must gain significantly from their union, and both must face high exit costs for ending it.

Despite dramatic changes in law and public policy that have erased (or minimized) distinctions between men and women, married life has changed more slowly and subtly. In the last four decades, the percentage of married women in the paid labor force increased from 32 percent to 59 percent, and the number of hours that wives commit to paid labor increased apace. While men do not appear to be doing much more housework today than they did two decades ago, women are doing less in response to their commitments to paid labor. Women did 2.5 times as much household labor as their husbands in 1975. By 1987, the ratio was 1.9. Wives' share of total (reported) household income increased marginally, from 35 percent in 1975–1980 to 38 percent in 1986–1991. In such small ways, husbands and wives are increasingly similar. Still, marriages are hardly genderless arrangements. My research for *Marriage in Men's Lives* showed that most marriages in America resemble a traditional model, with husbands as heads of households, and wives who do most housework and child care. Given the pace at which gender distinctions have been, or are being, eliminated from laws, work, school, religion, politics, and other institutions, the family appears to be curiously out of step.

One reason gender is still a central motif in marriage is because masculinity (and possibly, femininity) are defined by, and displayed in marriage. As the title of Sara Berk's book proclaimed, the family is *The Gender Factory.* Consider the consequences of unemployment for husbands. If spouses were economically rational, then the unemployed (or lower-paid) partner would assume responsibility for housework. Sociologist Julie Brines found just the opposite. After a few months of unemployment, husbands actually *reduced* their housework efforts. The reason is that housework is much more than an economic matter. It is also symbolic. "Men's" work means providing for the family and being a "breadwinner," whereas "women's" work means caring for the home and children. Such assumptions are part of our cultural beliefs. Doing housework, earning a living, providing for the family, and caring for children are ways of demonstrating masculinity or femininity. When wives are economically dependent on their husbands, doing housework is consistent with traditional assumptions about marriage. Such women conform to cultural understandings about what it means to be a wife, or a woman. However, a dependent husband departs from customary assumptions about marriage and men. Were he to respond by doing more housework, his deviance would be even greater. Marriage is still the venue in which masculinity and femininity are displayed.

The husband and wife who construct a new model of marriage that doesn't include gender as a primary organizing principle will face challenges. The husband who decides to be the primary child-care provider or the wife who elects to be the sole wage earner will find these unusual marital roles difficult but not impossible to sustain. Relationships with parents may be awkward. Friends may struggle to understand the arrangement if it differs from their own. Employers may also find such an arrangement difficult to understand and accept. Yet as difficult as it may be to forge a new model of marriage, it seems certain that some change is necessary if marriage is to endure.

The title of Goldscheider and Waite's recent book asks a stark question: *New Families, or No Families?* In "new" families, husbands and wives will share family economic responsibilities *and* domestic tasks equitably. The alternative, according to these authors, is "no" families. Quite simply, marriage must be redefined to reflect the greater gender equality found everywhere else, or women will not marry. Or those who do will face increasing difficulties reconciling their public and private lives. Indeed, young women today value marriage less than young men do. Three in four (75 percent) never-married men under age 30 described getting married as important for their lives in 1993. A smaller percentage (66 percent) of comparable women replied that way (1993 General Social Survey).

Research confirms that most women who marry today desire marriages that differ importantly from those of their grandmothers because women's lives have changed in so many other ways in recent decades. However, though the options available to women have expanded in other respects, the basic pattern of marriage is pretty much the same as it has been for decades. The revolution in gender has not yet touched women's marriages. Part of the reason is that men have been excluded from the gender revolution. While almost any young woman today will notice enormous differences between her life options and those of her great-grandmother, the differences between men and their great-grandmother, are minimal, at most. The script for men in America has not changed. In short, despite enormous changes in what it means to be a woman, marriage does not yet incorporate those changes. Neither men or women have yet figured out how to fashion "new" families.

Many of our problems are better seen as the result of institutional change than of individual moral decline. The *personal problems* that lead to family decline are also legitimate *public issues.* Institutions like the family are bigger than any individual. So when large numbers of people create new patterns of family life, we should consider the collective forces behind such novel arrangements. And if some of those innovations are harmful to adults or children, fixing them will require more than a call for stronger moral habits (though there is certainly nothing wrong with such advice). Fixing them will require restructuring some basic social arrangements.

Since *A Call to Civil Society* focuses on *institutional* decline, I want to consider the meaning of an institution. A society is a cluster of social institutions, and institutions are clusters of *shared* ideals. Only when people agree about how some core dimension of life should be organized is there a social institution. The family is a good example.

Although individual families differ in detail, collectively they share common features as a result of common problems and tested solutions. In resolving and coping with the routine challenges of family life such as child care, the division of household labor, or relations with relatives, individuals draw on conventional (i.e., shared) ideals. As disparate individuals rely on shared answers to questions about family life, typical patterns emerge that are understood and recognized—mother, father, son, daughter, husband, and wife. To the extent that such ideals are widely shared, the family is a social institution. Were individuals left completely on their own to resolve the recurring problems of domestic life, there would be much less similarity among families. Alternatively, were there no conventional values and beliefs to rely on, the family would not be an institution. The family, as an institution, differs in *form* from one culture to another. Yet everywhere, it consists of patterned (i.e., shared and accepted) solutions to the problems of dependency (of partners, children, and the elderly).

The problem today is that an assumption that was once central to all social institutions is no longer so compelling. Beliefs about gender have long been an organizing template that guided behaviors in both public and private. Yet while gender has become increasingly unimportant in public, neither women nor men have fully adjusted to these changes in their private married lives. If men and women are supposed to be indistinguishable at work or school, does the same standard apply in marriage? Americans have not yet agreed about the answer. As a result, the institution of the family (the assumptions about how married life should be organized) no longer complements other social arrangements. Increasingly, the family is viewed as a problem for people because the assumptions about domestic life no longer agree with those in other settings. When husbands and wives return home from work, school, church, or synagogue, they often struggle with traditional ideals about marriage that do not apply in these other areas. No matter what her responsibilities at work, the married mother will probably be responsible for almost all child care at home, for instance. Responsibilities at work are unlikely to be dictated by whether the person is male or female. But responsibilities at home are. This contradiction helps explain the trends identified in *A Call to Civil Society.*

High rates of divorce, cohabitation, and unmarried childbearing are documented facts, and all have clearly increased this century. Do such trends suggest that the family is losing its institutional anchor? In fact, the traditional arrangements that constitute the family are less compelling today than in the past. In this respect, the institution of the family is weaker. To understand why, I now consider the traditional basis of the family, legal marriage.

LEGAL MARRIAGE AND THE INSTITUTION OF THE FAMILY

The extent to which the family *based on legal marriage* is an institution becomes obvious when one considers an alternative way that adult couples arrange their intimate lives. Certainly there is no reason to believe that two people cannot enjoy a harmonious and happy life without the benefit of legal marriage. In fact, growing numbers of Americans appear to believe that unmarried *cohabitation* offers something that marriage does not. One thing that cohabitation offers is freedom from the rules of marriage because there are no widely accepted and approved boundaries around cohabitation. Unmarried partners have tremendous freedom to decide how they will arrange their legal and other relationships. Each partner must decide how to deal with the other's parents, for example. Parents, in turn, may define a cohabiting child's relationship as different from a married child's. Couples must decide whether vacations will be taken together or separately. Money may be pooled or held in separate accounts. If children are born, cohabiting parents must decide about the appropriate (non-legal) obligations each incurred as a result. In such small ways, cohabiting couples and their associates must *create* a relationship. Married couples may also face decisions about some of these matters. However, married spouses have a pattern to follow. For most matters of domestic life, marriage supplies a template. This is what cohabiting couples lack. They are exempt from the vast range of marriage norms and laws in our society.

A man can say to his wife: "I am your husband. You are my wife. I am expected to do certain things for you, and you likewise. We have pledged our faithfulness. We have promised to care for one another in times of sickness. We have sworn to forego others. We have made a commitment to our children. We have a responsibility and obligation to our close relatives, as they have to us." These statements are not simply personal pledges. They are also enforceable. Others will expect these things of the couple. Laws, religion, and customs bolster this contract. When this man says to someone, "I would like you to meet my wife," this simple declaration says a great deal.

Compare this to an unmarried couple living happily together. What, if any, are the conventional assumptions that can be made? What are the limits to behavior? To whom is each obligated? Who can this couple count on for help in times of need? And how do you complete this introduction of one's cohabiting partner: "I would like you to meet my . . ."? The lack of a word for such a partner is a clear indication of how little such relationships are governed by convention. Alternatively, we may say that such a relationship is *not* an institution, marriage is. I believe this helps explain why cohabiting couples are less happy, and less satisfied with their relationships than married couples.

Almost all worrisome social trends in regard to the *family* are actually problems related to *marriage:* declining rates of marriage, non-marital fertility, unmarried cohabitation, and divorce. Any understanding of the family must begin with a consideration of marriage. I now offer a *normative definition of marriage;* a statement of what Americans agree it *should* be, the assumptions and taken-for-granted notions involved. In so doing, I will lay the foundation for an explanation of family decline.

In *Marriage in Men's Lives,* I developed the details of normative marriage by consulting three diverse sources. First, I examined large national surveys conducted repeatedly over the past two decades. Second, I read domestic relations law, including state and federal appellate decisions. Finally, I consulted sources of religious doctrine, especially the Bible. Throughout, my goal was to identify all aspects of marriage that are widely shared, accepted as legitimate, and broadly viewed as compelling.

A normative definition of marriage draws attention to the central idea that marriage is more than the sum of two spouses. As an institution, marriage includes rules that originate outside the particular union that establish boundaries around the relationship. Those boundaries are the understood limits of behavior that distinguish marriage from all other relationships. Married couples have something that all other couples lack; they are heirs to a system of shared principles that help organize their lives. If we want to assess changes in the family, the starting point is an examination of the institutional foundation of marriage. Six ideals define legal marriage in America.

1) *Individual Free Choice.* Mate selection is based on romantic love. In the course of a century, parents have come to play a smaller and smaller role in the choice of married partners. Dating supplanted courtship as compatibility, attractiveness, and love replaced other bases for matrimony. National surveys show that "falling in love" is the most frequently cited reason for marrying one's spouse, and that the most important traits in successful marriages are thought to be "satisfying one another's needs" and "being deeply in love." Western religious ceremonies admonish partners to love one another until death, and every state permits a legal divorce when love fails ("incompatibility," "irreconcilable differences" or similar justifications). Love is associated with feelings of security, comfort, companionship, erotic attraction, overlooking faults, and persistence. Since love and marriage are so closely related, people expect all such feelings from their marriages.

2) *Maturity.* Domestic relations law defines an age at which persons may marry. Throughout the U.S., the minimum is 18, though marriage may be permitted with approval by parents or the court at earlier ages. Parental responsibilities for children end with legal *emancipation* at age 18. Thus, marriage may occur once parents are released from their legal obligations to children, when children are legally assumed to be mature enough to enter binding contracts, and once children are assumed able to become self-sufficient and able to provide for offspring. Traditional religious wedding ceremonies celebrate a new form of maturity as Genesis states: "A man leaves his father and mother and cleaves to his wife and they become one flesh."

3) *Heterosexuality.* Traditionally, and legally, the only acceptable form of sex has been with one's spouse. Sex outside of marriage (fornication or adultery) is still illegal in half of U.S. states. And sexual expression within marriage has traditionally been legally restricted to vaginal intercourse (sodomy laws). Though such laws are rarely enforced even where they exist, they remind us of the very close

association of marriage and conventional forms of heterosexual sexuality. Recent efforts to legalize homosexual marriage have been strenuously resisted. Since the full-faith-and-credit clause of the Constitution requires that marriages conducted legally in one state be recognized as legal in others, the possibility of legal homosexual marriage in Hawaii prompted an unprecedented federal "Defense of Marriage Act" in September 1996. This law will allow states to declare homosexual Hawaiian marriages void in their jurisdiction. Despite growing acceptance of homosexuality, there is very little support for homosexual marriages. The 1990 General Social Survey showed that only 12 percent of Americans believe homosexuals should be allowed to marry.

4) *Husband as Head.* Though Americans generally endorse equality between the sexes, men and women still occupy different roles in their marriages. Even if more and more couples are interested in egalitarian marriages, large numbers of people aren't. The 1994 General Social Survey shows that adults are almost evenly divided about whether both spouses should contribute to family income (57 percent approve of wives working, and in fact, 61 percent of wives are employed). Four in ten adults endorse a very traditional division of roles, where the wife takes care of the home and family, and the husband earns all the income. Traditional religious wedding ceremonies ask women to "honor and obey" their husbands. In reality, most husbands have more authority than their wives do. The spouse who is primarily responsible for income enjoys more authority, and in the overwhelming majority of American marriages, that is the husband. Demands made of husbands at work are translated into legitimate demands made on the family. So most husbands have more authority than their wives do, regardless of professed beliefs.

5) *Fidelity and Monogamy.* In law, sexual exclusivity is the symbolic core of marriage, defining it in more obvious ways than any other. Husbands and wives have a legal right to engage in (consensual) sex with one another. Other people may not have a legal right to engage in sex with either married partner. Adultery is viewed as sufficiently threatening to marriages that homicides provoked by the discovery of it may be treated as manslaughter. Extramarital sex is viewed as more reprehensible than any other form, including sex between young teenagers, premarital sex, and homosexual sex. Eight in ten adults in the 1994 General Social Survey described extramarital sex as "always wrong." Adultery, in fact, is rare. Recent research reported by Laumann and his colleagues in *The Social Organization of Sexuality* revealed that only 15 percent of married men and 5 percent of married women have ever had extramarital sex. Among divorced people these percentages are only slightly higher, 25 percent for men and 10 percent for women. Monogamy is closely related to fidelity because it restricts all sexual expression to one married partner. With the exception of Utah where some Mormons practice polygamy (against the canons of the Mormon Church), monogamy has gone largely unchallenged in the United States since 1878 when the U.S. Supreme Court upheld the conviction of a Mormon who practiced polygamy.

6) *Parenthood.* With rare exceptions, married people become parents. Despite high rates of unmarried fertility, there is little to suggest that married couples are less likely to have, or desire to have children today than they were several decades

ago. Only 13 percent of ever-married women aged 34 to 45 are childless today. Two decades ago, the comparable figure was 7 percent. The six-point difference, however, is due to delayed fertility, rather than higher childlessness. Overall completed cohort fertility (i.e., the total number of children born to women in their lifetime) has remained stable since the end of the Baby Boom. And while the legal disabilities once suffered by illegitimate children have been declared unconstitutional, marital and nonmarital fertility differ in important respects. Unmarried fathers may, through legal means, obtain the custody of their children, although few do. Indeed, a vast legal apparatus exists to enforce the parental obligations of men who do not voluntarily assume them. On the other hand, married men are automatically presumed to be the legal father of their wife's children, and nothing (except the absence of "unfitness") is required of them to establish custody.

CHALLENGES TO NORMATIVE MARRIAGE

This ensemble of behaviors and beliefs describes how most Americans understand marriage. Even if particular marriages depart in some ways from this model, this should be the starting point when attempting to assess family change. If "the family" has declined, then such change will be obvious in one or more of the foregoing dimensions. Widespread attempts to change normative marriage, or wholesale departures from it, are evidence that Americans do not agree about the institution. I now briefly review three obvious challenges to the normative model of marriage just outlined. High divorce rates, late ages at marriage, and declining rates of remarriage are a reflection of an underlying theme in such challenges. That theme is the importance of gender in marriage.

The increasingly common practice of unmarried *cohabitation* is an example of a challenge to normative marriage. In 1997, 4.1 million opposite-sex cohabiting couples were counted by the U.S. Bureau of the Census, the majority of whom (58 percent) have never married, and one in five of whom (22 percent) is under 25 years old. Research on cohabiting partners has identified a central theme in such relationships. Cohabiting individuals are more focused on gender equality in economic and other matters than married spouses. They are also less likely to have a gender-based division of tasks in most forms of household labor except the care of infants.

Yet another challenge to normative marriage is unmarried childbearing. One in three children in America is born to an unmarried woman; six in ten of those children were conceived outside of marriage (the balance were conceived prior to a divorce or separation). The historical connection between sexual intercourse and marriage weakened once effective contraception and abortion became available. Without contraception, married women became pregnant if fecund. There was no reason to ask a married woman why, or when she would have a child. Parenthood in an era of universal contraception, however, is a choice. It is now possible to ask *why* someone had a child. Since childbearing is thought to be a choice, it is viewed as a decision made chiefly by women. And the type of woman who chooses to have children differs from the childless woman because motherhood now competes with many other legitimate roles in a woman's life. Research has shown that women who choose motherhood give occupational and income considerations lower priority than childless women.

By the late 1960s, feminists who argued that wives should not be completely dependent on husbands joined this critique of the exclusive breadwinner role of husbands. Just as the exclusive breadwinner role for men was criticized, the exclusive homemaker-mother role for married women was identified as oppressive. And, of course, such women are the statistical exception today. Maternity must be balanced with many other adult roles in women's lives, and traditional marriage is faulted as creating a "second shift" for women who return home from work only to assume responsibility for their households with little help from their husbands.

All significant challenges to marriage focus on various aspects of gender. The traditional assignment of marital roles based on sex (i.e., husband as head of household) is the core problem in marriage. Other dimensions of normative marriage are less troublesome. There is little evidence of widespread disagreement about the ideas of free choice of spouses, fidelity, monogamy, heterosexual marriage, maturity, or parenthood.

Whether Americans are now less committed to common moral beliefs than in the past is an empirical question. Values (i.e., moral beliefs) are researchable issues, and it would be possible to investigate their role in matters of family life. Increases in divorce, cohabitation, illegitimacy, or premarital sex are certainly evidence that some beliefs are changing. However, social scientists have yet to identify all the various causes of these family trends. Undoubtedly there are many, including demographic (e.g., longer life, lower fertility), technological (e.g., contraception, public health) and cultural (e.g., shifting patterns of immigration). Changing values about gender are but one cause of family change. Still, I suspect they are the most important. When something as basic and fundamental as what it means to be a man or woman changes, virtually everything else must change accordingly. Now we must incorporate such new ideas into the institution of marriage.

31

Playing Fair

BARBARA RISMAN

The research of Berkeley sociologist Arlie Hochschild in the 1980s on the "second shift" discouraged many American women when they realized they were not alone in having entered the labor market full-time only to be shouldering the brunt of the responsibility for a second full-time job running the household and family. Men offered help, to varying degrees, but essentially failed to assume ownership of the tasks and roles in this domain. Risman searched to find the groups in society where this inequality did not exist, where husbands and wives comanaged this realm equally. She found this pattern primarily among "educationally elite" American households whose members worked in white-collar professions. A phenomenon arising predominantly in the late 1980s, equal sharing and labor management occurred where men's incomes were too low to sustain the family lifestyle on their own and the wife's paycheck had real meaning. Risman examines the commonalities and different patterns characterizing these couples, how they moved into "playing fair," and the very high marital satisfaction they display. Her findings offer hope, in contrast to Nock's analysis, for the future of gender roles in marriage. How equitable do you want your relationship with your spouse to be? Is Risman's forecast only applicable to the "educationally elite" that she studied, or are these people on the forefront for how most husbands and wives will organize their marriages in the future? If we move more toward the egalitarian marriages described by Risman, will other parts of society have to change to accommodate this?

Much research on contemporary families focuses on the stalled revolution in most American homes: even when women spend as many hours in the paid labor force as their husbands do, they retain primary responsibility for homemaking and childrearing. The power of gender as a social structure is apparent in these typical families. To understand how and when the gender structure changes we must consider not only the typical family but also families who live on the cutting edge of social change. In this chapter we look at

From Barbara Risman, *Gender Vertigo*, pp. 93–94, 100–101, 106–108, 110–113, 114.

a statistically rare phenomenon—"fair" role-sharing families, those in which husbands and wives occupy breadwinner and nurturer roles equally.

There is a scarcity of research on couples who have redistributed family work equitably and without regard for the gendered division of responsibilities. There is a good reason for this dearth of research: such families are very rare in a statistical sense, and social scientists prefer to study more mainstream families. All of the research in this field is made up of qualitative studies of recruited volunteer families because there are not enough such families to survey in a random sample. The research that does exist leads me to believe that there have been major changes in egalitarian families in the past two decades. The most striking change is that now some such families actually exist. . . .

THE "FAIR FAMILY" STUDY

We sent a survey to the seventy-five families who had volunteered to participate and seemed eligible based on our screening conversation. The survey worked well: only once did our screening device let through a family in which the woman did more than her share. The problem was, however, that this screened group of seventy-five families yielded us only fifteen who met our criteria for inclusion. Only one of five families who identified themselves as equitable on the telephone, or 20 percent, actually shared the household labor in a 40/60 or better split, agreed that they shared equally the responsibility for breadwinning and childrearing, and felt that their relationship was fair. That alone tells us much about the strength of our gender structure for creating inequitable marriage. Yet I do not want to discredit the tendency toward egalitarianism in the rest of the families who volunteered to participate. The husbands and wives in the families we did not include in our sample were trying to share equally—they were challenging the gender structure. Even if they were not entirely successful they clearly are part of the massive social change that feminism has inspired. Still, as a social analyst, I find it remarkable that the taken-for-granted nature of female responsibility for family work hides a gendered division of labor even from many fair-minded couples themselves.

The fifteen families who met our criteria were educationally elite. More than half of the parents (eight men and nine women) had a Ph.D. or an M.D. Another eight parents had master's degrees (usually in education or business). Three fathers and one mother had a bachelor's degree only, and one mother had never completed college. These families were not necessarily rich, as many had given up high-paying jobs or occupations in order to care for their children equally. Others were college professors, a group with notoriously high educational attainment relative to their incomes. Nevertheless, these were the only families who met our criteria. . . .

Four relationships leading to an equitable division of household labor were identified: dual-career couples, dual-nurturer couples, post-traditionals, and those pushed by external circumstances. In dual-career marriages both partners had always been interested in their own career growth and success, as well as in co-parenting their children. Dual-nurturer couples were more child-centered than work centered, with both parents organizing their work lives almost exclusively around their parental responsibilities. Post-traditional couples had spent at least part of their adult lives in husband-breadwinner and wife-nurturer roles and had consciously rejected that model. And two couples had been pushed into "fair" relationships by circumstances beyond their control. In one such family, the wife's job was the organizing principle for the family's life because she earned nearly twice as much as her husband, who was not very career oriented. In the other family the wife's chronic illness was at least partly responsible for the husband's domestic labor.

DUAL-CAREER COUPLES

The most common route to peer marriage—and the one I had expected because it mirrors my own experience—was the partnership of two career-oriented professionals who both held egalitarian values. Two-thirds of the couples in this study (ten of fifteen) divided their labor "fairly" because both partners were career and equity oriented. Both parents compromised work goals to balance family and career priorities, but both remained committed to their careers as well as their childrearing responsibilities. In all but one of these ten couples, both partners had always studied or worked full-time, shared household responsibilities before parenthood, and never considered any childrearing style other than co-parenting. In one couple, the wife had stayed home for a year after the birth of their daughter while her husband held a temporary position; she remembered the experience as atypical and miserable. She was lonely and unhappy with no work other than mothering. These are families in which both husband and wife simply assume that a fulfilling life involves both paid and family work. No husband or wife articulated, without some probing, an ideological justification for this assumption; they simply had adopted the culturally available feminist view on equality as their own taken-for-granted reality.

A remarkable finding is the absence of gender expectations used as a basis for organizing labor. Not one of these couples mentioned that they had ever considered that the wife devote herself exclusively to family work. No husband or wife complained about consistently doing more than a fair share of housework, nor did the men wish that their wives were more traditionally domestic. Only two of the ten dual-career couples reported any serious conflict over division of labor, usually long in the past. Often the couples never had negotiated at all. One mother, a mathematician married to a public policy analyst explained,

> We didn't tend to do the "You'll do half of the time, I'll do half of the time." We tended to divide tasks up, [decide] who would do them. So Karl liked to cook. He cooked for himself before we ever married. I didn't cook for myself at all before we were married. I ate out or in the cafeteria where I worked, so we didn't really have to adapt ourselves in that way. If he had wanted to share the cooking because he didn't much like it, I probably would've divided it in

> some way, half and half, but we tended to sort jobs. He didn't like to do laundry, I like to do laundry. I didn't mind cleaning up, I do the cleaning up . . . the dishes and wiping up counters and stuff. So we found a way that we considered equitable . . . but we did talk about it and we did consciously do it.

In another household the family tried to divide the cooking 50:50 but quickly reverted to sorting tasks by preference—here the mother cooked and father cleaned up afterward. . . . The one way that gender did manifest itself in the family labor of dual-career couples was that the wives sometimes came to the relationship with higher standards of cleanliness. But unlike more traditional families, these couples did not use this difference to justify an extra burden for the wife. Rather, differing standards were seen as a problem to be worked out equitably.

DUAL-NURTURER COUPLES

Two couples were dual nurturers, oriented to home, family, and lifestyle rather than career. They worked for pay so that they could spend time together and with their families. Only one of these couples was unequivocally dual nurturing, however. In the Woods family, neither parent has worked full-time consistently for years, and neither wants to do so. The father, an aspiring sculptor, works half-time as an editor; the mother is an accountant who sees clients three days a week. They try to organize their work schedules so that both are not working during the same day, even though their baby is in day care and their other child is in school and after-school programs. They feel that the evenings are too hectic if both arrive home at the same time after a full day away. Their work schedules and choice of jobs have varied with family needs. In many ways this looked like a home with two mothers in that neither parent was strongly attached to the labor force. This couple was particularly focused on the quality of their lives rather than material acquisition or career development. Yet the couple did not appear to be suffering economically from their career decisions. They lived in a modern, energy-efficient home they had designed themselves. Original artwork, painting, and pottery were in evidence in their rural, picturesque setting.

POST-TRADITIONAL COUPLES

The third route to a division of equitable labor was dissatisfaction with more traditional arrangements. Two of the couples in this study were post-traditional families. In the case of the Germanes, both partners' previous marriages had been organized around traditional gendered expectations and responsibilities. The woman stated clearly that she had left a previous marriage because her husband did not meet her needs for an equal partner. She had been married to a career diplomat and had moved every three years of her adult life. As a bookkeeper in government service she found it easy to relocate, and she had never been unemployed more than three months in her entire working life because of the "work ethic I grew up with." Paid work was important to her sense of self. When asked about her maternity leave, she answered, "I do feel that three months out of work—I don't care who I'm taking care of—is a lot of time." She cut back her work to thirty-two hours a week during her daughter's infancy, but only because her husband was on an assignment that involved much travel. Her current husband, who also had been a diplomat, had returned to school for another college degree when their daughter was in the early elementary grades. Ms. Germane enthusiastically recalled, "I liked the role reversal when he was the housewife. . . . He walked [our daughter] to school and I swear he knew every lady in the school. . . . He became a PTA mom." Both husband and wife reported having always shared work equally; the wife was very much aware that they seemed to share things "more in the middle" than other people. Even now that the husband is again working full-time in business management the couple continue to share labor equally. Both husband and wife had experienced less satisfying relationships, and they were, in Schwartz's (1994) language, very deep friends. Both wanted to keep things fair to protect their precious friendship.

The other post-traditional couple had renegotiated their gender-based roles when the youngest of their three boys entered kindergarten. It took this family three to four years to transform a decade-long male-breadwinner/female-homemaker pattern into a fair relationship. The wife was clearly the moving force behind this transformation. Ms. Potadman told us about her resentment at spending Saturdays doing housework after she returned to nursing full-time. Eventually her family—at her urging—divided the household tasks with a scheduling system. The mother explains it this way: "I had the idea, let's just list the tasks and we'll divide them up, and whenever he [her husband] gets his done, he gets them done." This wife still holds primary responsibility for the scheduling sessions. The entire family talked at length about the scheduling charts they used: each son cooked one day a week, the mother twice, and the father once, with dinner out every Friday. The sons were assigned cleanup tasks on days they did not cook.

COUPLES PUSHED INTO SHARING

The final route to a fair relationship was to be pushed by external forces. In one family the wife had a considerably better-paying job and a much less flexible schedule. In another family the wife was chronically ill. In both of these families gender equality was the conceptual framework that helped them make sense of their lives. In many ways they saw these external constraints in gender-neutral terms. In the Cody family the wife earned two-thirds of the household income. The husband had been passed over for a promotion that he had expected, and quit his job and changed careers. Mr. Cody was no longer work-focused and was pleased to have a flexible schedule. He enjoyed the freedom of owning a small business without the economic burden of supporting the family on his income.

> When we decided that we were going to have children Marilyn's job situation was such that . . . she worked in a reasonably structured environment, and in that case you can't take time off to bring somebody to gymnastics and, you know, go to the school for plays and . . . give parties and do that kind of stuff. So we made a decision that I would do that, and what it does basically is it takes me out of, you know, being the high-powered career person and, you know, . . . I just do something that brings in some money, and [parenting] gives me satisfaction that I can do it. . . .

In summary, not one of the thirty parents interviewed even suggested that the husband's paid work was more important than the wife's. In fact, four of these families had moved to North Carolina because of the wife's work, and not one husband or wife mentioned this as unusual. Another four had relocated to North Carolina because of the husband's job; in two of these families the wife's ability to transfer was mentioned as a prerequisite for relocation. Another three families relocated because both partners found positions in the area. The other four couples were living there when they met. Not one of the thirty parents suggested that caring for their babies was or should have been more the mother's responsibility. Not one believed that housework should be the wife's responsibility, although half did admit to conflict at some point in their relationship about standards of cleanliness. It is to how conflict is negotiated and power exercised in fair families that I now turn.

Control refers to the cultural lines of authority and coercion, the status differential between men and women that is at the root of our gender structure. There are countless overt and subtle means by which male privilege is constructed in daily life. Women's fear of rape constrains their mobility in ways that men never experience. Fear of ridicule and social disapproval keeps women, from girlhood on, worried about their weight and physical attractiveness. The wage gap and sexual harassment in the workplace also reinforce male dominance. The normative belief that husbands should head the household also explicitly reinforces dominance and provides cultural authority for men to control their wives. Similarly, the continuing expectation that wifehood involves domestic service is yet another way that male privilege and female subordination is re-created in daily life. While such daily inequity appears to pale in comparison with fear of rape, the expectation that women provide domestic services to husbands and primary care for children not only creates inequitable marriages but also disadvantages women in the world of work.

Yet there are families—fair families—in which the marital control issues that reinforce male privilege appear to be moot. These women and men are both pioneers in and beneficiaries of the women's liberation movement.

REFERENCES

Blumstein, Phillip and Pepper Schwartz. 1983. *American Couples.* New York: William Morrow.

Schwartz, Pepper. 1994. *Peer Marriage.* New York: Free Press.

32

Responsibilities of Intensive Mothering

SHARON HAYS

In an excerpt from her exceptionally well-received book on motherhood, Hays outlines the culture of intensive mothering that has become a part of the dominant middle-class culture. This is part of the squeeze in which contemporary women find themselves, as they are driven to succeed in marriage, at work, in the gym, and with their children. Far from losing its emphasis as women are working long hours and supporting households, motherhood has maintained its transcendent position as the guardians of our "sacred children." While men have adopted some dimensions of household labor, in the final analysis it is women who are considered ultimately responsible for the care and emotional development of their children. Their moral obligation is to place their children's needs before their own, to find greater satisfaction in giving than receiving, or else they risk social condemnation. Women who sacrifice themselves for their children are considered particularly admirable. Like men, they accept these obligations, feeling inadequate should these be unfulfilled. The good mother is responsible for giving her children unconditional love and ensuring that they grow up bathed in the protection and innocence of this sacred period. Women who do not experience this maternal instinct should feel shame. These expectations, another spoke in women's social responsibilities, contribute to the simultaneous well-being of children, greater freedom of men, and role strain of women. How does the society that Hays describes compare and contrast with the one described by Risman? Do you think there is such a thing as maternal instinct or is this just something that is imposed on women as an expectation? How might American mothering patterns differ from those found in other cultures?

The ideology of intensive mothering and the extent to which mothers attempt to live up to it is responsible for the cultural contradictions of motherhood. The same society that disseminates an ideology urging mothers to give unselfishly of their time, money, and love on behalf of sacred

From Sharon Hays, *The Cultural Contradictions of Motherhood*, pp. 97–110, 108–109, 111, 112–113, 115, 121–122. Reprinted by permission of Yale University Press.

children simultaneously valorizes a set of ideas that runs directly counter to it, one emphasizing impersonal relations between isolated individuals efficiently pursuing their personal profit. In other words, the cultural model of a rationalized market society coexists in tension with the cultural model of intensive motherhood. And this tension increasingly influences the lives of individual mothers as more and more of them enter the paid workforce and therefore participate directly in the world of the rationalized market. Yet, the history of child-rearing ideas and the words of today's child-rearing advisers seem to demonstrate that the more powerful and all encompassing the rationalized market becomes, the more powerful becomes its ideological opposition in the logic of intensive mothering. . . .

In the following pages I attempt to tease out the logic of that ideology as contemporary mothers understand it.[1] As you listen to these mothers, it may seem at times that they are simply speaking in clichés, trite truisms, and all too well-worn phrases. But clichés and truisms should not be underestimated or discounted—they often highlight recurring cultural themes. Ultimately, our familiarity with many of the phrases used by these mothers is a measure of the deep and pervasive power of the ideology of intensive mothering and the extent to which all of us recognize at least portions of its logic.

THE RESPONSIBILITY OF INDIVIDUAL MOTHERS

Christina is a working-class mother with two young children. In an arrangement that is an exception to the rule, her husband stays at home to care for the kids while Christina goes out to work at a full-time job. Yet even though she is the sole breadwinner, Christina is sure that the children are still primarily her responsibility.

Christina started working outside the home because her husband had quit his job in a fit of anger and had trouble finding another one. She was able to get a job that paid "a lot of money," so, as she puts it, "I really didn't have much of a choice; I didn't feel as if I had any other options." In fact, she admits, the time she spent as a stay-at-home mother nearly drove her "out of her mind"; she was quite happy to go back to work, to talk to people, and to use her brain. But she also emphasizes that she would have stayed home indefinitely if her husband hadn't left his job. She agrees with her husband that staying home to care for the kids is basically her obligation. She explains:

> If anyone cooks, I cook. And if anyone cleans, it's me, although I insist that the kids help me. So I still have all of the traditional roles, I just don't do a very good job at them 'cause I don't care. [She laughs.] I'm just doing what [my husband] expects.

Christina not only does all the cooking and cleaning but also visits the kids on her lunch break each day and takes responsibility for them in the evenings and on weekends. Although she clearly works a double shift, she tells me she is happy with this arrangement. It keeps her husband from feeling overburdened, it provides the family with medical benefits and some financial security, and it allows her to get out of the house. Nonetheless, she tells me, "The only reason I am working is *money.* Because, if that weren't so, I would have the obligation to stay home with my kids. That's just the way it is." Women's first obligation is to the children and the family, Christina believes; mothers should work for pay only if family finances absolutely require it.

Although Christina's situation is unusual and Christina herself maintains what seems to be, by most mothers' standards, an exaggerated sense of her own responsibility, her story points to the widely shared belief that women are primarily responsible for the care of home and children. In mothers' own accounts, they take primary responsibility for *every* child-rearing duty—including watching, feeding, disciplining, cleaning up after, and playing with children—in well over half the families in my sample. Mothers take primary responsibility for feeding and cleaning up after the children in four-fifths of these households. Those duties not performed primarily by the mother are generally reported either as "shared equally" with male partners[2] or as done by paid workers or other household members. There is not a single household in which fathers or male adults take responsibility for all child-rearing tasks, and men rarely take primary responsibility for any single child-rearing duty. Even nannies, housekeepers, au pairs, day-care providers, and others are more likely than male household members to be the main persons looking after the child. These findings become even more striking when it is noted that the mothers in my sample, on average, spend *four times* the hours men do as the primary caregivers. Specifically, the average number of hours that these mothers said they take primary responsibility for watching over the kids each weekday was 8.9. Their male partners, on the other hand, do the same for an average of 1.9 hours each day. One national study of married couples confirms this disparity: on average, mothers' caregiving accounts for 74 percent of the total hours spent in direct child care (excluding the housework and planning associated with child care), and fathers' caregiving accounts for the remaining 26 percent (Ishii-Kuntz and Coltraine 1992).[3]

Although nonparental caregivers are more likely than fathers to take responsibility for child rearing, few mothers consider such arrangements an acceptable solution. Older siblings, who in other times and places served as central alternate caregivers, are not considered viable, since most mothers feel older children should be at least fourteen before being entrusted with the care of younger children for any significant period. In addition, by the time these siblings are teenagers, most of them are at school much of the day and spend their remaining time (appropriately, their mothers say) in leisure activities, doing homework, or working for pay.[4] Adult relatives are sometimes acceptable substitute caregivers, but mothers do not always consider relatives competent caregivers, and also worry about asking them to do the job for free at the same time they fear they would transgress implicit rules for appropriate familial behavior if they paid them.

Nannies, au pairs, women who provide family day care, and employees of day-care centers are not considered wholly suitable substitutes for mothers either.[5] While in many cases such caregivers spend nearly as many hours with the children as mothers do, and many mothers think this is fine, almost all mothers also feel the need to compensate by ensuring that they themselves spend a good deal of "quality time" with their children (as Jacqueline and Rachel, for instance, testify). More important, as several mothers told me, paid caregivers can never *fully* substitute for mothers because they simply don't have the commitment to and love for the children that mothers have. As one mother put it,

> Yeah, I think I'd be the best one to take care of them. Even with their dad, I think I'm the best one. I think I'm the only one that would be able to give them a fair yes or no and be able to explain why yes and why no. As far as anybody else, *it's not their blood,* it's not theirs, their children, in order to take out the time to provide the yeses or the nos or the maybes or the playtime or the time playing when you're feeding them or choosing what they're going to wear to make them look nice, you know.
>
> I mean, you can pay somebody to do it, but to them it's a job, and it's like, "as long as they're clean." I know people that have taken care of kids and leave them in a corner watching TV all day. *They're not worried about the child, they're worried about the money.* So it's like, I think I'm the best one to take care of them. (Emphasis mine)

Blood and bonding, it seems, are thicker than money.

Mothers tell me that the only person other than the mother who *potentially* has the necessary commitment, love, and time for the children is the father. Yet not one of the mothers in my sample, including Christina, suggested that fathers should give up a paying job to stay home with the children. Most mothers *do,* on the other hand, argue that fathers should take more responsibility for caring for the kids than they currently do.[6] At the same time, however, most mothers also believe that there are problems with asking men to do more. . . .

But, more often than not, the primary reason mothers hesitate to ask their partners to do much more of the child rearing is not so much that they consider these men dangerous but that they consider them incompetent.[7] A lot of men, it seems, can't handle the simplest required child-rearing tasks: "The way [my husband] takes care of [our son] is, he's like, I think he's *mindless,* to tell you the truth. He won't wipe his face off, he won't wash his face, he won't put his shoes on. I have to tell him to feed him or whatever." Sometimes, after a mother has asked a man to watch the kids, she discovers that his way of watching them is far from what she had in mind. For instance:

> Sometimes when I come home I can tell that [the kids have] been doing things that they weren't supposed to, like they've been in the bathroom, and things that they've gotten into, they've gotten hold of a pen and I see coloring on the chair, or somewhere. [Their father] wasn't watching them, maybe involved in a program or something like that. And that scares me because if he's not watching them right here, what if they go outside or on the street or, you know? . . .

In all this, mothers indicate that they understand very well this aspect of the ideology of intensive motherhood. Although fathers, paid caregivers, and others may help out, in the end it is mothers who are held responsible and who understand themselves as accountable not only for keeping the kids fed and housed but also for shaping the kinds of adults those children will become. And in keeping with this sense of responsibility are the methods mothers use to ensure the child's proper development.

INTENSIVE METHODS

No other question I asked of these mothers evoked as deep an emotional response as the question "How would you feel if you never had children?" None of my prior research prepared me for the intensity of feeling mothers expressed at the mere thought of not having children. Nearly one-quarter of the women I talked to actually cried when I asked them this question.[8] And the answers of nearly all mothers—"lonely," "empty," "missing something"—were stunningly consistent.

Elaborating on their feelings of emptiness, of loneliness, of missing something, mothers point to a number of factors. Sometimes mothers associate these feelings with their desire to "make a mark on the earth" and to "create an extension of themselves and their love." More often, these feelings are said to be connected to the importance of creating a family. For instance:

> I've always wanted to be a mother and I don't think the thought has ever crossed my mind of not having children. And I think, if I didn't, something would be missing. I don't know, but I feel like that was such a big part of my life as a child, always watching my brothers and sisters and always a baby in the house. . . . Just that sense of family, it's very important to me. . . .

Just as the child gives nonjudgmental, noncontingent, absolutely accepting, unconditional love, a child deserves and requires this sort of love in return. The giving of this love, mothers tell me, can be very time-consuming:

> Sometimes I just hold [my daughter] all day. And I would think, "Boy, this is the most important thing I have to do today and it's really important and I'm not going to get to the post office and we're going to have pizza for dinner. This child really needs this and that's what I'm going to give her right now." And it was a big switch from coming from working [at a paid job] and feeling like I could achieve twenty-five things in a day and [instead] all I did was hold my baby that day. But I came to realize how important that was. I think children really have a right to that.

Constant nurture, if that is what the child needs, is therefore the child's right—even if it means the mother must temporarily put her own life on hold. . . .

The love that mothers feel for their children and the commitment that love engenders are crucial foundations for another basic tenet of intensive mothering: the child-centered nature of socially appropriate child rearing.

Child rearing is child-centered, first, in that the child is the center of attention and the child's needs come first:

> I pay a lot of attention to my daughter. I make it very clear to her that she's very important. If we're in the middle of something and the phone rings, I don't always answer the phone. I take time to be with her and to listen to her and to do what she wants to be doing. . . .

In your youth, you are allowed to be selfish. But when you become a mother, you know that the children's needs come first.

Making the child's needs a priority is connected to the second way in which child rearing is child-centered. A good mother will, as Spock (1985: 200) puts it, "follow the baby's lead." Following the baby's lead, mothers argue, is (at least) a three-step process. First, it involves recognizing that children are people too:

> The idea is that your child is a person, totally deserving and capable of respect, giving and receiving, and that you talk to your child and you treat your child, [and] you explain to your child as you would any other person.

Not only do you need to respect your children as human beings, you also need to respect them as *individuals*. This means that a good mother will take the time to get to know the particular interests and desires of her own unique child (as one mother put it, "You learn your kid"). Above all, this process involves listening to the child. . . .

But respecting them, listening to them, and determining their interests and desires are not enough. As the third step in the process, the good mother responds to and acts upon what the child seems to be requesting. This means that the mother allows the child to control the process of child rearing in line with his or her needs and desire. . . .

The child-centered nature of socially appropriate child rearing is what makes it so labor-intensive. Of course, raising a child necessarily requires work; ministering to the physical and emotional needs of a small and dependent being takes time. But the ideology of intensive child rearing requires a good deal more of the mother than simply ensuring that the minimal requirements for affection and physical sustenance are met. Just as mothers understand the development of child-centered, permissive methods as the creation of recent generations, so too do they recognize that child rearing is more labor-intensive than it once was. . . .

Mothers share much of the ideology that makes the process of raising a child labor-intensive. Working-class, poor, professional-class, and affluent mothers alike nearly all believe that child rearing is appropriately child-centered and emotionally absorbing. And, practically speaking, this common attitude means they understand that good child rearing requires the day-to-day labor of nurturing the child, listening to the child, attempting to decipher the child's needs and desires, struggling to meet the child's wishes, and placing the child's well-being ahead of their own convenience. All this is clearly time-consuming and labor-intensive. But, as noted in the previous chapter, it is middle- and upper-middle class mothers who are the most likely to take the labor-intensive tenet of intensive mothering to the extreme. In negotiating with the child, explaining to the child, reasoning with the child, apologizing to the child, and using methods meant to ensure internalized self-discipline, these mothers take the lead. . . .[9]

Finally, according to the ideology of intensive mothering, appropriate child rearing is expensive. First, for those who can afford them, there are the costs of the tumbling classes, the swimming and judo and piano and dancing lessons, the child psychologists, and the special child-centered outings and vacations. For many, there are also the costs of the child's education, from preschool through college. Then there is the expense of child-maintenance accessories—the toys, the books, and the designer fashions—all of which go well beyond the already high price of simply providing for the child's physical needs.[10] It is, in short, quite expensive to "follow the child's lead," to provide them with what they desire, and to purchase all the goods necessary to ensure what the experts understand as healthy development.

One must also take into account the lost wages involved. The majority of stay-at-home mothers in my sample left their previous jobs specifically because they felt it was in their children's best interests for them to do so. A large number of paid working mothers stayed at home or cut back their paid work hours during the time their children were infants. A few mothers have managed to dovetail their schedules with their husbands' (as Spock suggests), so that at least one parent is home with the child for the maximum time possible. Over half the professional-class, paid working mothers in my study have reduced the number of hours they devote to their careers in order to have more time with their children. All of this is costly. And when you add to this the cost of hiring someone to care for the child while the mother is away plus the cost of assuring that the child receives the best paid care possible, the high price of appropriate child rearing becomes clear.[11]

What all this adds up to, of course, is child-rearing methods that are child-centered, expert-guided, emotionally absorbing, labor-intensive, and financially expensive. And it is the individual mother who is ultimately held responsible for assuring that such methods are used.

NOTES

1. I wish to make it clear that my emphasis in this chapter is on *ideas* about mothering rather than on practices. What mothers actually do may be quite different from what they say.

2. In many cases, the primary male adult in the household is not the husband of the mother or the biological father of the children. He may be a father, brother, or other relative or a friend or lover of the mother. But in almost all the households in my sample, the male adult present was the husband of the mother and took the role of father to the children. Given this, I try to alternate their titles and hope the reader will bear with me.

3. One international study by the High/Scope Educational Research Foundation found an even more dramatic difference, with American mothers, on average, taking charge of watching over children for 10.7 hours each day, compared to fathers at .7 hours, both parents at .9 hours, and care by others at 3.7 hours. These American mothers spent more hours as primary caregivers than mothers in any of the other countries in the study, which included Belgium, China, Finland, Germany, Hong Kong, Nigeria, Portugal, Spain, and Thailand (Owen 1995).

Another indirect indicator of gender differences in responsibility for child care is the fact that of the one in five United States households managed by a single-parent, 85 to 90 percent are headed by women (Sorrentino 1990).

4. Although the increasing number of teenagers who work for pay seems to be an indication that older children are taking on more responsibility and that parents may be requiring more of such children, the fact is that the majority of these teenagers do not contribute the money they earn to the family economy. Instead, they use it to buy clothes and cars and to pay for leisure activities (Waldman and Springen 1992). The higher incidence of teenage employment, in this sense, may simply be an indication that teenagers are becoming increasingly feverish consumers.

5. For the uninitiated, mothers distinguish between commercial day-care centers, which employ a staff of caregivers and occupy a physical space used exclusively for child care, and family day-care arrangements, in which the caregiver (usually a mother herself) attends to other people's children in her own home. Nelson (1994) provides an interesting analysis of the tensions faced by family day-care providers who are simultaneously mothers and paid workers in a setting that is simultaneously home and the site of work.

6. Two-thirds of the mothers in my small sample said that they wanted their husbands to do more; the rest said they did not. The paid working mothers were actually much less likely than the stay-at-home moms to say that their husbands should do more. Although this might be partially explained by the fact that the husbands of employed working moms generally spend more time caring for their children than the husbands of mothers who stay at home, none of these husbands actually do half the work. If we calculate the work hours of a mother who has paid employment against a mother who stays at home with the kids, the former is the one who still suffers the larger "leisure gap" (see Hochschild 1989).

7. The notion that fathers can be dangerous is related to the belief that men in general tend to be more dangerous than women. When asked about their gender preference in making the choice of alternative caregivers, four-fifths of the mothers said they would prefer a woman and the rest said they had "no preference"; *none* said they would prefer a man. Although some seemed to base this answer on their belief that men are incompetent, more often than not it was connected to the notion that men can be dangerous. One mother explained her preference for a woman caregiver in this way: "Because there's so much in the news about men doing stuff to little girls. . . . It's not that I don't think that a man wouldn't be good with kids. It's just that I don't need to have that extra worry of "Is he a weirdo?' " In other words, men are perceived as possible child molesters. It was never suggested that a woman might be a pedophile.

8. This emotional response becomes even more striking when one realizes that this question was asked just five minutes into the interview. Although I had talked to these mothers by phone several times before our meeting, I was still a stranger to them, and we had had little chance to talk about kids and child rearing. Yet their feelings about life without children were so strong that, as I say, a number of them cried spontaneously in front of me, a stranger. As many of them told me, no childless person can fully understand how deeply they feel this.

9. Thus, the quotes and stories in the following discussion of labor-intensive techniques (up to, but not including, the treatment of paid caregivers) are all taken from such mothers. However, a number of working-class and poor mothers also mentioned the use of reasoning, negotiation, explanation, distraction, and time-out. The difference is the extent to which these two groups of women expounded upon such methods; it is therefore primarily a difference in degree rather than in kind.

10. The average couple whose oldest child is under six years old spends 10 percent more overall than the average couple with no children (Exter 1992). By one estimate, which excludes the cost of childbirth and college tuition, raising a child from birth through high school graduation will cost from $151,170 to $293,400, depending on one's income level. Thirty years ago, economists estimated that parents would spend between $13,408 and $69,333 (Wright 1991). Even after adjusting for inflation, this is a dramatic increase.

For discussions of the booming industry in providing children with designer toys, clothes, and accessories, see Groves (1990), Lawson (1990), and Strom (1991). Experts in the child-products industry estimate that sales rose by at least 10 percent per year from 1988 to 1991, while the nation's birthrate (after years of decline) was rising only 3 to 4 percent (Strom 1991). Working-class and poor women are less able to afford such fashions and accessories, but this does not mean that they are always able to ignore their children's demands for them.

11. Even though, as I have noted, hired caregivers are very poorly paid, child care eats up a good percentage of family income, especially, of course, in poor and working-class families. On average, those whose family income is under $15,000 a year spend 23 percent of it on child care; those making $15,000 to $49,999 spend between 12 percent and 7 percent, with the percentage falling as the income rises; and those earning over $50,000 spend 6 percent of their income on child care ("Who's Minding the Children?" 1993).

In fact, more than one mother noted that the cost of child care eats up most of her income from paid work. It is also significant that these mothers tend to speak as if such costs appropriately come out of their salaries rather than their husbands' (e.g., "after child care costs, I hardly make anything").

REFERENCES

Exter, Thomas G. 1992. "Big Spending on Little Ones." *American Demographics* 14 (2):6.

Groves, Martha. 1990. "Mothers of Invention." *Los Angeles Times,* September 5.

Hochschild, Arlie. 1989. *The Second Shift.* New York: Viking.

Ishii-Kuntz, Masako and Scott Coltraine. 1992. "Predicting the Sharing of Household Labor: Are Parenting and Housework Distinct? *Sociological Perspectives* 35(4):629–47.

Lawson, Carol. 1990. "For Stylish Babies: Zebra-Skin Diapers and Neon Strollers." *New York Times,* October 11.

Nelson, Margaret K. 1994. "Family Day Care Providers: Dilemmas of Daily Practice." Pp. 181–210 in *Mothering: Ideology, Experience Agency,* edited by Evelyn Nakano Glenn, Grace Chang, and Linda Rennie Forcey. New York: Routledge.

Owen, Kelly. 1995. "U.S. Dads Lag in Child-Care Duties, Global Study Finds." *Los Angeles Times,* February 19.

Sorrentino, Constance. 1990. "The Changing Family in International Perspective." *Monthly Labor Review* 113(3):41–56.

Spock, Benjamin. 1985. *Dr. Spock's Baby and Child Care.* New York: Pocket Books.

Strom, Stephanie. 1991. "Creating the Well-Groomed Child." *New York Times,* July 6.

Waldman, Steven and Karen Springen. 1992. "Too Old, Too Fast? Millions of American Teenagers Work, but Many are Squandering their Futures." *Newsweek,* November 16, p. 80.

Wright, Jeanne. 1991. "The High Cost of Kids." *Los Angeles Times,* November 12.

"Who's Minding the Children?" 1993. *New York Times,* January 28.

33

African-American Mothers and Gang Daughters

WILLIAM B. BROWN

Family forms differ between the dominant culture and alternative groups in society, and Brown examines some of the challenges faced by inner-city African American mothers trying to raise their gang-affiliated daughters. The difficulties these mothers must communicate to their daughters about what it means to grow up in America as African American women, doubly disadvantaged, are poignant, as is the necessity they must assume of shouldering the responsibilities for themselves and their children, unable to rely on the men from their community who father their children. While they clearly depart from the model assumed for the white middle-class intensive mothers, these women are shown to be strong and vibrant role models for their daughters, forging meaningful maternal bonds. How does the problem of being doubly disadvantaged, as is the case for many African American women born into poverty, negatively affect life chances? Is there anything that can be done about this? Do the mothers in this study give you any ideas for social change?

The present study, which centers on the relationships between African-American mothers and their gang-affiliated daughters, alludes to individual cases, and these cases are analyzed in their larger social context.[1] One of the purposes for writing this paper is to provide these young African-American women and their mothers with an opportunity to share some of their life experiences, hopes, dreams, and concerns about the future with interested readers. This paper will explore the relationships between these young women and their mothers, and demonstrate how these young women's gang affiliation affects those relationships. It will also show how these mother-daughter relationships impact the daughters' gang activities and their individual behavior within the context of the gang.

THE NEIGHBORHOOD

The participants of this study live in a neighborhood which occupies a four-block square in east Detroit located about one and one-half miles south of Eight-Mile Road. For those who are not familiar with the Metropolitan Detroit area, Eight-Mile Road—sometimes dubbed the Mason-Dixon-Line of Michigan—is the northern boundary that separates the city of Detroit, which is populated mostly by African-Americans, from its mostly white suburban neighbors. Although many African-Americans have moved out of Detroit and into some of the outlying suburbs, along with most of their White counterparts, Eight-Mile Road continues to be "guarded" by suburban police departments with a tactical agenda that appears to be designed to maintain social class and racial integrity for their respective suburbs.[2] Thus, the northern migration of African-Americans is, in part, curtailed through harassment and other forms of Constitutional rights violations (e.g., suburban police detaining African-Americans without probable cause or suspicion beyond racial bias) by social control agents. For many African-Americans, and indicative of the 1940s, movement into neighborhoods that are predominantly White continues to be stymied by predominantly White police.

Like many other inner city neighborhoods in America, this neighborhood is characterized by poverty, high unemployment rates, underemployment, deteriorating homes and businesses, vacant lots and abandoned buildings, and underfunded, mismanaged, and often neglected schools. Frequently, one can see graffiti on buildings, houses, stop signs, and streetlight poles. Some call it evidence of criminal or delinquent behavior, but many of the youngsters in this neighborhood refer to it as a way of expression. One youngster referred to the graffiti in this neighborhood as "us poor people's newspaper. It tells us what's going on and what to look out for." On one boarded-up building, just north of Seven-Mile Road, this "newspaper," written in red paint on a plywood "canvass," proclaims, "REVOLUTION IS THE ONLY HOPE FOR THE HOPELESS." . . .

MOTHER-DAUGHTER RELATIONSHIPS AND THE GANG

The young women in the present study are all gang members. Their ages range from 14–17 years. They are not just self-proclaimed gang members. They also are not gang member "wannabees." They are "hardcore" gang members who chose to be beat into the gang rather than to be sexed into the gang. This was a conscious decision made by each of these young women, since part of what they all were seeking through their gang affiliation was respect. Tanya,[3] who is 17 years old, explains, "Any bitch can give it up to become accepted. But then her commitment is as loose as she is. I was beat into the family[4] so that my commitment mean something." Watisha, who is 16 years old, says, "I got beat into the family because, well, I wanted to be a family member, not a family whore."

Some of the more common characteristics these girls share, outside of their gang affiliation bonds include: (1) Socioeconomic status: All of the girls live in families that are situated near the poverty line. Sometimes these families qualify as officially poor, while other times they are on the margin of poverty. (2) Education: All of the girls are currently in school. They all attend the same high school in Northeast Detroit which has a predominantly African-American student body. (3) All of the girls have experienced some negative contact with law enforcement. Several girls have been arrested several times for offenses such as vandalism, truancy, theft, and alcohol and drug violations. (4) Each of these girls share a history with all African-American women; they all must contend with a society that devalues women in general, and women of color even more. Thus, these girls are forced, in order to survive, to learn how to exist in "interlocking structures of race, class, and gender oppression while rejecting those same structures" (Collins 1991: 124). This is the historical linkage between these girls and their mothers. Like many of the participants in Alex Kotlowitz's (1991) study of gangs in Chicago, these young women, and their mothers, are survivors in the "other America."

The mothers in this study can be defined as simply mothers. They are women who, for whatever reasons, find themselves situated, socially and economically, in dire straights. They all work and several also work part-time jobs in addition to their full-time jobs. These mothers attempt to maintain households, which include preparing meals, doing the laundry, paying the household bills, and trying to provide direction for their children and, all but one mother accomplish this alone. All of the mothers in this study share a common bond. They are, by relative poverty standards, poor. They are living illustrations of the "working poor" in America. . . .

The relationships between these girls and their mothers are strong. Mutual respect and understanding exist between mother and daughter in all instances, although, to be certain, there are disagreements and, in some cases, tempers are revealed during the course of their interactions. All of the mothers explained that they try to raise their daughters much the same as they were raised by their mothers. They all have tried to convey to their daughters the urgency of establishing a strong sense of independence. According to Campbell (1984), the importance of this independence is linked to the girl's relationships with males, but only one mother in the present study alluded to this notion. Rather, most of the mothers try to stress upon their daughters that independence is necessary for their daughter's survival in a "society" that does not look favorably on women of color, and reaches beyond female-male relationships. As one mother notes,

> Any Black woman can have her a relationship with a man if she wants to. She needs to keep her independence though, because Black men have a lot of problems trying to support a family . . . I've had several so I know. No, nowadays a woman has got to be independent so that she is free do to what she's got to do. She has to be free to go to work, and that ain't always easy. She's got to learn how to work out there on her own in all kinds of relationships with other people. She's got to try to keep her independence when she has to deal with lots of White people cause sometimes they can hurt her as much as a Black man . . . I've been hurt by both, so I know what I'm talking about. . . . It ain't always easy, but she's got to do it to survive. . . .

The mothers in the present study are aware of their daughters' gang affiliations, but do not encourage them to become members of a gang, nor do they turn their heads and condone their daughter's illegal activities within, or outside, the context of the gang. All of the mothers are frightened over the potential dangers of their daughters gang involvement, and in all cases, there is a fear that someday they will receive a telephone call informing them that something has happened to their daughter. Denise's mother, Carrie, points out, "I am terrified that someday that phone is going to ring and I am going to find out that something has happened to her." Carrie tries to discourage Denise from becoming more involved in gang activities, but at the same time recognizes that the gang serves as a surrogate parent outside the home. Carrie understands that, on one hand, the gang provides a certain amount of protection for her daughter, but on the other hand, she expresses concern that, "The gang can also be the cause of something bad happening to Denise." The fear of something bad happening to her daughter is, as Carrie states, "always with me."

The mothers all concede that survival is enhanced, in many cases, through numbers and solidarity. Thus, the gang, which is often bound by solidarity, is a form of survival. Sophie's mother, Gladys, who also is concerned about her daughter's safety, is aware of Sophie's gang affiliation. Gladys readily admits that African-American women must be stronger than men and women who are members of other racial or ethnic groups. "I try to impress on Sophie that she has got to be strong and continue to fight and never give up. I can only pray that she has learned this," says Gladys, who quickly adds, "I think I have given her values that will help her, and God will have to guide and protect her." Gladys works long hours. She has problems with her back, but cannot seem to get out the "minimum-wage-rut." Four years ago Gladys attended a community college because, "I wanted to learn something so that I could get a job that didn't make my back hurt, but I had to stop because I couldn't afford to go to that school no more." Clearly, Gladys is the symbol of the strength for her daughter.

The girls express respect for their mothers' attempts to provide, as best they can, for their families. Respect is an important plank in these mother-daughter relationships. This does not, however, mean that the girls approve of the methods adopted by their mothers to support the family. Many of the girls expressed dissatisfaction that their mothers spend so much time working, and so little time with their families. But most seem to understand that this is their mother's way to insure survival. Latrese, a 15-year-old gang member, complains that her mother is rarely at home, and states, "I know my mother loves me, and all that, but she seems obsessed with working long hours. I hardly ever see her except on Sundays. Sundays she usually spends sleeping a lot." Charlene defends her long working hours, and points out that,

> I know what it is like to go without. I'll do whatever it takes to provide for Latrese. I know she gets upset with me for working so much, but that girl has to learn that it takes a lot for a Black woman to make it in this world. If I don't work there's nobody else that's going to pay the bills for us. She's got to learn to accept that. . . .

All of the girls indicate that they do not share with their mothers those experiences which involve illegal behavior. Several of the girls state that sharing those experiences would only hurt their mothers, while a couple of girls indicated that their mothers would not understand the rationale behind their illegal behavior. Similar to Padilla's (1993) findings in Chicago, several girls indicated that their illegal behavior tends to solidify their commitment to the gang. Like Campbell's (1984) findings among girls in New York's youth gangs, many of these girls believed that they had few opportunities in the legitimate world, largely due to the powerlessness associated with their underclass membership, which is compounded by being African-American. Angela, who is 15 years old, remarks, "Tanya says I got a good business head, and that's what I try to do. I try to do good business." Angela is one of three girls in the present study who has been arrested several times. Two of those arrests are linked to Angela's involvement with drugs (marijuana sales). "I sell weed to get some things I want, and to help my momma," says Angela. "Momma don't want me selling no weed," she states, "but I'm going to still sell weed because I have to." Responding to a question about the motivation for drug involvement, she says, "I am Black, and my family don't have nothing. Selling some weed lets me get some things that I want, and lets me help out at home. . . . Momma don't know I'm selling right now so don't you go and tell her." Sandra, Angela's mother, has a full-time and a part-time job. She is not aware of Angela's current involvement with drug sales. Kitchen (1995) explained aggressive behavior among female gang members as a residual of drug dealing. The behavior of many of the girls in the present study is best described as assertiveness rather than aggression, and this assertiveness appears to be linked to survival rather than extensive drug sales involvement.

Selling drugs is not a high-priority activity for most of the girls in the present study, yet most are quite assertive. As Sophie proclaims, "My mamma taught me to stand up for myself. I don't take no shit from nobody." Fishman (1988) found girls' independence, assertiveness, and risk-taking characteristics to be associated with their socialization process. In the present study, data support the notion that independence, assertiveness, and risk-taking are characteristics that originate within the mother-daughter relationships rather than from the girl's gang involvement. These characteristics ought not to be judged as either good or bad, rather, it is my contention that they are artifacts of survival within a society that devalues the existence of some of its members. The gang provides a testing-ground to pursue and develop these characteristics.

The young women's behavior, in the context of the gang, is greatly influenced by many of the lessons taught by their mothers. Those lessons include family responsibility, loyalty, survival, and love. Each of these young women has assumed the responsibility of watching out for fellow gang members, and their sense of loyalty to other gang members runs deep. In the process of applying these lessons they believe they are increasing their chances for survival. On the other hand, the gang affiliation shared by these young women place a strain on the relationships with their mothers. One strand of that strain rests in the obvious concern for the safety and well-being of their daughters. Another strand of tension for the mother-daughter relationships is in the daughter's insistence to keep much of their gang activities secret. Most of the young women indicated that they felt keeping things from their mothers violated the trust factor in their mother-daughter relationships. Many of the young women rationalized keeping many gang-related activities from their mothers in order to spare their mothers concern or worry. Other felt that keeping gang-related activities from their mothers simply reinforced some sense of individual independence.

THE MEANING OF FAMILY—AT HOME AND IN THE STREET

The definition of "family" in many African-American communities, a definition shared also by many Native-American communities, has, as Stephanie Coontz (1997: 120) points out, "traditionally included those who nurture and help to support a child, regardless of the household residence or degree of biological relatedness." Repeatedly, throughout the course of this study, participants introduced the notion that the foundation of a family is shored-up with trust and loyalty. The definitions of these concepts are summarized by Tanya, who says, "Trust is when you know the shit is going to hit, but you know someone is always there to be for you. Loyalty is like when you always put family first."

Denise, who is 16 years old and the oldest of three children in a family where her mother is the only adult present, says, "My mother is someone that I can always count on. Sometimes I can't talk to her about some of the things that I do, but I can always count on her. She has always been there for me, and I will always go down for her." Denise's mother, Carrie, strongly champions her daughter's words, "I know Denise is no angel, and I know that from time-to-time she is going to step out across the road, but if something happens to me, she will take care of her brother and sister because they are family." Tanya, the oldest female gang member represented in the present study (17 years old), is very close to her mother, Gloria. "My mother has had a hard life. She lost her husband, my father, ten years ago, and my brother was killed in this neighborhood almost five years ago." Continuing, Tanya says, "If something happens to my mother, well, then I will just have to take care of them kids my own self. My family is my responsibility." Gloria seems confident that Tanya will look after her younger brother and sister if something were to happen to her. . . .

The meaning of family transcends much of the popular belief about gangs and this meaning is adopted by most members of the gang. All of the gang members in the present study refer to their gang affiliation as a family affiliation. In fact, most participants reject the term gang completely, and argue that this term is used by various social institutions (e.g., mass media, criminal justice practitioners, and many researchers), because they want to focus only on the negative activities of the members. Sophie argues that most people who use the term gang, "want to have an excuse to fuck with us. A lot of them know that we love and protect each other, but they want to make us out as being something evil. Saying that we are a gang seems to give them an excuse to put us down." Chandra, the youngest member of this study, says "Being in the gang is like always being with family. One family is in my home, and the other family is in the streets. Sometimes I have a hard time seeing them as being different, because I am loved and protected in both families." All of the participants reveal that the associations and relationships within the context of the gang are, in most instances, replications of associations and relationships between family members in their own households. The gang provides both the opportunity for members to nurture and be nurtured.

NOTES

1. It must be clarified at the outset that this paper is about African-American mothers and their gang-affiliated daughters. Many of the mothers included in the present study maintain contact with their former husbands and the fathers of their children. Likewise, many of the girls in the present study have existing relationships with their fathers who no longer reside in the household. Many of the fathers no longer have significant influence over the activities of the mothers and their daughters. This should not be construed as any sort of cultural pathology, rather, it is typically a result of structural realities which, since the days of slavery, have devalued African-American males. Several of the more common structural explanations include high unemployment rates, underemployment, and a criminal justice system which isolate many of the African-American males through incarceration. However, in some cases, the husband/father of the families represented in the present study do make an effort to assist the family. This assistance is often in the form of watching children, helping with household chores, etc. The husband-fathers' economic situation, however, rarely provides for financial assistance which is often due to structural reasons mentioned previously. Elizabeth Higginbotham (1983), Derrick Bell (1987, 1992), Earl Ofari Hutchinson (1990), Zweigenhaft and Dornhoff (1991), Lewis R. Gordon (1995), and Robert C. Smith (1995) provide substantiated data to support the notion that structural factors, rather than culture, have had a profound and negative impact on the African-American male's role in the family.

2. In an unrelated study of suburban police activities, this author has witnessed, on numerous occasions, suburban police officers making routine "harassment" stops of vehicles driven by African-Americans. Usually, the driver was a young African-American male, and, in most cases, tickets not were issued. During one instance, I asked the officer why he had stopped a young Black male, he responded. "They [Blacks] know they don't belong here this time of night."

3. All of the participants' names are masked to assure confidentiality.

4. In most instances, the young women who participated in this study refer to their gang affiliations as family affiliations. In other words, they prefer to view themselves as family members rather than as gang members.

REFERENCES

Bell, D. 1987. *And We Are Not Saved.* New York: Basic Books.

Bell, D. 1992. *Faces at the Bottom of the Well.* New York: Basic Books.

Campbell, A. 1984. *The Girls in the Gang: A Report from New York City.* New York: Basil Blackwell Inc.

Collins, P. H. 1990. *Black Feminist Thought: Knowledge, Consciousness, and the Politics of Empowerment.* New York: Routledge.

Coontz, S. 1997. *The Way We Really Are: Coming to Terms with America's Changing Families.* New York: Basic Books.

Fishman, L.T. 1988. *The Vice Queens: An Ethnographic Study of Black Female Gang Members.* Paper presented at annual meeting of American Society of Criminology.

Gordon, L. R. 1995. *Bad Faith and Antiblack Racism.* New Jersey: Humanities Press.

Hacker, A. 1995. *Two Nations: Black and White, Separate, Hostile, Unequal* New York: Ballantine Books.

Hagedorn, J. M. 1988. *People and Folks: Gangs, Crime and the Underclass in a Rustbelt City.* Chicago: Lakeview Press.

Harrington, M. 1962. *The Other America: Poverty in the United States.* New York: The Macmillan Company.

Harrington, M. 1984. *The New American Poverty.* New York: Penguin Books.

Higginbotham, E. 1983. "Laid Bare by the System: Work and Survival for Black and Hispanic Women." In *Class, Race, and Sex: The Dynamic of Control,* edited by Amy Smerdlow and Hanna Lessinger, pp. 200–215. Boston: G. K. Hall.

Hutchinson, E. O. 1990. *The Mugging of Black America.* Chicago: African-American Images.

Kitchen, D. B. 1995. *Sisters in the Hood.* Ph.D. Dissertation, Western Michigan University.

Kotlowitz, A. 1991. *There Are No Children Here: The Story of Two Boys Growing Up in the Other America.* New York: Anchor Books.

Padilla, F. M. 1993. *The Gang as an American Enterprise.* New Brunswick, NJ.: Rutgers University Press.

Smith, R. C. 1995. *Racism in the Post-Civil Rights Era: Now You See It. Now You Don't.* New York: SUNY.

Zweigenhaft, R. L. and G. W. Domhoff. 1991. *Blacks in the White Establishment: A Study of Race and Class in America.* New Haven, CT: Yale University Press.

34

American Religion: The State, the Congregation, and the People

N. J. DEMERATH III

Several competing theories abound in America today about the state of contemporary religion. Some argue that we are experiencing an upswing in our level of piety, with greater adherence to religious membership and ideals among our leaders, greater levels of religious standards being applied to public policies, high levels of political activism in our churches, and a rise in the Charismatic, or Evangelical movement. Others feel that this religiosity does not penetrate much beyond the surface level, with public displays being morally correct, but private beliefs and behaviors advancing secularism. Our history as a nation founded by the Puritans and originally peopled by Protestants may be giving rise to greater national religious diversity, with the influx of immigrants from Catholic Latin America and Buddhist, Hindu, Moslem, and Islamic Asia. How we deal with this religious diversity will once again be a test of our country's cohesion and adaptability. Demerath addresses several of the pressing issues concerning religious observers today by looking at the different planes where religion is practiced. He considers our level of civil religion, current trends in the Protestant Church, and the level of religiosity actually felt and practiced by everyday Americans. Like many issues in religion, these touch on some controversial topics. You may want to reflect on how you feel about those discussed in the chapter and others mentioned here. What role does religion play in your life? Has your family, over the generations, become more or less religious? Why?

From *The Annals of the American Academy of Political & Social Science.* Reprinted by permission of Sage Publications, Inc.

AMERICAN CIVIL RELIGION: IS IT HOLDING?

One of the strongest identifiers of America as a religious society is the way its religion is publicly invoked and symbolically brandished. Our most important national holidays, such as July Fourth and Memorial Day, are religiously consecrated. Virtually every session of the nation's daily legislative business is prefaced by prayer. Both our coins and our politicians proclaim religious mottos. Our rites of passage—whether weddings, funerals, or presidential inaugurations—are marked with religious observance. Religious solace and supplication accompany every national crisis.

All of this is part of what Robert Bellah (1967) first termed America's "civil religion." In Bellah's view, this "Judeo-Christian" common denominator is a rich residue of a historical experience that has become a binding cultural force. The country is irretrievably religious at its roots and in its most luxuriant foliage. This is an important part of America's distinctiveness, since few other societies can boast such a natural melding of religion and nationhood.

Yet there are reasons to pause before accepting this portrayal. It is not clear whether this is an analysis of America's mythology or a contribution to it. On the one hand, our civil religion may not function quite the way it is depicted. On the other hand, other societies have their own versions of a civil religion, though some stretch the concept and its possibilities rather than merely illustrating it. Let me consider both complications in turn. . . .

Williams and I (1991) have explored the paradoxical tension between our heralded civil religion and our no less legendary "separation of church and state." Actually, these two seemingly inconsistent syndromes are strangely symbiotic. Each is a guard against the other's excesses, and each provides a countervailing assurance as a boost to the other's legitimacy. That is, we can indulge a symbolic civil religion precisely because there is a substantive separation of church and state in important matters of government policy; at the same time, our separation is never a total rupture because of the presence of overarching civil religious ceremonials. . . .

Meanwhile, America's civil religion has been pressured by more than its church-state separation. Our burgeoning pluralism includes increasing numbers of Muslims, Hindus, and Buddhists outside the Judeo-Christian tradition. Moreover, Richard Neuhaus (1984) has argued that religion has become conspicuously absent from our "naked public square," and Steven L. Carter (1992) has termed ours a "culture of disbelief"—though both are really lamenting the absence of more conservative religion in the national arena. In fact, Robert Wuthnow (1988) suggests that America now hosts not one but two civil religions—one liberal and one conservative—and this may vitiate the very point of civil religion as something both unitary and unifying.

James D. Hunter (1991) insists that religious cohesion has given way to "culture wars," although I have argued that this phrase is more incendiary than accurate given an American public huddled in the middle on key issues (Demerath and Yang 1997) and when one compares these expected abrasions of a democratic polity to the far more bloody conflicts of countries such as Guatemala, Northern Ireland, Israel, and India (Demerath and Straight 1997). Civil religion remains alive in the United States, though perhaps not well. . . .

AMERICAN RELIGIOUS ORGANIZATIONS AND CONGREGATIONALISM

Few images of the stereotypical American community fail to include at least one corner church with steeple. Whether the white, frame icons of New England congregationalism, the neo-Gothic edifices of the Catholics and Episcopalians, the contemporary "ring-a-ding God-boxes" of the Lutherans and Jews, the storefront sites of urban neighborhoods in transition, or the new suburban campuses of conservative Protestant groups, these structures dot the American landscape in profusion. They also loom large in the country's institutional sector. Religious congregations provide critical support mechanisms for their members as communities within communities. They account for the lion's share of the nation's charitable giving and community service, and they represent a major source of political influence and mobilization.

All of this reflects a distinctively American confluence of Protestantism's founding theology and our localized and democratic community ethos. When Protestantism broke from Catholicism, it did what every successful movement must, namely, stress a problem to which it had the solution. In this case, the problem was the individual's suddenly unmediated relationship to God and all of the loneliness and anxiety that this entailed. The solution was nothing less than the congregation itself as a source of both fellowship and reassurance. Of course, the congregation was especially important in the rural areas and small towns of Europe and America, where it was often the only gathering point. The United States' particular emphasis on frontier self-reliance and local democracy gave congregational life a further thrust that was later carried back to the new frontiers of urbanization.

It was especially important that the most sophisticated versions of congregational life occurred among the liberal Protestant denominations, with all of their higher-status membership resources. In fact, the very existence of a self-proclaimed liberal church is one of the truly exceptional characteristics of American mainline religion. Liberals in most other faiths around the world quickly move outside of their faith structures and shift all the way to full-blown secularism. Here, however, this shift was slowed by such liberal American denominations as the Congregationalists, Episcopalians, Presbyterians, and Unitarians. Much of the secret lay in organizational complexes that combined liberalized religion with an array of compelling social and service activities. Early on, the resulting organizational stability gave important standing and influence to the liberalism that these churches embodied. More recently, this same organizational legacy has helped to brake their slippage. . . .

DO PIOUS STATISTICS LIE?

It is not hard to find statistical evidence of the religious exceptionalism of individual Americans. Virtually any polling study of religious belief and behavior in various major countries will show the United States at or near the top in such matters as levels of church membership (close to 60 percent), weekly church attendance (better than 40 percent), belief in God (95 percent), and the experience of having encountered God (close to 75 percent) (cf. Ingelhart 1997).

Without actually pricking this balloon of national distinction, it is important to deflate it a bit. What is being compared in these studies is not just the religious penchants of differing individuals but the differing cultural expectations for religion and religiosity that apply from one society to the next. Thus it is paradoxical that participating in some form of American religion remains a generally compelling national norm even though we lack a formally established national religion to participate in. But disestablishment has put the onus on religious groups to develop at the grassroots level as expressions of community resolve molded to community needs. For many, church membership and church participation became an important part of community life itself. In addition, if there is no national establishment, there are surely local approximations, whether Catholic in the urban Northeast, Methodist in parts of the South, Baptist in much of the Southwest, and, of course, Mormon in Utah. Scholars disagree whether religious participation is highest in the context of a local religious monopoly or local religious competition (cf. Finke and Stark 1988; Blau, Redding, and Land 1997). However, there is little doubt that declaring oneself nonreligious is tantamount to an antisocial act in many circles.

Clearly, America's vaunted statistics of individual religiosity must be interpreted within this broader perspective. Without impugning the deep religious commitments of many American adherents, church membership does not necessarily entail regular church participation or personal religious commitment. Studies of belief in God are vulnerable to variations in the question and the context; in general, the levels of absolute belief decline when a respondent has more finely grained alternatives to choose from. One of the most significant, hence controversial, studies of individual religion in the last half-century involved a recent check to see whether people who claimed to have been in church on the previous Sunday could actually have showed up (Hadaway, Marler, and Chaves 1993). This study and subsequent refinements and replications show that actual levels of church attendance are less than half of those that are so widely cited. Are people lying? Perhaps a better way to put it is that many are telling the pollsters that they know what they ought to be doing and are anxious not to disappoint. Some respondents may also be wary of confessing nonbelief and nonparticipation to someone who may launch an evangelizing effort to enlist them to join the fold.

Just as context is important in America's putatively high levels of religious involvement, context is no less important in some of the low levels elsewhere. As we have already seen, the Judeo-Christian world is virtually unique in giving such emphasis to church or temple participation that it can become a central metric of religious behavior and devotion. To the extent that ritual and worship center on home and family in other faiths, they become more elusive. While one could always attempt to survey beliefs as opposed to behavior, this raises yet another source of troublesome variance. Christianity is very much a religion of the word; most other faiths are religions of the act where what one believes is far less important than what one does—not merely as a matter of ritual observance but also as part of one's overall ethical lifestyle.

Another problem in comparing levels of individual religiosity cross-nationally involves questions of religious identity. In many countries, religion has become more a matter of passive cultural heritage than active ongoing commitment. From Northern Ireland to Sweden and Poland and on to Israel, I found many respondents uneasy when I asked about their personal religiosity. However, when I offered the category of cultural Protestant, Catholic, or Jew, they brightened and clutched the label eagerly. Although they regarded themselves as beyond the reach of conventional religious rounds, their religious legacies resonated in other important ways.

There is no question that our combined principles of civil religion and church-state separationism still confer a stamp of distinctiveness. Much the same is true of our emphasis on religious congregational life and our stress on church and temple participation as a basis of individual piety.

REFERENCES

Bellah, Robert N. 1967. Civil Religion in America. *Daedalus* 96(Winter): 1–21.

Blau, Judith R., Kent Redding, and Kenneth C. Land. 1997. Ethnocultural Cleavages and the Growth of Church Membership in the United States, 1860–1930. In *Sacred Companies: Organizational Aspects of Religion and Religious Aspects of Organizations,* ed. N. J. Demerath III, Peter D. Hall, Terry Schmitt, and Rhys H. Williams. New York: Oxford University Press.

Carter, Steven L. 1992. *Culture of Disbelief.* New York: Basic Books.

Demerath, N. J., III and Karen Straight. 1997. Lambs Among the Lions: America's Culture Wars in Cross-Cultural Perspective. In *Cultural Wars in American Politics,* ed. Rhys H. Williams. Chicago: Aldine de Gruyter.

Demerath, N. J., III and Yonghe Yang. 1997. What American Culture War? A View from the Trenches as Opposed to the Command Posts and the Press Corps. In *Cultural Wars in American Politics,* ed. Rhys H. Williams. Chicago: Aldine de Gruyter.

Finke, Roger and Rodney Stark. 1988. Religious Economies and Sacred Canopies. *American Sociological Review 53*(Feb.):41–49.

Hadaway, C. Kirk, Penny Long Marler, and Mark Chaves. 1993. What the Polls Don't Show: A Closer Look at U.S. Church Attendance. *American Sociological Review* 58(Dec.):741–52.

Hunter, James D. 1991. *Culture Wars: The Struggle to Define America.* New York: Basic Books.

Ingelhart, Ronald. 1997. *Modernization and Postmodernization: Cultural, Economic, and Political Change in 43 Societies.* Princeton, NJ: Princeton University Press.

Neuhaus, Richard John. 1984. *The Naked Public Square: Religion and Democracy in America.* Grand Rapids, MI: Eerdmans.

Williams, Rhys H. and N. J. Demerath III. 1991. Religion and Political Process in an American City. *American Sociological Review* 56(Dec.):417–31.

Wuthnow, Robert. 1988. *The Restructuring of American Religion: Society and Faith Since World War II.* Princeton, NJ: Princeton University Press.

35

The Church, the Family, and the School in the African-American Community

ANDREW BILLINGSLEY AND CLEOPATRA HOWARD CALDWELL

Billingsley and Caldwell locate the church as one of the three most historically salient social institutions undergirding the African American community, in conjunction with the family and the school. These three function as expressions of the most basic African American cultural heritage and are responsible for the viability of the African American community. A major socializing agent, the church assists families in need and supports educational institutions. Thinking about some of the issues raised by Demerath in the previous article, why do you suppose the church remains a strong institution in the African American community? What are the ways that institutions are interconnected, and how does this operate in society in general?

This article focuses on the church as a social institution and examines how it interacts with and influences two other important social institutions in society (as identified by Moberg, 1962)—the family and the school. Historically, the church, the family, and the school are the three most critical institutions whose interactions have been responsible for the viability of the African American community (Roberts, 1980). The strengths of these three institutions are due in large measure to their function as expressions of the most basic values of the African American cultural heritage. These values include spirituality, high achievement aspirations, and commitment to family as enduring, flexible and adaptive functional mechanisms for survival (Billingsley, in press). . . .

From *The Journal of Negro Education*, 60(3), 1991. Reprinted by permission of Howard University.

The Black church continues to hold the allegiance of large numbers of African Americans and exerts great influence over their behavior. . . .

Why is the Black church so important? If the church is the institutionalized expression of the religious life of a people, as many sociologists generally believe, then the Black church is a powerful institution. Spirituality, according to Hill (1971), is one of the most distinctive features of African American culture. This is partially reflected in overt religious expression. A reanalysis of the data from the University of Michigan's National Survey of Black Americans (Billingsley, in press) shows that:

- 84% of African American adults considered themselves to be religious;
- 80% considered it very important to send their children to church;
- 78% indicated that they pray often;
- 76% said that the church was a very important institution in their early childhood socialization;
- 77% reported that the church was still very important;
- 71% attend church at least once a month; and
- nearly 70% were members of a church.

In the African American community the church is more than a religious institution. Lincoln (1989) describes the multiple functions of this institution as follows:

> Beyond its purely religious function, as critical as that has been, the Black church in its historical role as lyceum, conservatory, forum, social service center, political academy and financial institution, has been and is for Black America the mother of our culture, the champion of our freedom, the hallmark of our civilization. (p. 3) . . . .

THE AFRICAN AMERICAN FAMILY

The family is also a strong and functional institution in the African American community. As Hill (1971) explains, African American families are sustained by five major sources of strength including, notably, a strong religious orientation, flexibility of family roles, and a strong achievement orientation. Despite the strains on contemporary families these strengths remain evident. The majority of African Americans are part of family units. Indeed, as late as 1988, 70% of all African American households were family households (U.S. Census, 1988), albeit those households represented a highly diverse set of structures. Some of these were nuclear families, some were extended families, and some were augmented families. Further, some nuclear families were married-couple families without children, some were married couples with children, and some were single-parent families. Some Black households were headed by males and some were headed by females.

All of these variations are families in the sense that they are comprised of persons related to each other by blood, marriage, formal adoption, informal adoption, or by simple appropriation. The overwhelming majority of children, who constitute the major basis of family formation, live in family units. As late as 1988, 94% of African American children lived in families while less than 1% lived in nonfamily households (U.S. Census, 1988). . . .

THE SCHOOL

The school is also a highly valued institution for many African American people as expressed through high parental educational aspirations for their children (Willie, Garibaldi & Reed, 1990). For nearly half a century African American parents, their children, leaders, and organizations have been in the forefront of efforts to desegregate schools so that African American children could receive a quality education. In the process they have endured enormous hardships and sacrifices for limited gains. . . .

The importance of the school is not a modern phenomenon in the African American community. In a study covering the period 1865 to 1954, Johnson (1986) reports that African Americans historically have placed high value on education and on the schooling of their children:

> Black parents and the Black community from the era of the Reconstruction period in American history to the *Brown* decision (1954) have been interested in the education of Black children; have placed high value on their children's education; have maintained a belief that education is a method of benefitting from the opportunity of the American society; and that interest has been translated into action by providing schooling for Black children throughout the period covered by this study. (p. 199)

Shiloh Baptist Church (Washington, DC)

The Shiloh Baptist Church is a large inner-city church. Located in the center of the nation's capital, it is composed of more then 6,000 members representing a cross-section of working-class and middle-class city and suburban residents with a generous sprinkling of men (although the majority of its membership are women).

Like most Black churches, Shiloh is remarkably family-centered. This is reflected in both its privatistic functions in meeting the religious needs of its members and in its extensive communal functions in responding to the needs of the community. It celebrates the family at every opportunity and in all its infinite variety. On the religious side, special days and Sundays set aside to celebrate the family often constitute a week-long celebration. Mother's Day, Women's Day, Father's Day, Men's Day, Youth Week, Children's Day, and Family Day are only a few examples. The christening of children becomes at least a three-generation affair in which augmented family members are invited to take part. Single-parent families are not left out. The annual Single Person's Day recognizes that the church's singles organization is among the largest and most active organizations in the congregation. Moreover, the very name of the organization, "The Successful Singleness Movement," expresses its positive orientation. Like most other activities, this latter organization is not limited to church members, and its extensive community athletic clubs including tennis, bowling, and golf activities serve as social forums which often bring single people together.

However, it is Shiloh's community outreach programs directed at nonmembers that best reveal the family orientation of this church. To provide a home for many of its community activities as well as church organizations, the church, under the leadership of the late Reverend Henry C. Gregory, III, built the huge Shiloh Family Life Center at a total cost of $5 million. The center provides a wide range of activities and facilities for the whole family and for sectors of the family: a full-scale restaurant, full basketball court, racquetball courts, exercise rooms, banquet hall, meeting rooms, and a rooftop garden overlooking the neighborhood—all of which are open to the community. In addition Shiloh operates a nationally recognized child development center that has attracted numerous visitors including the president of the United States. It also conducts a parent-child mathematics and science learning center in addition to sponsoring a local branch of the Boys and Girls Clubs of America, the Boy Scouts, and the Girl Scouts. These programs provide opportunities for volunteers not only from the church membership but from nonmembers as well. They provide a major bridge between the social classes plus enormous opportunities for volunteers from local colleges, particularly students pursuing careers in education, social work, the ministry, or family studies.

Additionally, the Male Youth Health Enhancement Program the church sponsors has allowed the men of Shiloh to embrace about one hundred young males from the neighborhood, largely from low-income, single-parent families. The program, which has attracted foundation support, has a small staff. It is guided by a small advisory committee of parents, professionals, and lay persons, and is responsible to the board of directors of the Family Life Center, which is responsible to the senior deacon's board and the pastor. The male volunteers and the staff of the program work closely with the parents, who are organized into a parent's group that meets regularly. They keep in touch with the schools to monitor the progress of the boys. They also provide an after-school program each weekday, along with weekly seminars on life, drugs, sex, health, and African American history, and a wide range of local and regional field trips to scientific, educational, and cultural facilities.

The Congress of National Black Churches

Shiloh is not alone. Both in the Washington, DC, area and nationwide, myriad examples abound of the church, the family, and the school in dynamic interaction. The Congress of National Black Churches is an umbrella association of churches in the eight historically Black denominations. Under its leadership a number of outreach programs have been established. One is Project Spirit, a consortium of 15 Black churches in each of three cities, which was designed to develop a most innovative nexus among the church, family, and school. In each city—Oakland (CA), Indianapolis (IN), and Atlanta (GA)—5 churches of different denominations have come together to provide a program for parents and children after school and on weekends and holidays.

There are three components to each of these programs. The first component offers (a) after-school and Saturday tutorials, (b) skills development, and (c) life enhancement sessions for students of a local elementary school. Parental permission and participation are required for these activities. This component operates three to four hours daily and includes classes in social skills, reading, English, mathematics, science, and computers. Teachers are volunteers from the ranks of qualified and certified teachers. The second component is a parent education program which helps parents understand their children's behavior and includes emphasis on resolving behavior problems and other aspects of child rearing. A third component is a training program for pastors in family counseling and individual counseling. Each of the components serves about one hundred students and their parents.

With financial support from the Carnegie Corporation, the entire project is envisioned as a long-range, multi-year effort. Guided by a small advisory committee of ministers and scholars, Project Spirit is under the leadership of a social worker, Vanella Crawford, and is being evaluated by a psychologist, Harriet McAdoo of Howard University.

Church-Operated Schools

Finally, a number of other individual Black churches are responding more directly and aggressively to the educational aspirations of families in their communities. For example, in Brooklyn (NY), Concord Baptist Church, with its 10,000-plus membership, operates its own private elementary school. So, too, does the St. Paul Community Baptist Church, also in Brooklyn, and the Allen AME Church in nearby Queens. These and other Black churches are returning to the educational mission launched by the early Black church in creating schools. In the process many of these churches are following the example of the Catholic churches, which have a more extensive network of private schools as alternatives or supplements to the education available at public schools. This is a very constructive and promising development.

CONCLUSION

Nonetheless, it is in the public school system that the Black church will for some time make its greatest impact on the education of African American children and youth. This is the arena to which most African American families will continue to turn for the education of their children. The church can use its considerable and demonstrated influence to help African American families, and the African American community generally, extract from the public schools better performance in the education of African American children. This explains why efforts like Project Spirit hold such promise. The commitment to and dependence upon public schools by African American parents has caused them to place a great deal of hope in the promise of school desegregation, compensatory education, and Head Start. Thus far, however, all of these approaches have exhibited major shortcomings.

Indeed, the direction the Black community as a whole must take in this regard has been spelled out in a recent policy statement issued by a group of Black intellectuals under the leadership of the esteemed historian John Hope Franklin and sociologist Sarah Lawrence Lightfoot (Committee on Policy for Racial Justice, 1989). "What we must demand," they write, "is this: that the schools shift their focus from the supposed deficiencies of the black child—from the alleged inadequacies of black family life—to the barriers that stand in the way of academic success" (p. 2). As our ongoing research reveals, what better, more powerful, independent, numerous and resourceful ally can Black families count on to help them establish such relationships with schools than the Black church?

REFERENCES

Billingsley, A. (in press). *Climbing Jacob's ladder: The future of African American families.* New York: Simon & Schuster/Touchstone Books.

Committee on Policy for Racial Justice. (1989). *Visions of a better way.* Washington DC: Joint Center for Political Studies.

Hill, R. (1971). *The strengths of Black families.* New York: Emerson Hall.

Johnson, J. (1986). *An historical review of the role Black parents and the Black community played in providing school for Black children in the South: 1865–1954.* Unpublished doctoral dissertation, University of Massachusetts, School of Education.

Lincoln, C. E. (1989, April). *The Black church and Black self-determination.* Paper presented at the Association of Black Foundation Executives, Kansas City, MO.

Moberg, D. O. (1962). *The church as a social institution.* Englewood Cliffs, NJ: Prentice-Hall, Inc.

Roberts, J. D. (1980). *Roots of a Black future: Family and church.* Philadelphia: The Westminster Press.

Willie, C.V., Garibaldi, A. M., & Reed, W. L. (Eds.). (1990). *The education of African Americans,* Vol. 3. Boston: University of Massachusetts Press.

U.S. Bureau of the Census, Department of Commerce. (1988). *Current population reports: Population characteristics,* Series P-20, Nos. 388 and 437. Washington, DC: U.S. Government Printing Office.

36

Sport as Religion

MICHAEL NOVAK

Over the years, many sport sociologists have recognized strong parallels between the social institutions of sport and religion. Some might say, in fact, that the decline in church attendance has shifted concomitantly with the rise in athletic attendance, and that sport has significant elements of religion. Novak draws out some parallels between these two strong components in our society, and we leave you to see for yourself if you think the comparison works. Is Novak's point a useful one in explaining the incredible popularity of sports in the United States and throughout the world? If sport has replaced religion, at least in part, what might be the long-term effects of this?

A sport is not a religion in the same way that Methodism, Presbyterianism, or Catholicism is a religion. But these are not the only kinds of religion. There are secular religions, civil religions. The United States of America has sacred documents to guide and to inspire it: The Constitution, the Declaration of Independence, Washington's Farewell Address, Lincoln's Gettysburg Address, and other solemn presidential documents. The President of the United States is spoken to with respect, is expected to exert "moral leadership"; and when he walks among crowds, hands reach out to touch his garments. Citizens are expected to die for the nation, and our flag symbolizes vivid memories, from Fort Sumter to Iwo Jima, from the Indian Wars to Normandy: memories that moved hard-hats in New York to break up a march that was "desecrating" the flag. Citizens regard the American way of life as though it were somehow chosen by God, special, uniquely important to the history of the human race.

The institutions of the state generate a civil religion; so do the institutions of sport. The ancient Olympic games used to be both festivals in honor of the gods and festivals in honor of the state—and that has been the classical position of sports ever since. The ceremonies of sports overlap those of the state on one side,

Reprinted by permission of the author.

and those of the churches on the other. . . . Going to a stadium is half like going to a political rally, half like going to church. Even today, the Olympics are constructed around high ceremonies, rituals, and symbols. The Olympics are not barebones athletic events, but religion and politics as well. . . .

I am saying that sports flow outward into action from a deep natural impulse that is radically religious: an impulse of freedom, respect for ritual limits, a zest for symbolic meaning, and a longing for perfection. The athlete may of course be pagan, but sports are, as it were, natural religions. There are many ways to express this radical impulse: by the asceticism and dedication of preparation; by a sense of respect for the mysteries of one's own body and soul, and for powers not in one's own control; by a sense of awe for the place and time of competition; by a sense of fate; by a felt sense of comradeship and destiny; by a sense of participation in the rhythms and tides of nature itself.

Sports, in the second place, are organized and dramatized in a religious way. Not only do the origins of sports, like the origins of drama, lie in religious celebrations; not only are the rituals, vestments, and tremor of anticipation involved in sports events like those of religions. Even in our own secular age and for quite sophisticated and agnostic persons, the rituals of sports really work. They do serve a religious function: they feed a deep human hunger, place humans in touch with certain dimly perceived features of human life within this cosmos, and provide an experience of at least a pagan sense of godliness. . . .

Sports are religious in the sense that they are organized institutions, disciplines, and liturgies; and also in the sense that they teach religious qualities of heart and soul. In particular, they recreate symbols of cosmic struggle, in which human survival and moral courage are not assured. To this extent, they are not mere games, diversions, pastimes. Their power to exhilarate or depress is far greater than that. To say "It was only a game" is the psyche's best defense against the cosmic symbolic meaning of sports events. And it is partly true. For a game is a symbol; it is not precisely identified with what it symbolizes. To lose symbolizes death, and it certainly feels like dying; but it is not death. The same is true of religious symbols like Baptism or the Eucharist; in both, the communicants experience death, symbolically, and are reborn, symbolically. If you give your heart to the ritual, its effects upon your inner life can be far-reaching. . . .

Sports are not merely fun and games, not merely diversions, not merely entertainment. A ballpark is not a temple, but it isn't a fun house either. A baseball game is not an entertainment, and a ballplayer is considerably more than a paid performer. No one can explain the passion, commitment, discipline, and dedication involved in sports by evasions like these. . . .

The motive for regarding sports as entertainment is to take the magic, mystification, and falsehood out of sports. . . .

At a sports event, there may be spectators, just as some people come to church to hear the music. But a participant is not a spectator merely, even if he does not walk among the clergy. At a liturgy, elected representatives perform the formal acts, but all believers put their hearts into the ritual. It is considered inadequate, almost blasphemous, to be a mere spectator. Fans are not mere spectators. If they wanted no more than to pass the time, to find diversion, there are cheaper and less internally exhausting ways. Believers in sport do not go to sports to be entertained; to plays and dramas, maybe, but not to sports. Sports are far more serious than the dramatic arts, much closer to primal symbols, metaphors, and acts, much more ancient and more frightening. Sports are mysteries of youth and aging, perfect action and decay, fortune and misfortune, strategy and contingency. Sports are rituals concerning human survival on this planet: liturgical enactments of animal perfection and the struggles of the human spirit to prevail. . . .

In order to be entertained, I watch television: prime-time shows. They slide effortlessly by. I am amused, or distracted, or engrossed. Good or bad, they help to pass the time pleasantly enough. Watching football on television is totally different. I don't watch football to pass the time. The outcome of the games affects me. I care. Afterward, the emotion I have lived through continues to affect me. Football is not entertainment. It is far more important than that. If you observe the passivity of television viewers being entertained, and the animation of fans watching a game on television, the difference between entertainment and involvement of spirit becomes transparent. Sports are more like religion than like entertainment. Indeed, at a contest in the stadium, the "entertainment"—the bands, singers, comedians, balloons, floats, fireworks, jets screaming overhead—pales before the impact of the contest itself.

On Monday nights, when television carries football games, police officers around the nation know that crime rates will fall to low levels otherwise reached only on Mother's Day and Christmas. . . .

Sports, in a word, are a form of godliness. That is why the corruptions of sports in our day, by corporations and television and glib journalism and cheap public relations, are so hateful. If sports were entertainment, why should we care? They are far more than that. So when we see them abused, our natural response is the rise of vomit in the throat.

It may be useful to list some of the elements of religions, to see how they are imitated in the world of sports.

If our anthropologists discovered in some other culture the elements they can plainly see in our own world of sports, they would be obliged to write monographs on the religions of the tribes they were studying. Two experiments in thought may make this plain.

Imagine that you are walking near your home and come upon a colony of ants. They move in extraordinary busy lines, a trail of brown bodies across the whitish soil like a highway underneath the blades of grass. The lanes of ants abut on a constructed mudbank oval; there the ants gather, 100,000 strong, sitting in a circle. Down below, in a small open place, eleven ants on one side and eleven on the other contest bitterly between two lines. From time to time a buzz arises from the 100,000 ants gathered in their sacred oval. When the game is over, the long lines of ants begin their traffic-dense return to their colonies. In one observation, you didn't have time to discover the rules of their ritual. Or who made them up, or when. Or what they mean to the ants. Is the gathering mere "escape"? Does it mirror other facets in the life of ants? Do all ants everywhere take part? Do the ants "understand" what they are doing, or do they only do it by rote, one of the things that ants do on a lovely afternoon? Do ants practice, and stay in shape, and perfect their arts?

Or suppose you are an anthropologist from Mars. You come suddenly upon some wild, adolescent tribes living in territories called the "United States of America." You try to understand their way of life, but their society does not make sense to you. Flying over the land in a rocket, you notice great ovals near every city. You descend and observe. You learn that an oval is called a "stadium." It is used, roughly, once a week in certain seasons. Weekly, regularly, millions of citizens stream into these concrete doughnuts, pay handsomely, are alternately hushed and awed and outraged and screaming mad. (They demand from time to time that certain sacrificial personages be "killed.") You see that the figures in the rituals have trained themselves superbly for their performances. The combatants are dedicated. So are the dancers and musicians in tribal dress who occupy the arena before, during, and after the combat. You note that, in millions of homes, at corner shrines in every household's sacred room, other citizens are bound by invisible attraction to the same events. At critical moments, the most intense worshipers demand of the less attentive silence. Virtually an entire nation is united in a central public rite. Afterward, you note exultation or depression among hundreds of thousands, and animation almost everywhere.

Some of the elements of a religion may be enumerated. A religion, first of all, is organized and structured. Culture is built on cult. Accordingly, a religion begins with ceremonies. At these ceremonies, a few surrogates perform for all. They need not even believe what they are doing. As professionals, they may perform so often that they have lost all religious instinct; they may have less faith than any of the participants. In the official ceremonies, sacred vestments are employed and rituals are prescribed. Customs develop. Actions are highly formalized. Right ways and wrong ways are plainly marked out; illicit behaviors are distinguished from licit ones. Professional watchdogs supervise formal correctness. Moments of silence are observed. Concentration and intensity are indispensable. To attain them, drugs or special disciplines of spirit might be employed; ordinary humans, in the ordinary ups and downs of daily experience, cannot be expected to perform routinely at the highest levels of awareness. . . .

Religions also channel the feeling most humans have of danger, contingency, and chance—in a word, Fate. Human plans involve ironies. Our choices are made with so little insight into their eventual effects that what we desire is often not the path to what we want. The decisions we make with little attention turn out to be major turning points. What we prepare for with exquisite detail never happens. Religions place us in the presence of powers greater than ourselves, and seek to reconcile us to them. The rituals of religion give these powers almost human shape, forms that give these powers visibility and tangible effect. Sports events in baseball, basketball, and football are structured so that "the breaks" may intervene and become central components in the action.

Religions make explicit the almost nameless dreads of daily human life: aging, dying, failure under pressure, cowardice, betrayal, guilt. Competitive sports embody these in every combat. . . .

Religions consecrate certain days and hours. Sacred time is a block of time lifted out of everyday normal routines, a time that is different, in which different laws apply, a time within which one forgets ordinary time. Sacred time is intended to suggest an "eternal return," a fundamental repetition like the circulation of the human blood, or the eternal turning of the seasons, or the wheeling of the stars and planets in their cycles: the sense that things repeat themselves, over and over, and yet are always a little different. Sacred time is more like eternity than like history, more like cycles of recurrence than like progress, more like a celebration of repetition than like a celebration of novelty. Yet sacred time is full of exhilaration, excitement, and peace, as though it were more real and more joyous than the activities of everyday life—as though it were *really living* to be in sacred time (wrapped up in a close game during the last two minutes), and comparatively boring to suffer the daily jading of work, progress, history.

To have a religion, you need to have heroic forms to try to live up to: patterns of excellence so high that human beings live up to them only rarely, even when they strive to do so; and images of perfection so beautiful that, living up to them or seeing someone else live up to them, produces a kind of "*ah!*"

You need to have a pattern of symbols and myths that a person can grow old with, with a kind of resignation, wisdom, and illumination. Do what we will, the human body ages. Moves we once could make our minds will but our bodies cannot implement; disciplines we once endured with suppressed animal desire are no longer worth the effort; heroes that once seemed to us immortal now age, become enfeebled, die, just as we do. The "boys of summer" become the aging men of winter. A religion celebrates the passing of all things: youth, skill, grace, heroic deeds.

To have a religion, you need to have a way to exhilarate the human body, and desire, and will, and the sense of beauty, and a sense of oneness with the universe and other humans. You need chants and songs, the rhythm of bodies in unison, the indescribable feeling of many who together "will one thing" as if they were each members of a single body.

All these things you have in sports.

37

Savage Inequalities

JONATHAN KOZOL

Education has long been considered, rightly or wrongly, one of the dominant venues for upward mobility in the United States. We pride ourselves, as a nation, on having a system of public education that is available to all, one of the cornerstones of our value of equal opportunity. Over the years, we have seen many struggles in the educational arena that address this situation. For a long time students were bussed from their neighborhoods into other school systems to reduce segregation on the philosophy that separate education for different racial groups does not promote equality. We are currently in the midst of debate over the issue of school vouchers, discussing whether students who voluntarily withdraw from the public schools to attend private school should receive educational rebates to take to their new schools. In this selection Kozol shows us how unsuccessful we have been in trying to achieve equality of education in some situations. He compares an inner-city African American school with a white suburban school and shows us the dramatic differences. How are life chances affected by children who attend an inner-city school as compared to those who attend a suburban, well-funded one? Are there any solutions to the differentials that Kozol points out? Where does your school fall as compared to the ones described by Kozol? What is your opinion of the type of high school education you received?

"East of anywhere," writes a reporter for the *St. Louis Post-Dispatch,* "often evokes the other side of the tracks. But, for a first-time visitor suddenly deposited on its eerily empty streets, East St. Louis might suggest another world." The city, which is 98 percent black, has no obstetric services, no regular trash collection, and few jobs. Nearly a third of its families live on less than $7,500 a year; 75 percent of its population lives on welfare of some form. The U.S. Department of Housing and Urban Development describes it as "the most distressed small city in America."

Only three of the 13 buildings on Missouri Avenue, one of the city's major thoroughfares, are occupied. A 13-story office building, tallest in the city, has been boarded up. Outside, on the sidewalk, a pile of garbage fills a ten-foot crater.

The city, which by night and day is clouded by the fumes that pour from vents and smokestacks at the Pfizer and Monsanto chemical plants, has one of the highest rates of child asthma in America.

It is, according to a teacher at Southern Illinois University, "a repository for a nonwhite population that is now regarded as expendable." The *Post-Dispatch* describes it as "America's Soweto."

Fiscal shortages have forced the layoff of 1,170 of the city's 1,400 employees in the past 12 years. The city, which is often unable to buy heating fuel or toilet paper for the city hall, recently announced that it might have to cashier all but 10 percent of the remaining work force of 230. In 1989 the mayor announced that he might need to sell the city hall and all six fire stations to raise needed cash. Last year the plan had to be scrapped after the city lost its city hall in a court judgment to a creditor. East St. Louis is mortgaged into the next century but has the highest property-tax rate in the state. . . .

The dangers of exposure to raw sewage, which backs up repeatedly into the homes of residents in East St. Louis, were first noticed, in the spring of 1989, at a public housing project, Villa Griffin. Raw sewage, says the *Post-Dispatch,* overflowed into a playground just behind the housing project, which is home to 187 children, "forming an oozing lake of . . . tainted water." . . . A St. Louis health official voices her dismay that children live with waste in their backyards. "The development of working sewage systems made cities livable a hundred years ago," she notes. "Sewage systems separate us from the Third World." . . .

The sewage, which is flowing from collapsed pipes and dysfunctional pumping stations, has also flooded basements all over the city. The city's vacuum truck, which uses water and suction to unclog the city's sewers, cannot be used because it needs $5,000 in repairs. Even when it works, it sometimes can't be used because there isn't money to hire drivers. A single engineer now does the work that 14 others did before they were laid off. By April the pool of overflow behind the Villa Griffin project has expanded into a lagoon of sewage. Two million gallons of raw sewage lie outside the children's homes. . . .

The problems of the streets in urban areas, as teachers often note, frequently spill over into public schools. In the public schools of East St. Louis this is literally the case.

"Martin Luther King Junior High School," notes the *Post-Dispatch* in a story published in the early spring of 1989, "was evacuated Friday afternoon after sewage flowed into the kitchen. . . . The kitchen was closed and students were sent home." On Monday, the paper continues, "East St. Louis Senior High School was awash in sewage for the second time this year." The school had to be shut because of "fumes and backed-up toilets." Sewage flowed into the basement, through the floor, then up into the kitchen and the students' bathrooms. The backup, we read, "occurred in the food preparation areas."

School is resumed the following morning at the high school, but a few days later the overflow recurs. This time the entire system is affected, since the meals distributed to every student in the city are prepared in the two schools that have been flooded. School is called off for all 16,500 students in the district. The sewage backup, caused by the failure of two pumping stations, forces officials at the high school to shut down the furnaces.

At Martin Luther King, the parking lot and gym are also flooded. "It's a disaster," says a legislator. "The streets are under water; gaseous fumes are being emitted from the pipes under the schools," she says, "making people ill."

In the same week, the schools announce the layoff of 280 teachers, 166 cooks and cafeteria workers, 25 teacher aides, 16 custodians and 18 painters, electricians, engineers and plumbers. The president of the teachers' union says the cuts, which will bring the size of kindergarten and primary classes up to 30 students, and the size of fourth to twelfth grade classes up to 35, will have "an unimaginable impact" on the students. "If you have a high school teacher with five classes each day and between 150 and 175 students . . . , it's going to have a devastating effect." The school system, it is also noted, has been using more than 70 "permanent substitute teachers," who are paid only $10,000 yearly, as a way of saving money. . . .

East St. Louis, says the chairman of the state board, "is simply the worst possible place I can imagine to have a child brought up. . . . The community is in desperate circumstances." Sports and music, he observes, are, for many children here, "the only avenues of success." Sadly enough, no matter how it ratifies the stereotype, this is the truth; and there is a poignant aspect to the fact that, even with class size soaring and one quarter of the system's teachers being given their dismissal, the state board of education demonstrates its genuine but skewed compassion by attempting to leave sports and music untouched by the overall austerity.

Even sports facilities, however, are degrading by comparison with those found and expected at most high schools in America. The football field at East St. Louis High is missing almost everything—including goalposts. There are a couple of metal pipes—no crossbar, just the pipes. Bob Shannon, the football coach, who has to use his personal funds to purchase footballs and has had to cut and rake the football field himself, has dreams of having goalposts someday. He'd also like to let his students have new uniforms. The ones they wear are nine years old and held together somehow by a patchwork of repairs. Keeping them clean is a problem, too. The school cannot afford a washing machine. The uniforms are carted to a corner laundromat with fifteen dollars' worth of quarters. . . .

In the wing of the school that holds vocational classes, a damp, unpleasant odor fills the halls. The school has a machine shop, which cannot be used for lack of staff, and a woodworking shop. The only shop that's occupied this morning is the auto-body class. A man with long blond hair and wearing a white sweat suit swings a paddle to get children in their chairs. "What we need the most is new equipment," he reports. "I have equipment for alignment, for example, but we don't have money to install it. We also need a better form of egress. We bring the cars in through two other classes." Computerized equipment used in most repair shops, he reports, is far beyond the high school's budget. It looks like a very old gas station in an isolated rural town. . . .

The science labs at East St. Louis High are 30 to 50 years outdated. John McMillan, a soft-spoken man, teaches physics at the school. He shows me his lab. The six lab stations in the room have empty holes where pipes were once attached. "It would be great if we had water," says McMillan. . . .

Leaving the chemistry labs, I pass a double-sized classroom in which roughly 60 kids are sitting fairly still but doing nothing. "This is supervised study hall," a teacher tells me in the corridor. But when we step inside, he finds there is no teacher. "The teacher must be out today," he says.

Irl Solomon's history classes, which I visit next, have been described by journalists who cover East St. Louis as the highlight of the school. Solomon, a man of 54 whose reddish hair is turning white, has taught in urban schools for almost 30 years. A graduate of Brandeis University, he entered law school but was drawn away by a concern with civil rights. "After one semester, I decided that the law was not for me. I said, 'Go and find the toughest place there is to teach. See if you like it.' I'm still here. . . .

"I have four girls right now in my senior home room who are pregnant or have just had babies. When I ask them why this happens, I am told, 'Well, there's no reason not to have a baby. There's not much for me in public school.' The truth is, that's a pretty honest answer. A diploma from a ghetto high school doesn't count for much in the United States today. So, if this is really the last education that a person's going to get, she's probably perceptive in that statement. Ah, there's so much bitterness—unfairness—there, you know. Most of these pregnant girls are not the ones who have much self-esteem. . . .

"Very little education in the school would be considered academic in the suburbs. Maybe 10 to 15 percent of students are in truly academic programs. Of the 55 percent who graduate, 20 percent may go to four-year colleges: something like 10 percent of any entering class. Another 10 to 20 percent may get some other kind of higher education. An equal number join the military. . . .

"I don't go to physics class, because my lab has no equipment," says one student. "The typewriters in my typing class don't work. The women's toilets . . ." She makes a sour face. "I'll be honest," she says. "I just don't use the toilets. If I do, I come back into class and I feel dirty."

"I wanted to study Latin," says another student. "But we don't have Latin in this school."

"We lost our only Latin teacher," Solomon says.

A girl in a white jersey with the message DO THE RIGHT THING on the front raises her hand. "You visit other schools," she says. "Do you think the children in this school are getting what we'd get in a nice section of St. Louis?"

I note that we are in a different state and city.

"Are we citizens of East St. Louis or America?" she asks. . . .

In a seventh grade social studies class, the only book that bears some relevance to black concerns—its title is *The American Negro*—bears a publication date of 1967. The teacher invites me to ask the class some questions. Uncertain where to start, I ask the students what they've learned about the civil rights campaigns of recent decades.

A 14-year-old girl with short black curly hair says this: "Every year in February we are told to read the same old speech of Martin Luther King. We read it every year. 'I have a dream. . . .' It does begin to seem—what is the word?" She hesitates and then she finds the word: "perfunctory."

I ask her what she means.

"We have a school in East St. Louis named for Dr. King," she says. "The school is full of sewer water and the doors are locked with chains. Every student in that school is black. It's like a terrible joke on history."

It startles me to hear her words, but I am startled even more to think how seldom any press reporter has observed the irony of naming segregated schools for Martin Luther King. Children reach the heart of these hypocrisies much quicker than the grown-ups and the experts do. . . .

The train ride from Grand Central Station to suburban Rye, New York, takes 35 to 40 minutes. The high school is a short ride from the station. Built of handsome gray stone and set in a landscaped campus, it resembles a New England prep school. On a day in early June of 1990, I enter the school and am directed by a student to the office.

The principal, a relaxed, unhurried man who, unlike many urban principals, seems gratified to have me visit in his school, takes me in to see the auditorium, which, he says, was recently restored with private charitable funds ($400,000) raised by parents. The crenellated ceiling, which is white and spotless, and the polished dark-wood paneling contrast with the collapsing structure of the auditorium at [another school I visited]. The principal strikes his fist against the balcony: "They made this place extremely solid." Through a window, one can see the spreading branches of a beech tree in the central courtyard of the school.

In a student lounge, a dozen seniors are relaxing on a carpeted floor that is constructed with a number of tiers so that, as the principal explains, "they can stretch out and be comfortable while reading."

The library is wood-paneled, like the auditorium. Students, all of whom are white, are seated at private carrels, of which there are approximately 40. Some are doing homework; others are looking through the *New York Times.* Every student that I see during my visit to the school is white or Asian, though I later learn there are a number of Hispanic students and that 1 or 2 percent of students in the school are black.

According to the principal, the school has 96 computers for 546 children. The typical student, he says, studies a foreign language for four or five years, beginning in the junior high school, and a second foreign language (Latin is available) for two years. Of 140 seniors, 92 are now enrolled in AP [advanced placement] classes. Maximum teacher salary will soon reach $70,000. Per-pupil funding is above $12,000 at the time I visit.

The students I meet include eleventh and twelfth graders. The teacher tells me that the class is reading Robert Coles, Studs Terkel, Alice Walker. He tells me I will find them more than willing to engage me in debate, and this turns out to be correct. Primed for my visit, it appears, they arrow in directly on the dual questions of equality and race.

Three general positions soon emerge and seem to be accepted widely. The first is that the fiscal inequalities "do matter very much" in shaping what a school can offer ("That is obvious," one student says) and that any loss of funds in Rye, as a potential consequence of future equalizing, would be damaging to many things the town regards as quite essential.

The second position is that racial integration—for example, by the busing of black children from the city or a nonwhite suburb to this school—would meet with strong resistance, and the reason would not simply be the fear that certain standards might decline. The reason, several students say straightforwardly, is "racial" or, as others say it, "out-and-out racism" on the part of adults.

The third position voiced by many students, but not all, is that equity is basically a goal to be desired and should be pursued for moral reasons, but "will probably make no major difference" since poor children "still would lack the motivation" and "would probably fail in any case because of other problems."

At this point, I ask if they can truly say "it wouldn't make a difference" since it's never been attempted. Several students then seem to rethink their views and say that "it might work, but it would have to start with preschool and the elementary grades" and "it might be 20 years before we'd see a difference."

At this stage in the discussion, several students speak with some real feeling of the present inequalities, which, they say, are "obviously unfair," and one student goes a little further and proposes that "we need to change a lot more than the schools." Another says she'd favor racial integration "by whatever means—including busing—even if the parents disapprove." But a contradictory opinion also is expressed with a good deal of fervor and is stated by one student in a rather biting voice: "I don't see why we should do it. How could it be of benefit to us?"

Throughout the discussion, whatever the views the children voice, there is a degree of unreality about the whole exchange. The children are lucid and their language is well chosen and their arguments well made, but there is sense that they are dealing with an issue that does not feel very vivid, and that nothing that we say about it to each other really matters since it's "just a theoretical discussion." To a certain degree, the skillfulness and cleverness that they display seem to derive precisely from this sense of unreality. Questions of unfairness feel more like a geometric problem than a matter of humanity or conscience. A few of the students do break through the note of unreality, but, when they do, they cease to be so agile in their use of words and speak more awkwardly. Ethical challenges seem to threaten their effectiveness. There is the sense that they were skating over ice and that the issues we addressed were safely frozen underneath. When they stop to look beneath the ice they start to stumble. The verbal competence they have acquired here may have been gained by building walls around some regions of the heart.

"I don't think that busing students from their ghetto to a different school would do much good," one student says. "You can take them out of the environment, but you can't take the environment out of *them*. If someone grows up in the South Bronx, he's not going to be prone to learn." His name is Max and he has short black hair and speaks with confidence. "Busing didn't work when it was tried," he says. I ask him how he knows this and he says he saw a television movie about Boston.

"I agree that it's unfair the way it is," another student says. "We have AP courses and they don't. Our classes are much smaller." But, she says, "putting them in schools like ours is not the answer. Why not put some AP classes into their school? Fix the roof and paint the halls so it will not be so depressing."

The students know the term "separate but equal," but seem unaware of its historical associations. "Keep them where they are but make it equal," says a girl in the front row.

A student named Jennifer, whose manner of speech is somewhat less refined and polished than that of the others, tells me that her parents came here from New York. "My family is originally from the Bronx. Schools are hell there. That's one reason that we moved. I don't think it's our responsibility to pay our taxes to provide for *them*. I mean, my parents used to live there and they wanted to get out. There's no point in coming to a place like this, where schools are good, and then your taxes go back to the place where you began."

I bait her a bit: "Do you mean that, now that you are not in hell, you have no feeling for the people that you left behind?"

"It has to be the people in the area who want an education. If your parents just don't care, it won't do any good to spend a lot of money. Someone else can't want a good life for you. You have got to want it for yourself." Then she adds, however, "I agree that everyone should have a chance at taking the same courses. . . ."

I ask her if she'd think it fair to pay more taxes so that this was possible.

"I don't see how that benefits me," she says.

38

Student Participation in the College Classroom

DAVID KARP AND WILLIAM YOELS

When we talk about education it all comes down, in some ways, to what actually happens in the classroom. This is the environment where academic teaching and learning primarily occurs. Studies have shown that a variety of factors can impact on this educational process including the size of the classroom, the composition of the student body, the philosophy of the surrounding educational institution, and the dynamics of the interaction in the class. Karp and Yoels examine some of these in one northeastern university, focusing particularly on the latter factor. As you read this selection, you may want to compare your own classes to the classrooms they studied. Where do you like to sit in the classroom? Why? What do you think this says about your seriousness as a student? Can these small differences actually mean something about your academic inclinations?

Rarely have researchers attempted to consider the processes through which students and teachers formulate definitions of the classroom as a social setting. The problem of how students and teachers assign "meaning" to the classroom situation has been largely neglected in the various studies mentioned. Although writing about primary and secondary school classrooms, we would suggest that the following statement from Jackson's (1968:vii) *Life in Classrooms* holds true for college classrooms as well. He writes that:

> Classroom life . . . is too complex an affair to be viewed or talked about from any single perspective. Accordingly, as we try to grasp the meaning of what school is like for students and teachers, we must not hesitate to use all the ways of knowing at our disposal. This means we must read and look and listen and count things, and talk to people, and even muse introspectively.

Reprinted by permission of the authors.

The present study focuses on the meanings of student participation in the college classroom. Our examination of this problem will center on the way in which definitions of classrooms held by students and teachers relate to their actual behavior in the classroom.

Although we did not begin this study with any explicit hypotheses to be tested, we did begin with some general guiding questions. Most comprehensive among these, and of necessary importance from a symbolic interactionist perspective, was the question, "What is a college classroom?" We wanted to know how both students and teachers were defining the social setting, and how these definitions manifested themselves in the activity that goes on in the college classrooms. More specifically, we wanted to understand what it was about the definition of the situation held by students and teachers that led to, in most instances, rather little classroom interaction.

What knowledge, we might now ask, do students have of college classrooms that makes the decision not to talk a "realistic" decision? There would seem to be two factors of considerable importance as indicated by our data.

First, students believe that they can tell very early in the semester whether or not a professor really wants class discussion. Students are also well aware that there exists in college classrooms a rather distinctive "consolidation of responsibility." In any classroom there seems almost inevitably to be a small group of students who can be counted on to respond to questions asked by the professor or to generally have comments on virtually any issue raised in class. Our observational data indicated that on the average a very small number of students are responsible for the majority of all talk that occurs in class on any given day. The fact that this "consolidation of responsibility" looms large in students' consciousness is indicated by the fact, reported earlier, that more than 90 percent of the students strongly agreed or agreed with the statement "In most of my classes there are a small number of students who do most of the talking."

Once the group of "talkers" gets established and identified in a college classroom the remaining students develop a strong expectation that these "talkers" can be relied upon to answer questions and make comments. In fact, we have often noticed in our own classes that when a question is asked or an issue raised the "silent" students will even begin to orient their bodies towards and look at this coterie of talkers with the expectation, presumably, that they will shortly be speaking.

Our concept of the "consolidation of responsibility" is a modification of the idea put forth by Latane and Darley (1970) in *The Unresponsive Bystander.* In this volume Latane and Darley developed the concept of "the diffusion of responsibility" to explain why strangers are often reluctant to "get involved" in activities where they assist other strangers who may need help. They argue that the delegation of responsibility in such situations is quite unclear and, as a result, responsibility tends to get assigned to no one in particular—the end result being that no assistance at all is forthcoming. In the case of the classroom interaction, however, we are dealing with a situation in which the responsibility for talking gets assigned to a few who can be relied upon to carry the "verbal load"—thus the *consolidation of responsibility.* As a result, the majority of students play a relatively passive role in the classroom and see themselves as recorders of the teacher's information. This expectation is mutually supported by the professor's reluctance to directly call on *specific* students.

While students expect that only a few students will do most of the talking, and while these talkers are relied upon to respond in class, the situation is a bit more complicated than we have indicated to this point. It would appear that while these talkers are "doing their job" by carrying the discussion for the class as a whole, there is still a strong feeling on the part of many students that they ought not to talk *too much.*

More than 60 percent of the students responding to our questionnaire expressed annoyance with students who "talk too much in class." This is interesting to the extent that even those who talk very regularly in class still account for a very small percentage of total class time. While we have no systematic data on time spent talking in class, the comments of the observers indicate that generally a total of less than five minutes of class time (in a fifty-minute period) is accounted for by student talk in class.

A fine balance must be maintained in college classes. Some students are expected to do most of the talking, thus relieving the remainder of the students from the burdens of having to talk in class. At the same time, these talkers must not be "rate-busters." We are suggesting here that students see "intellectual work" in much the same way that factory workers define "piecework." Talking too much in class, or what might be called "linguistic rate-busting," upsets the normative arrangement of the classroom and, in the students' eyes, increases the probability of raising the professor's expectations vis-á-vis the participation of other students. It may be said, then, that a type of "restriction of verbal output" norm operates in college classrooms, in which those who engage in linguistic rate-busting or exhibit "overinvolvement" in the classroom get defined by other students as "brownnoses" and "apostates" from the student "team." Other students often indicate their annoyance with these "rate-busters" by smiling wryly at their efforts, audibly sighing, rattling their notebooks and, on occasion, openly snickering.

A second factor that insures in students' minds that it will be safe to refrain from talking is their knowledge that only in rare instances will they be directly called upon by teachers in a college classroom. Our data indicate that of all the interactions occurring in the classes under observation only about 10 percent were due to teachers calling directly upon a specific student. The unwillingness of teachers to call upon students would seem to stem from teachers' beliefs that the classroom situation is fraught with anxiety for students. It is important to note that teachers, unlike students themselves, viewed the possibility that "students might appear unintelligent in the eyes of other students" as a very important factor in keeping students from talking (Table 2). Unwilling to exacerbate the sense of risk which teachers believe is a part of student consciousness, they refrain from directly calling upon specific students.

The direct result of these two factors is that students feel no obligation or particular necessity for keeping up with reading assignments so as to be able to participate in class. Such a choice is made easier still by the fact that college students

Table 1 Percentage of Students Who Indicated That an Item Was an Important Factor in Why Students Would Choose Not to Talk in Class, by Sex of Student (in Rank Order)

Male			Female		
Rank	Item	%	Rank	Item	%
1.	I had not done the assigned reading	80.9	1.	The feeling that I don't know enough about the subject matter	84.8
2.	The feeling that I don't know enough about the subject matter	79.6	2.	I had not done the assigned reading	76.3
3.	The large size of the class	70.4	3.	The feeling that my ideas are not well enough formulated	71.1
4.	The feeling that my ideas are not well enough formulated	69.8	4.	The large size of the class	68.9
5.	The course simply isn't meaningful to me	67.3	5.	The course simply isn't meaningful to me	65.1
6.	The chance that I would appear unintelligent in the eyes of the teacher	43.2	6.	The chance that I would appear unintelligent in the eyes of other students	45.4
7.	The chance that I would appear unintelligent in the eyes of other students	42.9	7.	The chance that I would appear unintelligent in the eyes of the teacher	41.4
8.	The small size of the class	31.0	8.	The small size of the class	33.6
9.	The possibility that my comments might negatively affect my grade	29.6	9.	The possibility that my comments might negatively affect my grade	24.3
10.	The possibility that other students in the class would not respect my point of view	16.7	10.	The possibility that the teacher would not respect my point of view	21.1
11.	The possibility that the teacher would not respect my point of view	12.3	11.	The possibility that other students in the class would not respect my point of view	12.5

are generally tested infrequently. Unlike high school, where homework is the teacher's "daily insurance" that students are prepared for classroom participation, college is a situation in which the student feels quite safe in coming to class without having done the assigned reading and, not having done it, safe in the secure knowledge that one won't be called upon.[1] It is understandable, then, why such

Table 2 Percentage of Teachers Who Indicated That an Item Was an Important Factor in Why Students Would Choose Not to Talk in Class (in Rank Order)

Rank	Item	%
1.5	The large size of the class	80
1.5	The chance that I would appear unintelligent in the eyes of other students	80
4.0	The feeling that I don't know enough about the subject matter	70
4.0	The feeling that my ideas are not well enough formulated	70
4.0	The possibility that my comments might negatively affect my grade	70
6.0	The course simply isn't meaningful to me	50
7.5	I had not done the assigned reading	40
7.5	The chance that I would appear unintelligent in the eyes of the teacher	40
9.5	The possibility that the teacher would not respect my point of view	30
9.5	The possibility that other students in the class would not respect my point of view	30
11.5	The small size of the class	10

items as "not having done the assigned reading" and "the feeling that one does not know enough about the subject matter" would rank so high (Table 1) in students' minds as factors keeping them from talking in class.

In sum, we have isolated two factors relative to the way that classrooms actually operate that make it "practically" possible for students not to talk in class. These factors make it possible for the student to pragmatically abide by an early decision to be silent in class. We must now broach the somewhat more complicated question: what are the elements of students' definitions of the college classroom situation that prompt them to be silent in class? To answer this question we must examine how students perceive the teacher as well as their conceptions of what constitutes "intellectual work."

By the time that students have finished high school they have been imbued with the enormously strong belief that teachers are "experts" who possess the "truth." They have adopted, as Freire (1970) has noted, a "banking" model of education. The teacher represents the bank, the huge "fund" of "true" knowledge. As a student it is one's job to make weekly "withdrawals" from the fund, never any "deposits." His teachers, one is led to believe, and often led to believe it by the teachers themselves, are possessors of the truth. Teachers are in the classroom to ***teach,*** not to ***learn.***

If the above contains anything like a reasonable description of the way that students are socialized in secondary school, we should not find it strange or shocking that our students find our requests for criticism of ideas a bit alien. College students still cling to the idea that they are knowledge seekers and that faculty members are knowledge dispensers. Their view of intellectual work leaves little room for the notion that ideas themselves are open to negotiation. It is simply not part of their view of the classroom that ideas are generated out of dialogue, out of persons questioning and taking issue with one another, out of persons being *critical* of each other.

It comes as something of a shock to many of our students when we are willing to give them, at best, a "B" on a paper or exam that is "technically" proficient. When they inquire about their grade (and they do this rarely, believing strongly that our judgment is unquestionable), they want to know what they did "wrong." Intellectual work is for them dichotomous. It is either good or bad, correct or incorrect. They are genuinely surprised when we tell them that nothing is wrong, that they simply have not been critical enough and have not shown enough reflection on the ideas. Some even see such an evaluation as unfair. They claim a kind of incompetence at criticism. They often claim that it would be illegitimate for them to disagree with an author.

Students in class respond as uncritically to the thoughts of their professors as they do to the thoughts of those whom they read. Given this general attitude toward intellectual work, based in large part on students' socialization, and hence their definition of what should go on in classrooms, the notion of using the classroom as a place for generating ideas is a foreign one.

Part of students' conceptions of what they can and ought to do in classrooms is, then, a function of their understanding of how ideas are to be communicated. Students have expressed the idea that if they are to speak in class they ought to be able to articulate their point logically, systematically, and above all completely. The importance of this factor in keeping students from talking is borne out by the very high ranking given to the item (Table 1) "the feeling that my ideas are not well enough formulated."

In their view, if their ideas have not been fully formulated in advance, then the idea is not worth relating. They are simply unwilling to talk "off the top of their heads." They feel, particularly in an academic setting such as the college classroom, that there is a high premium placed on being articulate. This feeling is to a large degree prompted by the relative articulateness of the teacher. Students do not, it seems, take into account the fact that the teacher's coherent presentation is typically a function of the time spent preparing his/her ideas. The relative preparedness of the teacher leads to something of paradox vis-á-vis classroom discussion.

We have had students tell us that one of the reasons they find it difficult to respond in class involves the professor's preparedness; that is, students have told us that because the professor's ideas as presented in lectures are (in their view) so well formulated they could not add anything to those ideas. Herein lies something of a paradox. One might suggest that, to some degree at least, the better prepared a professor is for his/her class, the less likely are students to respond to the elements of his/her lecture.

We have both found that some of our liveliest classes have centered around those occasions when we have talked about research presently in progress. When it is clear to the student that we are ourselves struggling with a particular problem, that we cannot fully make sense of a phenomenon, the greater is the class participation. In most classroom instances, students read the teacher as the "expert,"[2] and once having cast the professor into that role it becomes extremely difficult for students to take issue with or amend his/her ideas.

It must also be noted that students' perceptions about their incapacity to be critical of their own and others' ideas leads to an important source of misunderstanding between college students and their teachers. In an open-ended question we asked students what characteristics they thought made for an "ideal" teacher. An impressionistic reading of these responses indicated that students were overwhelmingly uniform in their answers. They consensually found it important that a teacher "not put them down" and that a teacher "not flaunt his/her superior knowledge." In this regard the college classroom is a setting pregnant with possibilities for mutual misunderstanding. Teachers are working under one set of assumptions about "intellectual work" while students proceed under another. Our experiences as college teachers lead us to believe that teachers tend to value *critical* responses by students and tend to respond critically themselves to the comments and questions of college students. Students tend to perceive these critical comments as in some way an assault on their "selves" and find it difficult to separate a critique of their thoughts from a critique of themselves. Teachers are for the most part unaware of the way in which students interpret their comments.

The result is that when college teachers begin to critically question a student's statement, trying to get the student to be more critical and analytical about his/her assertions, this gets interpreted by students as a "put-down." The overall result is the beginning of a "vicious circle" of sorts. The more that teachers try to instill in students a critical attitude toward one's own ideas, the more students come to see faculty members as condescending, and the greater still becomes their reluctance to make known their "ill formulated" ideas in class. Like any other social situation where persons are defining the situation differently, there is bound to develop a host of interactional misunderstandings.

Before concluding this section, let us turn to a discussion of the differences in classroom participation rates of male versus female students. Given the fact that men and women students responded quite similarly to the *questionnaire items* reported here, much of our previous discussion holds for both male and female students. There are some important differences, however, in their *actual behavior* in the college classroom (as revealed by our observational data) that ought to be considered. Foremost among these differences is the fact that the sex of the teacher affects the likelihood of whether male or female teachers in these classes are "giving off expressions" that are being interpreted very differently by male and female students. Male students play a more active role in all observed classes regardless of the teacher's sex, but with female instructors the percentage of female participation sharply increases. Also of interest is the fact that the male instructors are more likely to directly call on male students than on female students (7.1 percent to 3.1 percent), whereas female instructors are just as likely to call on female students as on male students (12.5 percent to 12.8 percent). Possibly female students in female-taught classes interpret the instructor's responses as being more egalitarian than those of male professors and thus more sympathetic to the views of female students. With the growing [awareness] of women faculty and students [of women's issues] it may not be unreasonable to assume that female instructors are more sensitive to the problems of female students both inside and outside the college classroom.

With the small percentage of women faculty currently teaching in American universities it may well be that the college classroom is still defined by both male and female students as a setting "naturally" dominated by men. The presence of female professors, however, as our limited data suggest, may bring about some changes in these definitions of "natural" classroom behavior.

NOTES

1. We have no "hard" data concerning student failure to do the assigned reading other than our own observations of countless instances where we posed questions that went unanswered, when the slightest familiarity with the material would have been sufficient to answer them. We have also employed "pop" quizzes and the student performance on these tests indicated a woefully inadequate acquaintance with the readings assigned for that session. The reader may evaluate our claim by reflecting upon his/her own experience in the college classroom.

2. This attribution of power and authority to the teacher may be particularly exaggerated in the present study due to its setting in a Catholic university with a large number of students entering from Catholic high schools. Whether college students with different religious and socioeconomic characteristics attribute similar degrees of power and authority to professors is a subject worthy of future comparative empirical investigation.

REFERENCES

Freire, P. 1970. *Pedagogy of the Oppressed.* New York: Seabury Press.

Jackson, P. 1968. *Life in Classrooms.* New York: Holt, Rinehart and Winston.

Latane, B. and J. Darley. 1970. *The Unresponsive Bystander: Why Doesn't He Help?* New York: Appleton-Century-Crofts.

39

Conflict within the Ivory Tower

RUTH SIDEL

Beyond the factors occurring strictly inside the classroom, students learn many things from what goes on in the university. The battles that occur here attest to the importance people place on the institution of learning. Conflicts sometimes arise on college campuses, and these may interfere with students' ability to go to class and learn. While conflict and struggle is in some ways a natural part of life, when it escalates to the point of open hostilities, it may become unbearable. Sidel describes her experience with some highly destructive confrontations and the way these were experienced by members of the groups involved. Can you see the rumblings of similar patterns of conflict at your college or university? Are there underlying social problems that may bubble up to create hostilities? What conflicts are occurring in the arena of education today, and how have you seen them impact your school?

Admission to college or university is a first but crucial step in an individual's preparation for meaningful participation in the social, economic, and political life of postindustrial America. But admission is merely the first hurdle a student must clear in higher education. Financing college education, achieving academically, and maneuvering around the multitude of social, psychological, and political obstacles that impede the path to a bachelor's degree are often much higher hurdles than admission. Among the barriers that many students have had to face in recent years are virtually continuous clashes stemming from prejudice, ethnocentrism, and fear—fear of the unknown, of the stranger among us. At root these clashes are about entitlement and power, and about students' concerns with the precariousness of their position in the social structure.

Although colleges and universities have since the end of the Second World War been to a considerable extent transformed from elite bastions of privilege to increasingly open, heterogeneous communities, a wave of overt intolerance has recently swept over the academic community. There is little doubt that students today are more tolerant than their grandparents and their parents, yet clashes—some involving vocal or written assaults, some involving violence—continue to plague academic institutions and to shock observers. One of the reasons these so-called hate incidents are so shocking is the increasing unacceptability of overtly racist, sexist, anti-Semitic, and homophobic language and behavior in much of the wider society; another is the contrast between the violence of these incidents and the open expression of hatred and bigotry on the one hand, and the expectation of at least minimal civility in academic settings on the other.

Two relatively recent incidents that deal with the incendiary combination of race and gender point up the depth and pervasiveness of intergroup hostility on campuses all over the country. In the small rural town of Olivet, Michigan, at Olivet College, a school founded in 1844 by the abolitionist minister the Reverend John Shipherd as a "bastion of racial tolerance," a "racial brawl" involving approximately forty white students and twenty black students broke out one night in early April 1992. According to one report:

> Racial epithets were shouted at the black students as the two sides rumbled on the gray linoleum. Two students, one black, one white, were injured and briefly hospitalized.
>
> Afterward, blacks and whites who had crammed together for midterms and shared lunch money and dormitory rooms could not look each other in the face and were no longer on speaking terms.

This incident was the culmination of increasing hostility among black and white students at Olivet. In the months prior to the incident, white male students had become more openly resentful of black men dating white women. Then, on April 1, a white female student claimed she had been attacked by four black students and left unconscious in a field near the campus. She was not hospitalized, and, despite a police investigation, no arrests were made. College officials were said to be skeptical about her accusations. Nonetheless, word spread, and later that night two trash cans were set on fire outside the dormitory rooms of black student leaders.

The specific incident that precipitated the brawl occurred the next night and again involved a white female student and black male students. Three male students, two black and one white, knocked on a female student's door to ask about a paper she was typing for one of the black students. The men later described the conversation as "civil." The woman, a sorority member, called her brother fraternity for help, saying she was being harassed by some male students. Within a few minutes, about fifteen members of the white fraternity Phi Alpha Pi arrived and confronted the two black men. More whites joined in, and black female students called more black males to even the numbers. Who threw the first punch is unclear; black students claim it was a white fraternity member. "What is clear," according to one report of the incident, "is that instead of seeing a roommate or a fellow sophomore, the students saw race." Davonne Pierce, a dormitory resident assistant who is black, stated that his white friends shouted racial epithets at him as he was trying to break up the fight. He said to them, "How can you call me that when we were friends, when I let you borrow my notes?" But, he later recalled, "At that point, it was white against black. It was disgusting."

After the incident, most of the fifty-five black students, who said they feared for their safety, left the college and went home. They made up 9 percent of the student body. Davonne Pierce stayed on campus but stated, "Obviously they don't want us here." Dave Cook, a white junior who was one of the fraternity members involved in the fight, later said, "There were a lot of bonds that were broken that didn't need to be broken." He talked about his friendship with a black female student: "We would high-five each other and study for tests. But I don't know what she thinks about me. I don't know whether she's hating me or what. I didn't say one word to her, and she didn't say one word to me. Now she's gone."

Racist behavior on college campuses is, of course, not limited to students. An incident involving a campus in New York State reveals the deep-seated stereotyping and bigotry of some college administrators, police officers, and citizens in communities all over the country. In the early morning of September 4, 1992, a seventy-seven-year-old woman was attacked in the small town of Oneonta, New York. She told the state police that she thought her attacker was a black man who used a "stiletto-style" knife and that his hands and arms were cut when she fended him off. In response to a request from the police, the State University of New York at Oneonta gave the police a list of all of the black and Hispanic males registered at the college. Armed with the list, state and city police, along with campus security, tracked down the students "in their dormitories, at their jobs and in the shower." Each student was asked his whereabouts at the time the attack occurred, and each had to show his hands and his arms.

Michael Christian, the second of five children from a family headed by a single mother, grew up in the Bronx. His mother encouraged him to go to Oneonta to get him away from the problems of the city. Shortly after the attack, two state-police officers and representatives of campus security went to his dormitory room and woke him at 10:00 A.M. After asking him where he was at the time of the attack and demanding to see his hands, they said they wanted to question him downtown, and then they left. His roommate, Hopeton Gordon, a Jamaican student who had gone to high school in the Bronx with Mr. Christian, was questioned in front of other students from their dormitory. When the police asked to see his hands and he demanded their reasons, they responded, "Why? Do you have something to hide?" He said that he had felt humiliated in front of his suite mates and in front of female students.

This is not the first time black students, faculty, and administrators have been humiliated and have seen their civil rights trampled in Oneonta. Edward I. (Bo) Whaley, who went to the small town in upstate New York in 1968 as a student, remained, and is currently an instructor and counselor in the school's Educational Opportunity Program for disadvantaged students, recalls being followed by salespeople in Oneonta shops because they feared he would shoplift. He remembers the two minority ball players—one of whom he was trying to recruit—who were picked up as suspects in a rape case and had to pay for DNA testing even though someone else was convicted for the crime.

An admissions coordinator, Sheryl Champen, who is also black, was herself stopped by the state police the night of the attack. They demanded to see her identification before she could board a bus to New York City. It is unclear why the police questioned her and three other black women traveling with children, who also had to show identification before boarding the bus, since the attacked woman had reported that the person who assaulted her was a man. Their only common characteristic was race. When she heard about the treatment of the students of color, Ms. Champen said, "I was devastated, ashamed of being an admissions coordinator. Am I setting them up?" She feels that the behavior of the police was not an example of overeagerness to solve a crime. After recounting thirteen years of incidents that had begun when she was a first-year student at SUNY/Oneonta, she stated, "I know what it was. It was a chance to humiliate niggers."

The release of the names of the 125 black and Hispanic students not only violated their privacy (and their right to be presumed innocent until proven guilty) but also violated the Family Educational Rights and Privacy Act of 1974 (also known as the Buckley Amendment). Following the incident, the vice-president who authorized the release of the names was suspended for one month without pay and demoted. The president of SUNY/Oneonta called using the list in the investigation "an affront to individual dignity and human rights."

Though each of these events is unique and a product of the particular social environment, demographics, personalities, and stresses at the particular institution, during the late 1980s and early 1990s campuses were rife with similar episodes. A Brown student describes one incident at her university:

> It was April 25, 1989, the end of spring term. . . . Students . . . were preparing for Spring Weekend, an annual fling before final exams. That day, found scrawled in large letters across an elevator door in Andrews dormitory were the words, NIGGERS GO HOME. Over the next 24 hours, similar racial epithets were found on the doors of several women of color living in that hall; on the bathroom doors WOMEN was crossed out and replaced with NIGGERS, MEN was crossed out, replaced with WHITE. And in that same women's bathroom, a computer-printed flyer was found a day later which read: "Once upon a time Brown was a place where a white man could go to class without having to look at little black faces or little yellow faces or little brown faces except when he went to take his meals. Things have been going downhill since the kitchen help moved into the classroom. Keep white supremcy [sic] alive! Join the Brown Chapter of the KKK."

Seven years earlier, the *Dartmouth Review* had set the standard for racist denigration by publishing an article ridiculing black students. The article was entitled "Dis Sho' Ain't No Jive, Bro," and read in part: "Dese boys be saying that we be comin hee to Dartmut an' not takin' the classics. . . . We be culturally 'lightened, too. We be takin' hard courses in many subjects, like Afro-Am studies . . . and who bee mouthin' bout us not bein' good read?"

During the late 1980s, the University of Michigan experienced several racist incidents. One of the most infamous took place in 1988, when a poster mocking the slogan of the United Negro College Fund was hung in a classroom. It read "Support the K.K.K. College Fund. A mind is a terrible thing to waste—especially on a nigger."

Violent behavior has also been part of the cultural climate over the past decade. In February 1991, two black students from the University of Maine were allegedly assaulted by nine white men. The two students, both twenty-one, were driving in downtown Orono when approximately a dozen white men attacked their car and shouted, "Nigger, get out of here." When they got out of the car to see what was going on, the two men were kicked and beaten. Three students from the university were among those who attacked the students.

Incidents have not been limited to one kind of school, but have occurred at private as well as public, urban and rural, large and small, at Ivy League as well as less prestigious, little-known institutions. . . .

According to the Anti-Defamation League, anti-Semitic incidents on college campuses have risen sharply in recent years, from fifty-four in 1988 to over double that number, 114, in 1992. In February 1990, at American University in Washington, D.C., anti-Semitic graffiti were spray-painted on the main gate and on a residence hall. On the gate were painted a Star of David, an equal sign, and a swastika. On the dormitory was sprayed an expletive followed by "Israel Zionist." In 1991, at California State University at Northridge, a ceremonial hut used to celebrate the Jewish holiday of Sukkoth was vandalized with anti-Semitic writing. In addition to swastikas, "Hi' [sic] Hitler" and "Fuckin [sic] Jews" defaced the informative signs and flyers that decorated the hut. Two months earlier, Dr. Leonard Jeffries, Jr., then chair of the African-American Studies department at the City College of New York, delivered a speech at a black cultural festival in which he spoke of "a conspiracy, planned and plotted and programmed out of Hollywood" by "people called Greenberg and Weisberg and Trigliani." He went on to say that "Russian Jewry had a particular control over the movies and their financial partners, the Mafia, put together a financial system of destruction of black people."

Gay bashing has also been widely visible on college campuses. A Syracuse University fraternity, Alpha Chi Rho, was suspended by its national organization in 1991 for selling T-shirts with antihomosexual slogans, including one advocating violence against gays. On the front the shirts said "Homophobic and Proud of It!" and on the back, "Club Faggots Not Seals!" The picture illustrating the words was of a muscled crow, the fraternity's symbol, holding a club and standing over a faceless figure lying on the ground. Next to them is a seal hoisting a mug of beer.

During the same year, *Peninsula,* a conservative campus magazine at Harvard, published an issue entirely devoted to the subject of homosexuality. The magazine called homosexuality a "bad alternative" to heterosexuality and stated in its introduction that "homosexuality is bad for society." Within one hour of the magazine's distribution, the door of a gay student's room was defaced with antihomosexual words.

But, of course, discrimination does not need to be physical or perpetrated by students to wound, and to exclude some from mainstream college life. The football coach at the University of Colorado has called homosexuality "an abomination" and has supported a statewide group working to limit gay rights. In 1992, the governor of Alabama signed legislation prohibiting gay student groups from receiving public money or using buildings at state universities.

As we have seen, a variety of groups have been perceived and treated as "the Other"—in Patricia Hill Collins' words, "viewed as an object to be manipulated and controlled"—on college campuses over the past few years. Though many of the bias incidents have involved racial enmity and misunderstanding, anti-Semitism, homophobia, and blatant sexism have also been catalysts for hostile acts. Many academic institutions, concerned about overtly demeaning, sometimes violent behavior as well as the far more subtle denigration of women and other minority groups, have attempted to address these problems through a variety of measures: speech codes; orientation programs for entering students that stress respect for diversity and the importance of civility; curriculum changes that focus on multiculturalism; hiring policies whose goals are to increase the number of women and members of minority groups on the faculty and staff of the institution; and the recruitment of more students of color. These measures, often employed to counter the ignorance, ethnocentrism, and anger within the college community, have themselves become the subject of controversy and debate. Both academic and popular discourse have focused far more on political correctness, on affirmative action, and on changes in the curriculum than on the hate incidents and violence that continue to occur. Speech codes at the universities of Wisconsin and Michigan became front-page news; discussions of what and who was p.c. seemed ubiquitous; and the pros and cons of a multicultural curriculum have been debated in university governing bodies and editorial meetings across the country.

40

The Structure of the U.S. Health Care System

DEBRA M. MCPHEE

Contemporary health care offers high levels of promise, but does not deliver these equally to all. The system of social stratification is replicated in the system of health care delivery, as this is influenced by the economics of health care. In this selection McPhee takes an overview of current developments in American health care as it is related to structural elements in health care delivery. She reflects on some of the struggles we have gone through over universal health insurance, and what the implications are of our current system of coverage. How much we pay for our health care and whether we have access to it, including our freedom of choice to see a doctor of our preference, are clearly shown to be factors that differ significantly among various groups in our society. What some take for granted as a minimum necessity (i.e., seeing a doctor they like) may be out of the reach of others who are grateful to see any doctor at all. This inequality in our nation's health care has profound impacts on the life chances of our citizens and is one of the primary ways that social hierarchies are reproduced from one generation to the next. What kind of health care system do you use? Do you think it is effective for treating you or would you suggest some other form of delivering health care? How does one's social class affect access to health care? How does the health care system in the United States compare to the system in other countries? Which system do you prefer?

The word crisis has frequently been used to describe the current state of the American health care system. The majority of writers on the subject agree on the paramount crisis issues: escalating health care costs; inadequate coverage; relatively high levels of public dissatisfaction; and expensive, complex administrative requirements. There is, however, much less agreement on what should be done in the way of reform.

Proposals for health care reform generally fall into three categories: (1) private marketplace insurance; (2) employer-based health insurance; (3) centralized or single-payer health care models. This chapter will examine the current nature of the U.S. health care debate. Included is a critical review of the prominent health care reform proposals as they relate to issues of coverage, access to care, and the management of costs. The examination considers the influence of public opinion on the development of public health policy, and provides a critical analysis of some of the salient political issues likely to affect the future direction of the national health care debate in the United States.

THE STRUCTURE OF THE U.S. HEALTH CARE SYSTEM

Mapping the structure of the U.S. health care system is a challenging task. In theory, health care coverage is available to virtually all Americans through one of four routes: (1) Medicare for the elderly and disabled; (2) Medicaid for low-income/public assisted individuals and persons with certain disabilities; (3) employer-subsidized coverage in the workplace; or (4) self-purchased coverage available through private insurance companies (e.g., Blue Cross/Blue Shield). A few additional sources of health care coverage do exist for special category individuals which include medical services provided to members of the Armed Services, and Veterans Administration (VA) medical programs. Each form of provision differs regarding eligibility requirements and the specific care aspects that are covered. The financing and administration of the various forms of health care are handled by way of a complicated patchwork of federal, state, and local legislation. Thus, the health care delivery "system" as it presently operates is in fact a complex, multilayered industry—an industry which has been plagued by a decade of rising costs and increasingly restrictive governmental policies resulting in a limiting of resources and service availability.

NATURE OF THE U.S. HEALTH CARE CRISIS

Coverage

It is estimated that between 31 to 37 million Americans are currently lacking some form of private or public medical insurance coverage.[1] While in excess of 33 million persons are presently enrolled in the Medicare program, coverage often becomes problematic when the physician bills the patient for more than the Medicare approved charge, leaving the patient liable for the difference.[2] In addition, the minimum deductibles and co-insurance amounts that the Medicare beneficiary must pay out-of-pocket can be substantial. With specific reference to public health assistance the Medicaid program has increasingly failed to provide adequate coverage to the persons it was designed to protect. The number of persons covered by this program has risen over the past decade from 30 to 37 million at the same time that the percentage of the poor *not* covered by Medicaid has risen from about 35% to about 60%.[3]

Americans who do not qualify for health care coverage through either the Medicare or Medicaid programs, and are not covered by way of employer-assisted plans, are left with the option of securing adequate coverage for themselves and their families through one of the many private health insurance companies. Presently, consumers must choose from among the 1,500 private insurance companies operating throughout the United States. Each company is free to specify its own eligibility requirements, extent of coverage, and reimbursement rates. Insurance premiums are, of course, the responsibility of the individual, and vary widely from company to company.

As with any other consumer good, purchasing health insurance in a private marketplace system means one generally gets what one pays for. Insurance packages that provide anything above a minimum level of coverage are often costly and beyond the financial resources of most Americans. With health care costs on the rise, many private insurance companies have begun to institute strict eligibility requirements, restricting types and amounts of coverage offered, and have significantly increased their premiums. Consequently, even those persons who do manage to retain coverage may find themselves without enough coverage at a time of illness or injury. It is estimated that well over 50 million persons in the United States are inadequately covered by their present health insurance, meaning that a major illness will translate into almost certain financial ruin.[4]

As an alternate choice, Health Maintenance Organizations (HMOs) have been presented as the solution to the health care crisis in general and to the problems of the Medicare/Medicaid programs in particular. It has been argued that incorporating HMOs into Medicaid would improve the states' abilities to budget expenditures, simplify management, eliminate abusive practices directly linked to fee-for-service care, and contain costs.[5] It has further been argued that increasing the use of HMOs will increase general access to mainstream medicine.[6] But while alternative structures such as HMOs continue to enjoy support, they have also been severely criticized for illustrating poor program designs, unethical marketing practices, inadequate or underservicing, and excessive administrative costs.[7]

In the past, coverage for American workers, and often their dependents, was a cornerstone of health insurance. In recent years, however, employer-subsidized health insurance plans have suffered greatly. Since the early 1980s it is estimated that the percentage of corporate profits consumed by health care has risen from 29% to 49%.[8] In response to continually rising costs many employers have reduced the amount of health care coverage they offer or have canceled employee insurance benefits altogether. Even when employers do not restrict health care coverage, an increasing number of insurance companies are excluding individuals based on prior health conditions, and some exclude entire industries whose employees are considered "high risk." Consequently, employment alone does not necessarily guarantee greater access to health care coverage and many of the "working poor" across the United States do not receive adequate medical treatment. Ironically, the nonworking poor may, in some cases, receive better medical care than their employed counterparts. For example, an unemployed poor person may be eligible for Medicaid benefits or indigent care, whereas an employed poor person may be denied benefits or indigent care precisely because he or she is employed. In brief, the working poor fall between the cracks— they have too little income to purchase adequate health insurance and too much income to qualify for medical assistance. Thus, with respect to health care coverage there exists great disparity in the U.S. system. At one extreme, active duty military personnel receive complete medical coverage; at the other end, well over 14% of the population receive little or no medical coverage.

Access to Care

On a state-by-state comparison the proportion of uninsured varies depending on several factors, including the level of Medicaid coverage in the state, the demographics of the population, insurance practices, overall income, the nature of employment, and the individual state health policy. Of those clients who are covered by Medicaid, access to care is not necessarily guaranteed. While care is theoretically available, health care facilities are frequently so overloaded that access is unrealistic. Again, while Medicaid patients theoretically have the option of choosing their own physicians, in reality the choice is not the recipients'. More and more physicians have refused to accept Medicaid clients. Physicians' resistance to treating such patients has been ascribed to many causes, including low and delayed Medicaid payments, fears of malpractice litigation, paperwork, cultural or language problems, noncompliance, and other factors, including racial discrimination.[9] Undoubtedly, the prospect of low or nonexistent payment is a disincentive to most providers who, under the guidelines set down by the American Medical Association, have the right to refuse treatment if the patient cannot demonstrate an ability to pay up-front.[10] Thus, a complete health care screening in the United States includes a specific evaluation of a person's ability to pay *before* he or she can receive care. Many cannot measure up and as a result it is estimated that every year more than 1 million American citizens are turned away by health care practitioners and institutions because they lack financial means or medical coverage.[11]

Health Care Spending and the Issue of Cost

The issue of cost and the need for cost containment with respect to health care spending is most often the primary focus of the health care debate. Unquestionably the cost of health care in all the industrialized countries is high and rising fast. Since the early 1970s the percentage of the U.S. GNP consumed by health care has continued to rise to well over 12%.[12] Interestingly, the percentage of the GNP consumed by health care costs for both the United States and its closest neighbor, Canada, was virtually identical (around 9%) until 1971, when Canada adopted a centralized national health insurance plan.[13] Since that time Canadian percentages have stayed around 9% while the U.S. figures have continued to climb. In just one decade U.S. annual health care spending has increased from $248 billion to $650 billion.[14] This means the United States currently spends over

$3,000 per person, per year, in a system which still leaves one in four persons uninsured or without adequate health insurance coverage.[15]

Public health programs, and specifically the Medicaid program, have long been primary targets of governmental cost containment initiatives in the health care field. In the early 1980s under the Reagan administration, both the federal and state governments sought to control or reduce Medicaid expenditures in the face of tax cuts, growing costs, and reduced federal funds for the program. This led to freezes and reductions in both eligibility and provider payments. The result was a basically stable number of beneficiaries despite an increase in the poverty population.[16] Thus the health care cutbacks and cost containment measures of both the Reagan and Bush administrations resulted in decreased access and eligibility for persons in need, without significantly reducing the overall cost of health care delivery.

Social Implications

Although inadequate health care may bring personal tragedy to impoverished individuals, the long-term consequences may be even more costly to the society as a whole. The National Center for Health Statistics reports that in 1988, 17% of American children under the age of 18 years had neither private insurance nor Medicaid coverage.[17] Given the numerous studies validating the importance of preventative medical treatment and early intervention, particularly with respect to prenatal care, these rates are, at the very least, alarming.

Studies have consistently shown that babies born to mothers with no prenatal care are five times more likely to die and three times more likely to end up in neonatal intensive care units.[18] The United States, the same country that ranks number one in the world for health care spending, presently ranks twentieth in the world in infant mortality.[19] The maintenance of a system that fails to focus on prevention as a medical priority has a great deal to do with this situation. Statistics indicate that while 95% of Canadian women will receive prompt prenatal care, this is true for only about 75% of pregnant women in the United States.[20] Currently, a baby born in Detroit, Michigan is 38% more likely to die in the first year of life than a baby born just across the border in Windsor, Ontario.[21] Clearly preventative medical services do not come cheaply. However, while Canada may be ranked number two in the world in health care spending, it also boasts the second highest life expectancy rate at birth for both sexes and has the second lowest infant mortality rate, next to Japan.[22]

Although health care coverage in itself is not the sole determinant of health status, Medicaid data has repeatedly demonstrated that coverage is a key factor in overall health and well-being.[23] And although availability of care does not in and of itself guarantee improved health, medical indigence has been strongly associated with lack of care and poor health status. What is clear, is that presently the uninsured represent significant costs both economically and socially. While most of the uninsured will receive acute care services in necessary situations, the point of entry into the system is generally at the worst stage—through the hospital emergency room. Here it is often a case of too little too late with the results generally being high costs and poor outcomes. Furthermore, it is increasingly the case that preventable conditions are being treated as emergency cases in the least cost-effective manner, with hospitals having to provide primary care in their emergency settings.

Clearly the impact of the U.S. health care crisis can be felt in every sector, both public and private. The nature of the health care crisis for the individual American is essentially identical to that of the nation as a whole—restricted accessibility and the high and uncertain costs of obtaining medical care. In recent years pressure on the government to reform the health care system has escalated, coming mainly from the inadequately covered, employers, big business, and employee groups. Further, there is a growing consensus that the incremental reform efforts of the past are no longer sufficient for bringing about the level of fundamental change required. . . .

CONCLUSION

There is little doubt that an era of inevitability is upon the United States with respect to health care reform. Over the past decade pressure has been exerted from many directions for an overhaul of the present system. While support for fundamental reform appears to be growing among the public and the policy makers alike, there remains a lack of consensus as to what shape the new system should take. The governmental cost containment measures of the 1980s have more often than not been unsuccessful in their prime objective, and have served only to restrict further services to those most in need. In recent years there has been an increase in the number of advocates in favor of the United States adopting a plan of national health insurance, based on a centralized model. Yet there remains a significant number of powerful groups and individuals who oppose any scheme which would call for greater government involvement or restrictions which could potentially interfere with this enormously profitable industry. . . .

It has been demonstrated that there has yet to be developed a system without shortcomings or gaps in the "safety net." All countries utilizing some form of centralized universal health insurance also incorporate some means of rationing their resources. As costs continue to rise, countries supporting universal plans will need to be more efficient in *how* their medical care is delivered and received. The United States on the other hand, continues to focus on measures that propose being more selective about *who* receives medical care based on socioeconomic and/or employment status. The widespread negative consequences of this kind of selectivity have been clearly demonstrated with respect to the health status both of individual Americans and the nation as a whole. As social and economic conditions force Americans to rethink their current health delivery practices, government's success in securing universal health care for all citizens hinges on the extent to which it is willing to reject proposals which divide its populace into the "deserving" and the "undeserving."

NOTES

1. See Marmor, T. (1993). Commentary on Canadian Health Insurance: Lessons for the United States. *International Journal of Health Services,* 23(1), 45–62: Reinhardt, U. (1992). Commentary: Politics and the Health Care System. *The New England Journal of Medicine,* 372(11), 809–811; Woolhandler, S. et al. (1993). High Noon for U.S. Health Care Reform. *International Journal of Health Services,* 23(2), 193–211.

2. Renas, S., and Kinard, J. (1990). Importing the Canadian Plan. *Health Progress,* March, p. 23.

3. Inglehart, J. (1992). Health Policy Report. *The New England Journal of Medicine,* 327(20), 1467–1472.

4. Friedman, E. (1991). The Uninsured: From Dilemma to Crisis. *JAMA,* 19(May), p. 2492.

5. Spitz, B. (1979). When a Solution is Not a Solution: Medicaid and Health Maintenance Organizations. *Journal of Health Politics, Policy, and Law,* 3(4), 497–518.

6. Spitz, p. 498.

7. Spitz, p. 514.

8. Weisberg, R., and Mayers, L. (1990). Borderline Medicine. *PBS: Public Policy Productions.*

9. Fossett, J. W. et al. (1991). Medicaid Patients Access to Office Based Obstetricians. In Friedman, E. (1991), The Uninsured: From Dilemma to Crisis. *JAMA,* 19(May), p. 2492.

10. Relman, A., and Reinhardt, U. (1986). Debating For-Profit Health Care and the Ethics of Physicians. *Health Affairs* (Summer), p. 6.

11. Weisberg and Mayers.

12. Woolhandler, S. et al. (1993). High Noon for U.S. Health Care Reform. *International Journal of Health Services,* 23(2), 193–211.

13. Woolhandler et al.

14. Weisberg and Mayers.

15. Clinton, W. (1992). The Clinton Health Care Plan. *The New England Journal of Medicine,* 327(11), 804–807.

16. Friedman, p. 2492.

17. National Center for Health Services Research and Health Care Technology Assessment (1991). In Friedman, E. (1991), The Uninsured: From Dilemma to Crisis. *JAMA,* 19(May), p. 2491.

18. Weisberg and Mayers.

19. Weisberg and Mayers.

20. Weisberg and Mayers.

21. Weisberg and Mayers.

22. Mhatre, S., and Deber, R. (1992). From Equal Access to Health Care to Equitable Access to Health: A Review of Canadian Provincial Health Commissions and Reports. 22(4), p. 648.

23. Friedman, p. 2493.

41

Patient-Doctor Relations in the Era of Managed Care

HOWARD WAITZKIN

One of the great debates surrounding contemporary American health care involves the shift from a fee-for-service system of health care delivery to the health maintenance organizations (HMOs). These new plans have arisen from the need to control the rising cost of medicine and hold the promise of potentially making health care more affordable. Controversy has arisen surrounding several aspects of their functioning, however, as profit considerations have been increasingly seen to impinge on medical decision making. Doctors are pressured to spend less time with each patient and to refrain from ordering unnecessary tests that might be expensive. How this impinges on the relationships that develop between doctors and their patients is the focus of this selection by Waitzkin. Under the HMO system, each patient is assigned a primary care doctor, Waitzkin notes, who is responsible for coordinating that person's care. This frequently places doctors in administrative positions that they previously lacked, and strains the quality of health care delivery. Through illustrative case studies, Waitzkin examines the sometimes conflicting nature of the primary care physician's gatekeeper (purse-string manager) and patient advocate (medical manager) roles, showing how these undermine the quality of health care delivery and lead doctors to feel less good about the way they practice medicine. Think back to the last time you visited your doctor. Did you feel that you received personalized treatment or do you feel as if you were treated merely as a number? What aspects of the doctor-patient relationship do you think need improvement and what aspects do you think are good (given the massive numbers of people that doctors have to face daily)? How might the doctor-patient relationship be different in fee-for-service systems as compared to HMOs?

From Howard Waitzkin, *At The Frontlines.* Reprinted by permission of Rowman & Littlefield.

Alongside issues of access, cost, and the social origins of illness, there are important problems of clinical care that need to be addressed through reform. As managed care proliferates in the United States and other countries, it has transformed patient-doctor relationships in very fundamental ways. This transformation has occurred rapidly, with little preparation for either clinicians or patients. In any assessment of managed care's impact, the effects on patient-doctor relationships deserve attention. The debate so far has raised many potential problems, few of which have received serious enough analysis to exert an impact on health care policy, either within or outside the managed care industry.

The impact of managed care illustrates a more general principle: changes in the health-policy environment pattern changes experienced in patient-doctor relationships. Important constraints emerge from the economic structure of the health-policy environment. Such constraints include the financial and contractual arrangements that physicians and patients make with managed care organizations (MCOs).

Although problems in patient-physician relationships of course antedated the growth of managed care, the characteristics of managed care both exacerbate old problems and create new ones. There is no reason to idealize the patient-physician relationships of the past, but managed care has led to worrisome changes in these relationships. Also, while the characteristics of MCOs differ, structural features of managed care introduce strains in patient-physician relationships that cut across different types of MCOs. In this chapter, I focus on the connections between the structural features of managed care and the changes that have occurred in patient-physician relationships.

GATEKEEPER VERSUS DOUBLE AGENT

Advocates of managed care often claim that this method of organizing services can and should actually improve the patient-physician relationship. For instance, since MCOs generally assign patients to a single primary care physician, that person presumably can provide continuity and can help improve the coordination of services. The primary care provider can communicate about preventive services and encourage their utilization. Since managed care services are mostly paid in advance through monthly capitation payments, the predictability of copayments required for each outpatient visit may also reduce financial barriers to access for some patients.

Not a single research project, however, has conclusively demonstrated improved patient-physician communication processes or patient satisfaction in managed care systems. The limited studies comparing communication and satisfaction in managed care versus fee-for-service sectors have found either no difference or observations disfavoring managed care.[1] The adverse impact of managed care on communication processes and the patient-physician relationship also has received attention in influential editorials and position papers.[2] Further, the claimed advantages of managed care in the arenas of communication and interpersonal relationships have not been assessed in detail for growing subgroups of enrollees, including minorities, the poor, non-English speakers, the elderly, and the chronically ill.

Meanwhile, managed care refers to primary care practitioners as "gatekeepers." That is, such physicians tend the gate, keeping it closed for expensive procedures or referrals to specialists or emergency visits, and open the gate only when it is absolutely necessary for preservation of life or limb. The reason for tending the gate carefully is very clear: that is how physicians and their bosses keep enough of patients' capitation payments to break even or maybe come out a little ahead.

Double agent, as Marcia Angell and others have pointed out, has probably become a more cogent way to think about physicians' role as gatekeeper under managed care.[3] In essence, while continuing to pose as advocates for patients, physicians in actuality work as double agents for both patients and MCOs. The latter organizations hold interests that structurally are often diametrically opposed to patients'.

Perhaps "continuing to pose" as patient advocates may not convey the conflict adequately. Recently a primary care physician spent nearly a day advocating for a single patient, a 58-year-old psychologist with a displaced fracture of her elbow, to various Health Net bureaucrats, in order to convince them that she really did need an orthopedic appointment today rather than in three weeks, and also really needed surgery in three days rather than possibly in the indefinite future. Many clinician colleagues are burning out with the energy that such maneuvering takes, with such little apparent benefit for either patients or the physician "gatekeepers." At the very least, doctors find that such activities lead to rationing of services of inconvenience.[4] That is, when obtaining services for patients entails such inconvenience, an incentive arises not to pursue the matter vigorously, thus decreasing the probability that the patient will receive the services, even though needed.

More often, clinicians feel an inherent conflict—either ethical or financial or both—between patients' interests and those of the managed care systems that physicians represent as gatekeepers. This is the essence of physicians' work as double agents. Clinicians experience this conflict even when they supposedly benefit financially by keeping the gate closed.

ILLUSTRATIVE CASE SUMMARIES

These structural constraints of course also affect patients' experience of medicine in general and managed care in particular. The following encounters illustrate some generic issues that increasingly manifest themselves as managed care proliferates.

The first patient was a 58-year-old male engineering professor, insured by UC Care, the University of California's new self-insured managed care program. This patient developed severe substernal chest pain at 1 A.M. one morning during a Thanksgiving holiday weekend. The patient called the on-call primary care internist to approve a visit to an emergency room, as he had been instructed to do when he signed up for the plan. The on-call physician had covered more than thirty patients on the inpatient wards, in the intensive care units, and by phone that day for his colleagues. He was sleeping soundly when the patient called. He also was a little hard of hearing and did not wake up until a half hour after the patient called, when the answering service again tried to reach him. By that time, the patient had left for the emergency room because of continuing pain. The physician then called the emergency room to approve the visit. As soon as the patient arrived, his electrocardiogram revealed a large myocardial infarction. If the patient had waited for approval, as he was supposed to do, he would have arrived too late for treatment with streptokinase to help lyse the clot in his coronary artery, which was the standard of care for a patient with this type of heart attack.

The second patient, a 25-year-old, Spanish-speaking woman who worked as a hospital maintenance worker, was insured by Health Net, the MCO that workers for the University of California can join for the cheapest rate. She presented to the emergency room during her shift at 3 A.M. with sore throat and a fever of 101 degrees. Because she had not realized that she was supposed to call the on-call physician for permission, the emergency room staff called the physician and woke him up. Groggy and mad at being awoken solely for this bureaucratic reason, he asked to speak with the patient (fortunately, he could speak Spanish), realized the problem could wait until the daytime, and did not approve the emergency room visit. The patient complained that it would be difficult to come during the day because of child care responsibilities, but she could not persuade the physician to approve the emergency room visit, since he had been instructed not to approve such visits for minor outpatient problems.

In dealing with these cases, the physician experienced several feelings:

- guilt that he had not heard or responded quickly to the initial phone call from the patient having a heart attack;
- anger that the nature of managed care led to a critical delay in this patient's evaluation and treatment;
- sympathy for the patient whose visit he refused to approve because a later visit would prove inconvenient and because she hadn't understood the rules;
- annoyance that he had to perform mainly a gatekeeper role in both cases, using essentially none of his clinical skills;
- awareness that financial considerations underlay all these decisions: that he and his colleagues would receive a bonus at the end of the year if they could hold down emergency room utilization; that the paltry $7 per month that two of his colleagues received to cover all outpatient care for each of these patients would decrease even further if utilization became much higher; and that his own motivation to provide services subtly decreased with such managed care patients, since doing more work was not associated with more income. He also thought of Health Net's chief executive officer at the time, Roger Greaves, who reportedly received a salary of almost $3 million, plus bonuses and stock options, in 1994, and who seemed the main beneficiary of physicians' good work as gatekeepers.[5]
- frustration that all these emotions deviated enormously from those he had expected to have in medicine, a career he selected with the assumption that

he would mainly have the opportunity to serve those in need and to receive an adequate salary for doing so, not linked to patients' ability to pay or insurance coverage.

- most of all, awareness that his communication with managed care patients was becoming distorted by the structural nature of these payment arrangements, and that the openness and honesty he valued were becoming ever more tenuous.

Although the above situations were uniquely experienced by the author, many clinicians have had similar experiences under managed care. That is why some clinicians have struggled for suitable policies that would not require such conflictual activities.

The two patients in the above case summaries directly confronted the limitations that managed care imposes on individual discretion. In the first case, the patient—though knowledgeable about the rule that emergency room visits must be pre-approved to be paid for by the managed care plan—overrode that constraint through a judgment decision about the urgency of his symptoms. As it later became clear, his sophistication as a scientist and educator contributed to his choice to take action without the gatekeeper's permission—a decision that may have saved his life. The second patient acted from a position of ignorance about the structural constraints of the plan in which she had enrolled and could not present a convincing enough argument to accomplish her own preferences to be seen sooner rather than later. In neither case did the structural constraints imposed by managed care's gatekeeping principle conform with a close and trusting patient-physician encounter, in which the patient's preferences could guide the course of care in a predictable way.

FINANCIAL AND ORGANIZATIONAL ISSUES

Also from the patient's viewpoint, the structure of managed care constrains the very nature of the communicative process. Here, instead of exploring possible options with full participation by patients in decision making, the patient must make a case strong enough to be accepted by the gatekeeping physician. Financial interests, by which up to 80 percent of physicians' annual income may be at risk if they do not adequately restrict services under managed care contracts (Kuttner, 1998), reinforce the physician's skeptical appraisal. Under these conditions, especially as patients become more aware of these financial relationships, patients' trust of their physicians may seriously erode.

Further, the communicative process increasingly occurs under constraints of time. The on-call physician, unpaid for time spent on the phone in the middle of the night, is not disposed to lengthy and supportive conversation, especially with a patient who is a stranger. For more routine encounters during daylight hours, additional constraints on communication arise, as the productivity expectations of MCOs create standards that require physicians to see greater numbers of patients per unit of time. From the organizational viewpoint, the fixed capitation received per patient exerts pressure to maximize the number of patients seen by each salaried practitioner in each patient-care session. Since MCOs strive to fill physicians' schedules, patients may have to see other practitioners than their own, including physician substitutes like nurse practitioners and physician assistants; such organizations frequently employ mid-level practitioners to handle overflow from physicians' full schedules. Physicians employed by MCOs therefore enjoy little discretion in determining how much time to spend with each patient.

On the other hand, the structural constraints in the patient-physician relationship did not begin with managed care. Under the prior fee-for-service system, communication between patients and physicians suffered from a variety of problems, some tied to the financial underpinnings of that particular form of practice organization.[6] For instance, encounters between primary care practitioners and patients tended to be hurried, and little time was spent communicating information. Interruptions and dominance gestures by physicians commonly cut off patients' concerns. Further, exploration of issues in the social context of medical encounters, which patients experienced as important components of their lived experience of illness, tended to become marginalized in patient-physician encounters.

The financial structure of fee-for-service medicine created an incentive to maximize patients seen per unit of time, and to decrease the time devoted to in-depth exploration of patients' concerns. In contrast to managed care, this productivity constraint usually permitted the practitioners substantial discretion in choosing how much time to spend with a given patient. Patients' dissatisfaction with communication under the fee-for-service system nevertheless continued to rank among their most frequently voiced complaints about U.S. medical practice (Waitzkin, 1984). In recent years, even before the advent of managed care, many calls for improvements in physicians' communicative practices came to the surface (Roter and Hall, 1992; Lipkin, et al. 1995; Smith, 1996).

It therefore would be an error to see the fee-for-service structure as necessarily more conducive to favorable communication and relationships than managed care. The financial incentives of fee-for-service medicine also have created their own adverse effects. Yet the structure of managed care does little to improve those earlier problems and also introduces a new set of constraints that may prove even more contradictory and discouraging for patient-physician relationships.

NOTES

1. A. R. Davies. 1986. "Consumer Acceptance of Prepaid and Fees-For-Service Medical Care: Results from a Randomized Controlled Trial." *Human Services Research* 21: 429–452; H. R. Rubin. 1993. "Patients Ratings of Outpatients' Visits in Different Practice Settings." *Journal of the American Medical Association* 270: 835–840; "Health Care in Crisis: Are HMOs the answer?" *Consumer Reports* 57: 519–531, August 1992. On the quality of care in for-profit managed cared organizations, see D. U. Himmelstein,

S. Woolander, I. Hellander, and S. M. Wolfe. 1999. "Quality of care in Investor-Owned vs. Not-For-Profit HMOs." *Journal of American Medical Association* 282(2): 159–163 (July 14).

2. M. Angell. 1993. "The Doctor as Double Agent." *Kennedy Institute for Ethics Journal* 3: 279–286; E. J. Emanual and N. N. Dubler, "Preserving the Physician-Patient Relationship in the Era of Managed Care." 1995. *Journal of the American Medical Association* 273: 323–329; J. Balint and W. Shelton. 1996. "Regaining the Initiative: Forging a New Model of the Patient-Physician Relationship." *Journal of the American Medical Association* 275: 887–891.

3. See Angell, "The Doctor as Double Agent."

4. G. W. Grumet. 1989. "Health Care Rationing through Inconvenience." *New England Journal of Medicine* 321: 607–611.

5. W. Coyle and G. Hostetter. 1996. "Kaiser Leader in Care Payout." *Fresno Bee,* p. E1. February 13.

6. H. Waitzkin. 1984. "Doctor-Patient Communication: Clinical Implications of Social Scientific Research." *Journal of the American Medical Association* 252: 2441–2446; H. Waitzkin. 1985. "Information Giving in Medical Care." *Journal of Health and Social Behavior* 26: 81–101; H. Waitzkin. 1991. *The Politics of Medical Encounters: How Patients and Doctors Deal with Social Problems.* New Haven: Yale University Press; D. L. Roter and J. A. Hall. 1992. *Doctors Talking with Patients/Patients Talking with Doctors.* Westport, CT: Auburn House.

REFERENCES

Kuttner, R. 1998. "Must Good HMOs Go Bad?" *New England Journal of Medicine* 338: 1558–1563, 1635–1639.

Lipkin, Jr., M., S. M. Putnam, and A. Lazare. (eds.) 1995. *The Medical Interview: Clinical Care, Education and Research.* New York: Springer.

Roter, D. L. and J. A. Hall. 1992. *Doctors Talking with Patients/ Patients Talking with Doctors.* Westport, CT: Auburn House.

Smith, R. C. 1996. *The Patient's Story: Integrated Patient-Doctor Interviewing.* Boston: Little, Brown.

Waitzkin, H. 1984. "Doctor-Patient Communication: Clinical Implications of Social Scientific Research." *Journal of the American Medical Association* 252: 2441–2446.

42

Millions for Viagara, Pennies for Diseases of the Poor

KEN SILVERSTEIN

This highly readable selection brings the sphere of international medicine into focus by examining the role of global economics in the treatment of disease. Drug companies are located in the wealthy, industrial countries, and operate within their free market economy on a for-profit basis. As a result, the economic laws of supply and demand govern medical decision making about drug research and development. Not only do richer groups within a society have access to greater health care, as Waitzkin showed us earlier, but richer countries dominate the creation and dissemination of drugs. Poor countries that cannot afford to pay for medicine frequently see their health needs overlooked while rich countries see even relatively trivial medical needs pampered. As a result the "lifestyle" needs of the developed nations (impotence, baldness, toenail fungus, face wrinkles, pet care) have seen millions of dollars invested in medical research and drug development, while fatal diseases on the African continent go unchecked. This economic system is not significantly ameliorated by international social welfare concerns by either the industrialized or Third World countries. Should these differences described by Silverstein be expected given the large economic differences between the Western world and many third world countries? How would you feel if some of the medical research dollars, for say, acne treatment, were funnelled toward alleviation of basic illnesses in other countries? What is the large impact of the conditions that Silverstein describes?

Almost three times as many people, most of them in tropical countries of the Third World, die of preventable, curable diseases as die of AIDS. Malaria, tuberculosis, acute lower-respiratory infections—in 1998, these

Reprinted with permission from the July 19, 1999, issue of *The Nation*.

claimed 6.1 million lives. People died because the drugs to treat those illnesses are nonexistent or are no longer effective. They died because it doesn't pay to keep them alive.

Only 1 percent of all new medicines brought to market by multinational pharmaceutical companies between 1975 and 1997 were designed specifically to treat tropical diseases plaguing the Third World. In numbers, that means thirteen out of 1,223 medications. Only four of those thirteen resulted from research by the industry that was designed specifically to combat tropical ailments. The others, according to a study by the French group Doctors Without Borders, were either updated versions of existing drugs, products of military research, accidental discoveries made during veterinary research or, in one case, a medical breakthrough in China.

Certainly, the majority of the other 1,210 new drugs help relieve suffering and prevent premature death, but some of the hottest preparations, the ones that, as the *New York Times* put it, drug companies "can't seem to roll . . . out fast enough," have absolutely nothing to do with matters of life and death. They are what have come to be called lifestyle drugs—remedies that may one day free the world from the scourge of toenail fungus, obesity, baldness, face wrinkles and impotence. The market for such drugs is worth billions of dollars a year and is one of the fastest-growing product lines in the industry.

The drug industry's calculus in apportioning its resources is cold-blooded, but there's no disputing that one old, fat, bald, fungus-ridden rich man who can't get it up counts for more than half a billion people who are vulnerable to malaria but too poor to buy the remedies they need.

Western interest in tropical diseases was historically linked to colonization and war, specifically the desire to protect settlers and soldiers. Yellow fever became a target of biomedical research only after it began interfering with European attempts to control parts of Africa. "So obvious was this deterrence . . . that it was celebrated in song and verse by people from Sudan to Senegal," Laurie Garrett recounts in her extraordinary book *The Coming Plague*. "Well into the 1980s schoolchildren in Ibo areas of Nigeria still sang the praises of mosquitoes and the diseases they gave to French and British colonialists."

US military researchers have discovered virtually all important malaria drugs. Chloroquine was synthesized in 1941 after quinine, until then the primary drug to treat the disease, became scarce following Japan's occupation of Indonesia. The discovery of Mefloquine, the next advance, came about during the Vietnam War, in which malaria was second only to combat wounds in sending US troops to the hospital. With the end of a ground-based US military strategy came the end of innovation in malaria medicine.

The Pharmaceutical Research and Manufacturers of America (PhRMA) claimed in newspaper ads early this year that its goal is to "set every last disease on the path to extinction." Jeff Trewhitt, a PhRMA spokesman, says US drug companies will spend $24 billion on research this year and that a number of firms are looking for cures for tropical diseases. Some companies also provide existing drugs free to poor countries, he says, "Our members are involved. There's not an absolute void."

The void is certainly at hand. Neither PhRMA nor individual firms will reveal how much money the companies spend on any given disease—that's proprietary information, they say—but on malaria alone, a recent survey of the twenty-four biggest drug companies found that not a single one maintains an in-house research program, and only two expressed even minimal interest in primary research on the disease. "The pipeline of available drugs is almost empty," says Dyann Wirth of the Harvard School of Public Health, who conducted the study. "It takes five to ten years to develop a new drug, so we could soon face [a strain of] malaria resistant to every drug in the world." A 1996 study presented in *Cashiers Santé,* a French scientific journal, found that of forty-one important medicines used to treat major tropical diseases, none were discovered in the nineties and all but six were discovered before 1985.

Contributing to this trend is the wave of mergers that has swept the industry over the past decade. Merck alone now controls almost 10 percent of the world market. "The bigger they grow, the more they decide that their research should be focused on the most profitable diseases and conditions," one industry watcher says. "The only thing the companies think about on a daily basis is the price of their stocks; and announcing that you've discovered a drug [for a tropical disease] won't do much for your share price."

That comment came from a public health advocate, but it's essentially seconded by industry. "A corporation with stockholders can't stoke up a laboratory that will focus on Third World diseases, because it will go broke," says Roy Vagelos, the former head of Merck. "That's a social problem, and industry shouldn't be expected to solve it."

Drug companies, however, are hardly struggling to beat back the wolves of bankruptcy. The pharmaceutical sector racks up the largest legal profits of any industry, and it is expected to grow by an average of 16 to 18 percent over the next four years, about three times more than the average for the Fortune 500. Profits are especially high in the United States, which alone among First World nations does not control drug prices. As a result, prices here are about twice as high as they are in the European Union and nearly four times higher than in Japan.

"It's obvious that some of the industry's surplus profits could be going into research for tropical diseases," says a retired drug company executive, who wishes to remain anonymous. "Instead, it's going to stockholders." Also to promotion: In 1998, the industry unbuckled $10.8 billion on advertising. And to politics: In 1997, American drug companies spent $74.8 million to lobby the federal government, more than any other industry; last year they spent nearly $12 million on campaign contributions.

Just forty-five years ago, the discovery of new drugs and pesticides led the World Health Organization (WHO) to predict that malaria would soon be eradicated. By 1959, Garrett writes in *The Coming Plague,* the Harvard School of Public Health was so certain that the disease was passé that its curriculum didn't offer a single course on the subject.

Resistance to existing medicines—along with cutbacks in healthcare budgets, civil war and the breakdown of the state—has led to a revival of malaria in Africa, Latin America, Southeast Asia and, most recently, Armenia and Tajikistan. The WHO describes the disease as a leading cause of global suffering and says that by "undermining the health and capacity to work of hundreds of millions of people, it is closely linked to poverty and contributes significantly to stunting social and economic development."

Total global expenditures for malaria research in 1993, including government programs, came to $84 million. That's paltry when you consider that one B-2 bomber costs $2 billion, the equivalent of what, at current levels, will be spent on all malaria research over twenty years. In that period, some 40 million Africans alone will die from the disease. In the United States, the Pentagon budgets $9 million per year for malaria programs, about one-fifth the amount it set aside this year to supply the troops with Viagra. For the drug companies, the meager purchasing power of malaria's victims leaves the disease off the radar screen. As Neil Sweig, an industry analyst at Southeast Research Partners, puts it wearily, "It's not worth the effort or the while of the large pharmaceutical companies to get involved in enormously expensive research to conquer the Anopheles mosquito."

The same companies that are indifferent to malaria are enormously troubled by the plight of dysfunctional First World pets. John Keeling, a spokesman for the Washington, DC–based Animal Health Institute, says the "companion animal" drug market is exploding, with US sales for 1998 estimated at about $1 billion. On January 5, the FDA approved the use of Clomicalm, produced by Novartis, to treat dogs that suffer from separation anxiety (warning signs: barking or whining, "excessive greeting" and chewing on furniture). "At Last, Hope For Millions of Suffering Canines Worldwide," reads the company's press release announcing the drug's rollout. "I can't emphasize enough how dogs are suffering and that their behavior is not tolerable to owners," says Guy Tebbitt, vice president for research and development for Novartis Animal Health.

Also on January 5 the FDA gave the thumbs up to Pfizer's Anipryl, the first drug approved for doggie Alzheimer's. Pfizer sells a canine pain reliever and arthritis treatment as well, and late last year it announced an R&D program for medications that help pets with anxiety and dementia.

Another big player in the companion-animal field is Heska, a biotechnology firm based in Colorado that strives to increase the "quality of life" for cats and dogs. Its products include medicines for allergies and anxiety, as well as an antibiotic that fights periodontal disease. The company's Web site features a "spokesdog" named Perio Pooch and, like old "shock" movies from high school driver's-ed classes, a photograph of a diseased doggie mouth to demonstrate what can happen if teeth and gums are not treated carefully. No one wants pets to be in pain, and Heska also makes drugs for animal cancer, but it is a measure of priorities that US companies and their subsidiaries spend almost nothing on tropical diseases while, according to an industry source, they spend about half a billion dollars for R&D on animal health.

Although "companion animal" treatments are an extreme case—that half-billion-dollar figure covers "food animals" as well, and most veterinary drugs emerge from research on human medications—consider a few examples from the brave new world of human lifestyle drugs. Here, the pharmaceutical companies are scrambling to eradicate:

- *Impotence.* Pfizer invested vast sums to find a cure for what Bob Dole and other industry spokesmen delicately refer to as "erectile dysfunction." The company hit the jackpot with Viagra, which racked up more than $1 billion in sales in its first year on the market. Two other companies, Schering-Plough and Abbott Laboratories, are already rushing out competing drugs.
- *Baldness.* The top two drugs in the field, Merck's Propecia and Pharmacia & Upjohn's Rogaine (the latter sold over the counter), had combined sales of about $180 million in 1998. "Some lifestyle drugs are used for relatively serious problems, but even in the best cases we're talking about very different products from penicillin," says the retired drug company executive. "In cases like baldness therapy, we're not even talking about healthcare."
- *Toenail fungus.* With the slogan "Let your feet get naked!" as its battle cry, pharmaceutical giant Novartis recently unveiled a lavish advertising campaign for Lamisil, a drug that promises relief for sufferers of this unsightly malady. It's a hot one, the war against fungus, pitting Lamisil against Janssen Pharmaceutical's Sporanox and Pfizer's Diflucan for shares in a market estimated to be worth hundreds of millions of dollars a year.
- *Face wrinkles.* Allergan earned $90 million in 1997 from sales of its "miracle" drug Botox. Injected between the eyebrows at a cost of about $1,000 for three annual treatments, Botox makes crow's feet and wrinkles disappear. "Every 7 1/2 seconds someone is turning 50," a wrinkle expert told the *Dallas Morning News* in an article about Botox last year. "You're looking at this vast population that doesn't want frown lines."

Meanwhile, acute lower respiratory infections go untreated, claiming about 3.5 million victims per year, overwhelmingly children in poor nations. Such infections are third on the chart of the biggest killers in the world; the number of lives they take is almost half the total reaped by the number-one killer, heart disease, which usually strikes the elderly. "The development of new antibiotics," wrote drug company researcher A.J. Slater in a 1989 paper published in the Royal Society of Tropical Medicine and Hygiene's *Transactions,* "is very costly and their provision to Third World countries alone can never be financially rewarding."

In some cases, older medications thought to be unnecessary in the First World and commercially unviable in the Third have simply been pulled from the market. This created a crisis recently when TB re-emerged with a vengeance in US inner cities, since not a single company was still manufacturing Streptomycin after mid-1991. The FDA set up a task force to deal with the situation, but it was two years before it prodded Pfizer back into the field.

In 1990 Marion Merrell Dow (which was bought by German giant Hoechst in 1995) announced that it would manufacture Ornidyl, the first new medicine in forty years that was effective in treating African sleeping sickness. Despite the benign sounding name, the disease leads to coma and death, and kills about 40,000 people a year. Unlike earlier remedies for sleeping sickness, Ornidyl had few side effects. In field trials, it saved the lives of more than 600 patients, most of whom were near death. Yet Ornidyl was pulled from production; apparently company bean-counters determined that saving lives offered no return.

Because AIDS also plagues the First World, it is the one disease ravaging Third World countries that is the object of substantial drug company research. In many African countries, AIDS has wiped out a half-century of gains in child survival rates. In Botswana—a country that is not at war and has a relatively stable society—life expectancy rates fell by twenty years over a period of just five. In South Africa, the Health Ministry recently issued a report saying that 1,500 of the country's people are infected with HIV every day and predicting that the annual deathrate will climb to 500,000 within the next decade.

Yet available treatments and research initiatives offer little hope for poor people. A year's supply of the highly recommended multidrug cocktail of three AIDS medicines costs about $15,000 a year. That's exorbitant in any part of the world, but prohibitive in countries like Uganda, where per capita income stands at $330. Moreover, different viral "families" of AIDS, with distinct immunological properties, appear in different parts of the world. About 85 percent of people with HIV live in the Third World, but industry research to develop an AIDS vaccine focuses only on the First World. "Without research dedicated to the specific viral strains that are prevalent in developing countries, vaccines for those countries will be very slow in coming," says Dr. Amir Attaran, an international expert who directs the Washington-based Malaria Project.

All the blame for the neglect of tropical diseases can't be laid at the feet of industry. Many Third World governments invest little in healthcare, and First World countries have slashed both foreign aid and domestic research programs. Meanwhile, the US government aggressively champions the interests of the drug industry abroad, a stance that often undermines healthcare needs in developing countries.

In one case where a drug company put Third World health before profit—Merck's manufacture of Ivermectin—governmental inertia nearly scuttled the good deed. It was the early eighties, and a Pakistani researcher at Merck discovered that the drug, until then used only in veterinary medicine, performed miracles in combating river blindness disease. With one dose per year of Ivermectin, people were fully protected from river blindness, which is carried by flies and, at the time, threatened hundreds of millions of people in West Africa.

Merck soon found that it would be impossible to market Ivermectin profitably, so in an unprecedented action the company decided to provide it free of charge to the WHO. (Vagelos, then chairman of Merck, said the company was worried about taking the step, "as we feared it would discourage companies from doing research relevant to the Third World, since they might be expected to follow suit.") Even then, the program nearly failed. The WHO claimed it didn't have the money needed to cover distribution costs, and Vagelos was unable to win financial support from the Reagan Administration. A decade after Ivermectin's discovery, only 3 million of 120 million people at risk of river blindness had received the drug. During the past few years, the WHO, the World Bank and private philanthropists have finally put up the money for the program, and it now appears that river blindness will become the second disease, after smallpox, to be eradicated.

Given the industry's profitability, it's clear that the companies could do far more. It's equally clear that they won't unless they are forced to. The success of ACT UP in pushing drug companies to respond to the AIDS crisis in America is emblematic of how crucial but also how difficult it is to get the industry to budge. In late 1997, a coalition of public health organizations approached a group of major drug companies, including Glaxo-Wellcome and Roche, and asked them to fund a project that would dedicate itself to developing new treatments for major tropical diseases. Although the companies would have been required to put up no more than $2 million a year, they walked away from the table. Since there's no organized pressure—either from the grassroots or from governments—they haven't come back. "There [were] a number of problems at the business level," Harvey Bale, director of the Geneva-based International Federation of Pharmaceutical Manufacturers' Association, told *Science* magazine. "The cost of the project is high for some companies."

While the industry's political clout currently insures against any radical government action, even minor reforms could go a long way. The retired drug company executive points to public hospitals, which historically were guaranteed relatively high profit margins but were obligated to provide free care to the poor in return. There's also the example of phone companies, which charge businesses higher rates in order to subsidize universal service. "Society has tolerated high profit levels up until now, but society has the right to expect something back," he says. "Right now, it's not getting it."

The US government already lavishly subsidizes industry research and allows companies to market discoveries made by the National Institutes of Health and other federal agencies. "All the government needs to do is start attaching some strings," says the Malaria Project's Attaran. "If a company wants to market another billion-dollar blockbuster, fine, but in exchange it will have to push through a new malaria drug. It will cost them some money, but it's not going to bankrupt them."

Another type of "string" would be a "reasonable pricing" provision for drugs developed at federal laboratories. By way of explanation, Attaran recounted that the vaccine for hepatitis A was largely developed by researchers at the Walter Reed Army Institute. At the end of the day, the government gave the marketing rights to SmithKline Beecham and Merck. The current market for the vaccine, which sells for about $60 per person, is $300 million a year. The only thing Walter Reed's researchers got in exchange for their efforts was a plaque that hangs in their offices. "I'll say one thing for the companies," says Attaran. "They didn't skimp on the plaque; it's a nice one. But either the companies should have paid for part of the government's research, or they should have been required to sell the vaccine at a much lower price."

At the beginning of this year, Doctors Without Borders unveiled a campaign calling for increased access to drugs needed in Third World countries. The group is exploring ideas ranging from tax breaks for smaller firms engaged in research in the field, to creative use of international trade agreements, to increased donations of drugs from the multinational companies. Dr. Bernard Pécoul, an organizer of the campaign, says that different approaches are required for different diseases. In the case of those plaguing only the Southern Hemisphere—sleeping sickness, for example—market mechanisms won't work because there simply is no market to speak of. Hence, he suggests that if multinational firms are not willing to manufacture a given drug, they transfer the relevant technology to a Third World producer that is.

Drugs already exist for diseases that ravage the North as well as the South—AIDS and TB, for example—but they are often too expensive for people in the Third World. For twenty-five years, the WHO has used funding from member governments to purchase and distribute vaccines to poor countries; Pécoul proposes a similar model for drugs for tropical diseases. Another solution he points to: In the event of a major health emergency, state or private producers in the South would be allowed to produce generic versions of needed medications in exchange for a small royalty paid to the multinational license holder. "If we can't change the markets, we have to humanize them," Pécoul says. "Drugs save lives. They can't be treated as normal products."

PART V

Social Change

Throughout this book we have integrated social change into each section, but now we address it directly. So many changes are shaping the world in which we live, that we found it important to offer you a contemporary view of the current issues our society is dealing with and the controversies that arise from them. Social change occurs as new technologies arise to transform our lifestyles, both liberating us and capturing us with their advances. New ideas foster change as well, and these have led to corresponding transformations in the structures and forms of society, such as gender, the family, religion, and medicine. All of society's institutions, organizations, statuses, and roles, in fact, are subject to the influence of new trends and patterns that occur, some sweeping through rapidly, while others evolve slowly and gradually, almost imperceptibly. Hierarchies may slowly and subtly shift, or they may suffer more sudden reversals and upheavals. Changes may be generated by forces outside of society or through developments that arise from within, altering social values, norms, institutions, social relationships, and self-identities. These may result from environmental events, warfare and invasion, cross-cultural and subcultural contact, and the diffusion of ideas from one society or group to another. New innovations may arise in people's lifestyles, inventions, marketing, healing, educating, spending, voting, and communicating that may radically transform the way they think and live. The rate of social change is astounding many people today. Less than a generation ago we did not have computers, cellular phones, answering machines, banking machines, the

Internet, or a host of other "modern conveniences" that we could now no longer imagine living without. These have all transformed our major social institutions and the way they operate.

In our first section we consider some of the changes that have occurred to our **Visions of Society.** This section looks at broad trends influencing the shape of society. The world that we encounter has been altered in ways that we interact with daily on a commercial basis. Our sociological odyssey takes us to look at two related themes, George Ritzer's concept of the "McDonaldization," or bureaucratization, of society, and Alan Bryman's related observations about its "Disneyization," or theming. We can see these trends influencing the shape of our entire culture, its organizations, and the roles, relationships, and behavior people enact within them. McDonald's standardization and Disney's specially constructed imaging are now pervasive forces in nearly all aspects of social and commercial life. Look for these all around you when you shop, eat, select entertainment, or even in the way that you choose your college courses.

In some ways, the most important issue we face as a nation at the millennium is the challenge of **Community.** What once surrounded and nurtured us has eroded, and we worry about the quality of our lives and our society as a result. The *gemeinschaft* of small, traditional communities where people related to each other in primary relationships and across generational borders have been replaced by the *gesellschaft* of contractual relations, instrumental behavior, larger-scale, and urban societies where people do not have as much in common. The traditional forms in which we used to recognize community have changed. Just as the "Leave it to Beaver" families from the 1950s have become rarer, so have the neighborhoods in which they lived. Developers still create suburban tracts and people flock to them, but the feeling of neighborliness is not the same. People are not rooted in their local social institutions the way they used to be. Robert Putnam expresses this well when he writes of the decline of the traditional civic institutions and questions what has become of community. We are forced to wonder if community is a necessary environment to nurture people and their development, and what component parts, if any, are the most critical. We face the challenge of social change and the maintenance of our core values. Andrew Shapiro offers one small answer to this question in his discussion of cyber-communities. Virtual communities and interest communities, those forged around people's more central preoccupations and selves, are arising to take the place of what is gone. You must decide what you think of this trend, and how you evaluate chat rooms and other interest groups. This is your society, and the decisions and values you adopt will help shape it into the next millennium.

43

The McDonaldization of Society

GEORGE RITZER

The success of fast food chains is used by Ritzer as a metaphor for some general trends characterizing contemporary American society. We have become a nation driven by concerns for rationality, speed, and efficiency that are so well illustrated by the McDonalds' style of operation. Food, packaging, and service are designed to move quickly and cheaply through and out of these restaurants, giving customers the most modern eating experience. Speed, convenience, and standardization have replaced the flair of design and creation in cooking, the comfort of relationships in serving, and the variety available in choice. McDonaldization has become so pervasive that one can travel to nearly any city or town in America and find familiar chain-style restaurants, shops, hotels, and other avenues for commercial exchange. This has fostered the homogenization of American culture and life, streamlined along a set of rational, efficient, and impersonal principles. How has the McDonaldization phenomenon affected your life? What types of commercial exchanges are affected by this process? What are the benefits of this for society? What are some of the detriments that you see?

A wide-ranging process of *rationalization* is occurring across American society and is having an increasingly powerful impact in many other parts of the world. It encompasses such disparate phenomena as fast-food restaurants, TV dinners, packaged tours, industrial robots, plea bargaining, and open-heart surgery on an assembly-line basis. As widespread and as important as these developments are, it is clear that we have barely begun a process that promises even more extraordinary changes (e.g. genetic engineering) in the years to come. We can think of rationalization as a historical process and rationality as the end result of that development. As a historical process, rationalization has distinctive roots in the western world. Writing in the late nineteenth and early twentieth centuries, the great German sociologist Max Weber saw his society as the center

From the *Journal of American Culture,* V. 6, No. 1, 1983, pp. 100–107. Reprinted by permission of the publisher.

of the ongoing process of rationalization and the bureaucracy as its paradigm case. The model of rationalization, at least in contemporary America, is no longer the bureaucracy, but might be better thought of as the fast-food restaurant. As a result, our concern here is with what might be termed the "McDonaldization of Society." While the fast-food restaurant is not the ultimate expression of rationality, it is the current exemplar for future developments in rationalization.

A society characterized by rationality is one which emphasizes ***efficiency, predictability, calculability, substitution of nonhuman for human technology,*** and ***control over uncertainty.*** In discussing the various dimensions of rationalization, we will be little concerned with the gains already made, and yet to be realized, by greater rationalization. These advantages are widely discussed in schools and in the mass media. In fact, we are in danger of being seduced by the innumerable advantages already offered, and promised in the future, by rationalization. The glitter of these accomplishments and promises has served to distract most people from the grave dangers posed by progressive rationalization. In other words, we are ultimately concerned here with the irrational consequences that often flow from rational systems. Thus, the second major theme of this essay might be termed "the irrationality of rationality." . . .

EFFICIENCY

The process of rationalization leads to a society in which a great deal of emphasis is placed on finding the best or optimum means to any given end. Whatever a group of people define as an end, and everything they so define, is to be pursued by attempting to find the best means to achieve the end. Thus, in the Germany of Weber's day, the bureaucracy was seen as the most efficient means of handling a wide array of administrative tasks. Somewhat later, the Nazis came to develop the concentration camp, its ovens, and other devices as the optimum method of collecting and murdering millions of Jews and other people. The efficiency that Weber described in turn-of-the-century Germany, and which later came to characterize many Nazi activities, has become a basic principle of life in virtually every sector of a rational society.

The modern American family, often with two wage earners, has little time to prepare elaborate meals. For the relatively few who still cook such meals, there is likely to be great reliance on cookbooks that make cooking from scratch much more efficient. However, such cooking is relatively rare today. Most families take as their objective quickly and easily prepared meals. To this end, much use is made of prepackaged meals and frozen TV dinners.

For many modern families, the TV dinner is no longer efficient enough. To many people, eating out, particularly in a fast-food restaurant, is a far more efficient way of obtaining their meals. Fast-food restaurants capitalize on this by being organized so that diners are fed as efficiently as possible. They offer a limited, simple menu that can be cooked and served in an assembly-line fashion. The latest development in fast-food restaurants, the addition of drive-through windows, constitutes an effort to increase still further the efficiency of the dining experience. The family now can simply drive through, pick up its order, and eat it while driving to the next, undoubtedly efficiently organized, activity. The success of the fast-food restaurant has come full circle with frozen food manufacturers now touting products for the home modeled after those served in fast-food restaurants.

Increasingly, efficiently organized food production and distribution systems lie at the base of the ability of people to eat their food efficiently at home, in the fast-food restaurant, or in their cars. Farms, groves, ranches, slaughterhouses, warehouses, transportation systems, and retailers are all oriented toward increasing efficiency. A notable example is chicken production where they are mass-bred, force-fed (often with many chemicals), slaughtered on an assembly line, iced or fast frozen, and shipped to all parts of the country. Some may argue that such chickens do not taste as good as the fresh-killed, local variety, but their complaints are likely to be drowned in a flood of mass-produced chickens. Then there is bacon which is more efficiently shipped, stored, and sold when it is preserved by sodium nitrate, a chemical which is unfortunately thought by many to be carcinogenic. Whatever one may say about the quality or the danger of the products, the fact remains that they are all shaped by the drive for efficiency. . . .

One of the most interesting and important aspects of efficiency is that it often comes to be not a means but an end in itself. This "displacement of goals" is a major problem in a rationalizing society. We have, for example, the bureaucrats who slavishly follow the rules even though their inflexibility negatively affects the organization's ability to achieve its goals. Then there are the bureaucrats who are so concerned with efficiency that they lose sight of the ultimate goals the means are designed to achieve. A good example was the Nazi concentration camp officers who, in devoting so much attention to maximizing the efficiency of the camps' operation, lost sight of the fact that the ultimate purpose of the camps was the murder of millions of people.

PREDICTABILITY

A second component of rationalization involves the effort to ensure predictability from one place to another. In a rational society, people want to know what to expect when they enter a given setting or acquire some sort of commodity. They neither want nor expect surprises. They want to know that if they journey to another locale, the setting they enter or the commodity they buy will be essentially the same as the setting they entered or product they purchased earlier. Furthermore, people want to be sure that what they encounter is much like what they encountered at earlier times. In order to ensure predictability over time and place a rational society must emphasize such things as discipline, order, systemization, formalization, routine, consistency, and methodical operation.

One of the attractions of TV dinners for modern families is that they are highly predictable. The TV dinner composed of fried chicken, mashed potatoes, green peas, and peach cobbler is exactly the same from one time to another and one city to another. Home cooking from scratch is, conversely, a notoriously unpredictable enterprise with little assurance that dishes will taste the same time after time. However, the cookbook cannot eliminate all unpredictability. There are often simply too many ingredients and other variables involved. Thus the cookbook dish is far less predictable than the TV dinner or a wide array of other prepared dishes.

Fast-food restaurants rank very high on the dimension of predictability. In order to help ensure consistency, the fast-food restaurant offers only a limited menu. Predictable end products are made possible by the use of similar raw materials, technologies, and preparation and serving techniques. Not only the food is predictable; the physical structures, the logo, the "ambience," and even the personnel are as well.

The food that is shipped to our homes and our fast-food restaurants is itself affected by the process of increasing predictability. Thus our favorite white bread is indistinguishable from one place to another. In fact, food producers have made great efforts to ensure such predictability.

On packaged tours travelers can be fairly sure that the people they travel with will be much like themselves. The planes, buses, hotel accommodations, restaurants, and at least the way in which the sites are visited are very similar from one location to another. Many people go on packaged tours *because* they are far more predictable than travel undertaken on an individual basis.

Amusement parks used to be highly unpredictable affairs. People could never be sure, from one park to another, precisely what sorts of rides, events, foods, visitors, and employees they would encounter. All of that has changed in the era of the theme parks inspired by Disneyland. Such parks seek to ensure predictability in various ways. For example, a specific type of young person is hired in these parks, and they are all trained in much the same way, so that they have a robot-like predictability.

Other leisure-time activities have grown similarly predictable. Camping in the wild is loaded with uncertainties—bugs, bears, rain, cold, and the like. To make camping more predictable, organized grounds have sprung up around the country. Gone are many of the elements of unpredictability replaced by RVs, paved-over parking lots, sanitized campsites, fences and enclosed camp centers that provide laundry and food services, recreational activities, television, and video games. Sporting events, too, have in a variety of ways been made more predictable. The use of artificial turf in baseball makes for a more predictable bounce of a ball. . . .

CALCULABILITY OR QUANTITY RATHER THAN QUALITY

It could easily be argued that the emphasis on quantifiable measures, on things that can be counted, is *the* most defining characteristic of a rational society. Quality is notoriously difficult to evaluate. How do we assess the quality of a hamburger, or a physician, or a student? Instead of even trying, in an increasing number of cases, a rational society seeks to develop a series of quantifiable measures that it takes as surrogates for quality. This urge to quantify has given great impetus to the development of the computer and has, in turn, been spurred by the widespread use and increasing sophistication of the computer.

The fact is that many aspects of modern rational society, especially as far as calculable issues are concerned, are made possible and more widespread by the computer. We need not belabor the ability of the computer to handle large numbers of virtually anything, but somewhat less obvious is the use of the computer to give the illusion of personal attention in a world made increasingly impersonal in large part because of the computer's capacity to turn virtually everything into quantifiable dimensions. We have all now had many experiences where we open a letter personally addressed to us only to find a computer letter. We are aware that the names and addresses of millions of people have been stored on tape and that with the aid of a number of word processors a form letter has been sent to every name on the list. Although the computer is able to give a sense of personal attention, most people are nothing more than an item on a huge mailing list.

Our main concern here, though, is not with the computer, but with the emphasis on quantity rather than quality that it has helped foster. One of the most obvious examples in the university is the emphasis given to grades and cumulative grade point averages. With less and less contact between professor and student, there is little real effort to assess the quality of what students know, let alone the quality of their overall abilities. Instead, the sole measure of the quality of most college students is their grade in a given course and their grade point averages. Another blatant example is the emphasis on a variety of uniform exams such as SATs and GREs in which the essence of an applicant is reduced to a few simple scores and percentiles.

Within the educational institution, the importance of grades is well known, but somewhat less known is the way quantifiable factors have become an essential part of the process of evaluating college professors. For example, teaching ability is very hard to evaluate. Administrators have difficulty assessing teaching quality and thus substitute quantitative scores. Of course each score involves qualitative judgments, but this is conveniently ignored. Student opinion polls are taken and the scores are summed, averaged, and compared. Those who score well are deemed good teachers while those who don't are seen as poor teachers. There are many problems involved in relying on these scores such as the fact that easy teachers in "gut" courses may well obtain high ratings while rigorous teachers of difficult courses are likely to score poorly. . . .

In the workworld we find many examples of the effort to substitute quantity for quality. Scientific management was heavily oriented to turning everything work-related into quantifiable dimensions. Instead of relying on the "rule of thumb" of the operator, scientific management sought to develop precise measures of how much work was to be done by each and every motion of the worker. Everything that could be was reduced to numbers and all these numbers were then analyzable using a variety of mathematical formulae. The assembly line is similarly oriented to a variety of quantifiable dimensions such as optimizing the speed of the line, minimizing time for each task, lowering the price of the finished product, increasing sales and ultimately increasing profits. The divisional system pioneered by General Motors and thought to be one of the major reasons for its past success was oriented to the reduction of the performance of each division to a few, bottom-line numbers. By monitoring and comparing these numbers, General Motors was able to exercise control over the results without getting involved in the day-to-day activities of each division. . . .

Thus, the third dimension of rationalization, calculability or the emphasis on quantity rather than quality, has wide applicability to the social world. It is truly central, if not the central, component of a rationalizing society. To return to our favorite example, it is the case that McDonald's expends far more effort telling us how many billions of hamburgers it has sold than it does in telling us about the quality of those burgers. Relatedly, it touts the size of its product (the "Big Mac") more than the quality of the product (it is not the "Good Mac"). The bottom line in many settings is the number of customers processed, the speed with which they are processed, and the profits produced. Quality is secondary, if indeed there is any concern at all for it.

SUBSTITUTION OF NONHUMAN TECHNOLOGY

In spite of Herculean efforts, there are important limits to the ability to rationalize what human beings think and do. Seemingly no matter what one does, people still retain at least the ultimate capacity to think and act in a variety of unanticipated ways. Thus, in spite of great efforts to make human behavior more efficient, more predictable, more calculable, people continue to act in unforeseen ways. People continue to make home-cooked meals from scratch, to camp in tents in the wild, to eat in old-fashioned diners, and to sabotage the assembly lines. Because of these realities, there is great interest among those who foster increasing rationality in using rational technologies to limit individual independence and ultimately to replace human beings with machines and other technologies that lack the ability to think and act in unpredictable ways.

McDonald's does not yet have robots to serve us food, but it does have teenagers whose ability to act autonomously is almost completely eliminated by techniques, procedures, routines, and machines. There are numerous examples of this including rules which prescribe all the things a counterperson should do in dealing with a customer as well as a large variety of technologies which determine the actions of workers such as drink dispensers which shut themselves off when the cup is full; buzzers, lights, and bells which indicate when food (e.g., french fries) is done; and cash registers which have the prices of each item programmed in. One of the latest attempts to constrain individual action is Denny's use of pre-measured packages of dehydrated food that are "cooked" simply by putting them under the hot water tap. Because of such tools and machines, as well as the elaborate rules dictating worker behavior, people often feel like they are dealing with human robots when they relate to the personnel of a fast-food restaurant. When human robots are found, mechanical robots cannot be far behind. Once people are reduced to a few robot-like actions, it is a relatively easy step to replace them with mechanical robots. Thus, Burgerworld is reportedly opening a prototypical restaurant in which mechanical robots serve the food.

Much of the recent history of work, especially manual work, is a history of efforts to replace human technology with nonhuman technology. Scientific management was oriented to the development of an elaborate and rigid set of rules about how jobs were to be done. The workers were to blindly and obediently follow those rules and not to do the work the way they saw fit. The various skills needed to perform a task were carefully delineated and broken down into a series of routine steps that could be taught to all workers. The skills, in other words, were built into the routines rather than belonging to skilled craftspersons. Similar points can be made about the assembly line which is basically a set of nonhuman technologies that have the needed steps and skills built into them. The human worker is reduced to performing a limited number of simple, repetitive operations. However, the control of this technology over the individual worker is so great and omnipresent that individual workers have reacted negatively manifesting such things as tardiness, absenteeism, turnover, and even sabotage. We are now witnessing a new stage in this technological development with automated processes now totally replacing many workers with robots. With the coming of robots we have reached the ultimate stage in the replacement of humans with nonhuman technology.

Even religion and religious crusades have not been unaffected by the spread of nonhuman technologies. The growth of large religious organizations, the use of Madison Avenue techniques, and even drive-in churches all reflect the incursion of modern technology. But it is in the electronic church, religion through the TV screens, that replacement of human by nonhuman technology in religion is most visible and has its most important manifestation. . . .

CONTROL

This leads us to the fifth major dimension of rationalization—control. Rational systems are oriented toward, and structured to expedite, control in a variety of senses. At the most general level, we can say that rational systems are set up to allow for greater control over the uncertainties of life—birth, death, food production and distribution, housing, religious salvation, and many, many others. More specifically, rational systems are oriented to gaining greater control over the major source of uncertainty in social life—other people. Among other things, this means control over subordinates by superiors and control of clients and customers by workers.

There are many examples of rationalization oriented toward gaining greater control over the uncertainties of life. The burgeoning of the genetic engineering movement can be seen as being aimed at gaining better control over the production of life itself. Similarly, amniocentesis can be seen as a technique which will allow the parents to determine the kind of child they will have. The efforts to rationalize food production and distribution can be seen as being aimed at gaining greater control over the problems of hunger and starvation. A steady and regular supply of food can make life itself more certain for large numbers of people who today live under the threat of death from starvation.

At a more specific level, the rationalization of food preparation and serving at McDonald's gives it great control over its employees. The automobile assembly line has a similar impact. In fact, the vast majority of the structures of a rational society exert extraordinary control over the people who labor in them. But because of the limits that still exist on the degree of control that rational structures can exercise over individuals, many rationalizing employers are driven to seek to more fully rationalize their operations and totally eliminate the worker. The result is an automated, robot-like technology over which, barring some *2001* rebellion, there is almost total control.

In addition to control over employees, rational systems are also interested in controlling the customer/clients they serve. For example, the fast-food restaurant with its counter, the absence of waiters and waitresses, the limited seating, and the drive-through windows all tend to lead customers to do certain things and not to do others.

Irrationality of Rationality

Although not an inherent part of rationalization, the *irrationality of rationality* is a seemingly inevitable byproduct of the process. We can think of the irrationality of rationality in several ways. At the most general level it can simply be seen as an overarching label for all the negative effects of rationalization. More specifically, it can be seen as the opposite of rationality, at least in some of its senses. For example, there are the inefficiencies and unpredictabilities that are often produced by seemingly rational systems. Thus, although bureaucracies are constructed to bring about greater efficiency in organizational work, the fact is that there are notorious inefficiencies such as the "red tape" associated with the operation of most bureaucracies. Or, take the example of the arms race in which a focus on quantifiable aspects of nuclear weapons may well have made the occurrence of nuclear war more, rather than less, unpredictable.

Of greatest importance, however, is the variety of negative effects that rational systems have on the individuals who live, work, and are served by them. We might say that *rational systems are not reasonable systems.* As we've already discussed, rationality brings with it great dehumanization as people are reduced to acting like robots. Among the dehumanizing aspects of a rational society are large lecture classes, computer letters, pray TV, work on the automobile assembly line, and dining at a fast-food restaurant. Rationalization also tends to bring with it disenchantment leaving much of our lives without any mystery or excitement. Production by a hand craftsman is far more mysterious than an assembly-line technology where each worker does a single, very limited operation. Camping in an RV tends to suffer in comparison to the joys to be derived from camping in the wild. Overall a fully rational society would be a very bleak and uninteresting place.

CONCLUSION

Rationalization, with McDonald's as the paradigm case, is occurring throughout America, and, increasingly, other societies. In virtually every sector of society more and more emphasis is placed on efficiency, predictability, calculability, replacement of human by nonhuman technology, and control over uncertainty. Although progressive rationalization has brought with it innumerable advantages, it has also created a number of problems, the various irrationalities of rationality, which threaten to accelerate in the years to come. These problems, and their acceleration should not be taken as a case for the return to a less rational form of society. Such a return is not only impossible but also undesirable. What is needed is not a less rational society, but greater control over the process of rationalization involving, among other things, efforts to ameliorate its irrational consequences.

44

The Disneyization of Society

ALAN BRYMAN

Bryman, a British sociologist, proposes the idea of Disneyization as a complementary notion to McDonaldization in contemporary society. He outlines four trends that have been carried over from the popular Disney theme park motif to the larger world that influence the shape and content of our lives. The clearest is theming, where previously disparate elements become combined into one cohesive image to coordinate together and present an integrated, fun motif, such as is found in malls, resorts, and shopping centers. We also see the dedifferentiation of consumption, where consumption from disparate spheres becomes intertwined, so that theme parks sell merchandise and food in addition to rides, fast-food restaurants sell movie-related goods, and ostensible knowledge centers are equipped with marketing outlets. We have also seen the rise of new dimensions of merchandising, with goods being manufactured and promoted featuring copyrighted images and logos of athletic teams, movies, theme park characters, and other novelties, for the express purpose of manufacturing and sales (and even children's television shows designed around toys to increase their market demand). The final element is emotional labor, in which service work has been scripted to include the kind of cheerful friendliness that helps distract customers from realizing that they are being captivated into an artificially constructed themed environment and pitched the sale of goods. It may be instructive for you to compare Ritzer's and Bryman's article to see how much you think they represent the larger culture in which you live. Are there any other organizations that can metaphorically be used to illustrate larger trends in your society?

Ritzer's (1993) concept of McDonaldization represents a stimulating and important attempt to address large-scale issues concerning social change and the nature of modernity and to link these topics to some minutiae of everyday life. Ritzer is at pains to point out that McDonald's is merely a symbol of McDonaldization though it has undoubtedly been a major force behind the process. McDonaldization refers to "the process by which *the principles* of the

From *Sociological Review,* Vol. 47, No. 125–26, 29–30. Reprinted by permission of Blackwell Publishers.

fast-food restaurant are coming to dominate more and more sectors of American society as well as the rest of the world" (Ritzer, 1993: 3, emphasis added). This means that McDonaldization is not simply about the spread of McDonald's restaurants or of restaurants explicitly modelled on them; nor is it a process that can be specifically attributed to McDonald's alone, since the restaurants incorporate practices that were formulated long before the McDonald brothers started their first restaurant, such as scientific management, Fordism, and bureaucracy.

The purpose of this article is to propose that a similar case can be made for a process that I will call "Disneyization," by which I mean:

> the process by which *the principles* of the Disney theme parks are coming to dominate more and more sectors of American society as well as the rest of the world.

My view of Disneyization is meant to parallel Ritzer's notion of McDonaldization: it is meant to draw attention to the spread of principles exemplified by the Disney theme parks. Of course, the Disney theme parks are sites of McDonaldization too. A number of Ritzer's (1993) illustrations of the four dimensions of McDonaldization—efficiency, calculability, predictability, and control—are drawn from Disney parks and from theme parks that appear to have been influenced by them. There are, moreover, numerous parallels between McDonald's restaurants and the Disney parks (Bryman, 1995: 123; King, 1983). . . .

Further, we may well find that the McDonald's fast-food restaurants will be bearers of Disneyization, in much the same way that Disney theme parks are bearers of McDonaldization.

In the following account of Disneyization, four dimensions will be outlined. In each case, the meaning of the dimension and its operation in the context of the Disney parks will be outlined, its diffusion beyond the realms of the Disney parks will be indicated, and aspects of any of the dimensions which precede the opening of the first Disney theme park (Disneyland in California) in 1955 will be explored. The overall aim is to identify large-scale changes that are discernible in economy and culture that can be found in, and are symbolized by, the Disney parks. As with Ritzer's (1993) treatment of McDonald's in relation to McDonaldization, it is not suggested that the Disney parks *caused* these trends, though the parks' success may have hastened the assimilation of Disneyization.

The four trends are:

1. theming
2. dedifferentiation of consumption
3. merchandising
4. emotional labor

This list is probably not exhaustive, any more than McDonaldization's four dimensions can be so regarded. They are meant to be considered as four major trends which are discernible in and have implications for (late) modernity.

THEMING

Theming represents the most obvious dimension of Disneyization. More and more areas of economic life are becoming themed. There is now a veritable themed restaurant industry, which draws on such well-known and accessible cultural themes as rock and other kinds of music, sport, Hollywood and the film industry more generally, and geography and history (Beardsworth and Bryman, 1999). These themes find their expression in chains of themed restaurants, like Hard Rock Cafe, Planet Hollywood, All Sports Cafe, Harley-Davidson Cafe, Rainforest Cafe, Fashion Cafe, as well as one-off themed eating establishments. Diners are surrounded by sounds and sights that are constitutive of the themed environment, but which are incidental to the act of eating as such, though they are major reasons for such restaurants being sought out. In Britain, themed pubs are increasingly prominent and popular, while in the USA, bars themed on British pubs are big business too. Hotels are increasingly being themed and it is no coincidence that two of the more successful themed restaurant brands—Hard Rock Cafe and Planet Hollywood—are being deployed for such a purpose. Ritzer and Liska (1997) suggest that cruise ships are increasingly becoming themed. In Las Vegas, virtually every new hotel on the "strip" is heavily themed. The famous strip now contains such themes as Ancient Rome (Caesar's), Ancient Egypt (Luxor), ye olde England (Excalibur), the movies (MGM Grand), city life (New York New York), turn-of-the-century high life on the Mediterranean (Monte Carlo), the sea (Treasure Island), and so on. It seems quite likely that this penchant for themed hotels will proliferate though possibly not with the exotic facades that adorn the Las Vegas establishments. Certainly, the theming of hotel rooms as in the Madonna Inn near San Luis Obispo, California, and in the Fantasy Hotel in West Edmonton Mall seems to be becoming increasingly prominent (Eco, 1986; Hopkins, 1990).

Shopping in malls is increasingly being accomplished in themed environments. Mall of America in Minneapolis and West Edmonton Mall in Edmonton, Alberta exemplify this feature . . . Gottdiener (1997) suggests that airports are increasingly becoming themed environments. . . .

Theming accomplished at least two things in this connection. First, it established coherence to the various rides and attractions in Disneyland and the environments in which they were located. Secondly, in the design of rides and attractions, the accent was placed on their theming rather than on the thrill factor, which was the emphasis in traditional amusement parks. Indeed, Walt Disney initially did not plan for roller coaster rides in order to set his park apart from the amusement parks he loathed so much. Gradually, such rides have been incorporated as a result of pressure from younger visitors who found Disney fare too tame. However, when such rides were built they were in heavily themed form, for example, Big Thunder Mountain Railroad (themed on prospecting in the Wild West), Space Mountain (space travel) and Splash Mountain (*Song of the South*). By establishing coherence to rides and by placing an emphasis on the theme rather than on thrills, Walt Disney was able to differentiate Disneyland from the traditional amusement parks that he so disliked. . . .

DEDIFFERENTIATION OF CONSUMPTION

The term "dedifferentiation of consumption" denotes simply the general trend whereby the forms of consumption associated with different institutional spheres become interlocked with each other and increasingly difficult to distinguish. For one thing, there has been a tendency for the distinction between shopping and theme parks to be elided. Walt Disney realized at a very early stage that Disneyland had great potential as a vehicle for selling food and various goods. Main Street USA typified this in that its main purpose is not to house attractions but to act as a context for shopping. As Eco puts it: "The Main Street façades are presented to us as toy houses and invite us to enter them, but their interior is always a disguised supermarket, where you buy obsessively, believing that you are still playing" (1986: 43). Nowadays, the Disney theme parks are full of shops and restaurants to the extent that many writers argue that their main purpose increasingly is precisely the selling of a variety of goods and food. With many attractions, visitors are forced to go through a shop containing relevant merchandise in order to exit (e.g., a shop containing Star Wars merchandise as one leaves the Star Tours ride in the two American Disney parks and Disneyland Paris). In the EPCOT Center, a Disney World theme park which opened in 1982, there is an area called World Showcase which comprises representations of different nations. But one of the main ways in which the nations and their nationhood is revealed is through eating and shopping. Indeed, the buildings which iconically represent some of the countries do not contain attractions at all (e.g., Britain, Italy), or perhaps contain little more than a film about the country concerned (e.g., Canada, France). However, each "country" has at least one restaurant (some, like France, Mexico and China, have two) and at least one shop. It is not surprising, therefore, that for many commentators EPCOT and indeed the other parks are often portrayed as vehicles for selling goods and food. Thus, the Euro Disneyland share prospectus presented as one of the main management techniques associated with "the Disney theme park concept" the fact that "Disney has learned to optimise the mix of merchandise in stores within its theme parks, which consequently are highly profitable and achieve some of the highest sales per square metre for retail stores in the United States" (page 13). If we add hotels into this equation, the cases for dedifferentiation in the parks is even more compelling. At Disney World the number of hotels has grown enormously since Michael Eisner took the helm at the Walt Disney Company in 1984. In addition to being themed, there has been a clear attempt to ratchet up the number of guests staying in its hotels by emphasizing their advantages over non-Disney ones. For example, Disney guests are able to enter the parks earlier and can therefore get to the main attractions before the arrival of hordes of tourists. They are also able to secure tables for the sought after restaurants (especially the EPCOT ethnic ones) from their hotels rather than having to take a chance on their availability when they turn up at the parks. Also, for some time now Disney has been offering its hotel guests inclusive length-of-stay passes to the parks. It is striking that it was recognized during the days when Euro Disneyland's financial troubles were common knowledge that one of the reasons for its problems was not the number of visitors to the parks but the fact that they were not spending as much on food, souvenirs and Disney hotels as had been predicted (Bryman, 1995: 77). Thus, we see in the Disney parks a tendency for shopping, eating, hotel accommodation and theme park visiting to become inextricably interwoven. . . .

Las Vegas is possibly a better illustration than the Disney theme parks of Disneyization in the form of dedifferentiation. For a start, the hotels mentioned in the previous section could equally be described, and probably more accurately, as casinos. Each houses a massive casino, although they could equally be described as casinos with hotels attached. But in recent years, dedifferentiation has proceeded apace in Las Vegas. You may enter the Forum shops at Caesar's on the moving walkway but the only exit is to walk through the casino. More than this, in order to attract families and a wider range of clientele (Grossman, 1993), the casino/hotels have either built theme parks (e.g., MGM Grand, Circus Circus) or have incorporated theme park attractions (e.g., Luxor, Stratosphere, New York New York, Treasure Island, Excalibur). In the process, conventional distinctions between casinos, hotels, restaurants, shopping, and theme parks collapse. Crawford has written that "malls routinely entertain, while theme parks function as disguised marketplaces" (1992: 16), but current trends imply that even this comment does not capture the extent of dedifferentiation.

MERCHANDISING

In this discussion, I will use the term "merchandising" simply to refer to the promotion of goods in the form of or bearing copyright images and logos, including such products made under license. This is a realm in which Disney have been pre-eminent. Walt Disney's first animated star was arguably not Mickey Mouse, but Oswald the Lucky Rabbit, around which he and his studio had created a popular series of shorts in 1927. When he tried to negotiate a better financial deal over these shorts, Walt found that it was not he but the distributor that owned the rights to them. As a result, the studio had no rights to Oswald's name and therefore to the small range of merchandise that had begun to appear bearing the character's name and image. Thereafter, he zealously guarded his rights in this regard. A major factor may well have been the revenue-producing capability of merchandise bearing Oswald's image, including a pop up puppet, stencil set, celluloid figures and posters (Tumbusch, 1989: 28).

Merchandise and licensing proliferated, however, in the wake of Mickey's arrival in November 1928 (deCordova, 1994). A year later, Walt Disney Productions was transformed into four mini-companies, one of which dealt with merchandising and licensing. Deals were handled through first of all George Borgfeldt and from 1934 onwards by the flamboyant Kay Kamen. Walt Disney certainly did not create the idea of merchandising or even of merchandising animated cartoon characters. Felix the Cat was the subject of a large range of merchandise in the mid-1920s (Canemaker, 1991). What Walt Disney did realize was its immense profitability. In the years after Mickey's arrival, the company did not make large sums from its cartoons, because Walt Disney's incessant quest for improvements in the quality of animation cut deeply into the studio's profits. To a very large extent, he was able to finance expensive technical innovation and his unyielding insistence on quality by using profits from merchandise. Klein (1993) has suggested that about half of the studio's profits were attributable to merchandise (see also, Merritt and Kaufman, 1992: 144). Indeed, some writers have suggested that in later years, the design of cartoon characters, in particular their "cuteness," was at least in part motivated by a consideration of their capacity to be turned into merchandise (Bryman, 1995; Forgacs, 1992). It may also account for the changes in Mickey's increasingly less rodent-like appearance over the years (Gould, 1979).

The Disney theme parks have two points of significance in relation to merchandising as a component of Disneyization. Firstly, and most obviously, they provide sites for the selling of the vast array of Disney merchandise that has accumulated over the years: from pens to clothing, from books to sweets and from watches to plush toys. Sales from merchandise are a major contributor to profits from the parks. The parks are carefully designed to maximize the opportunity for and inclination of guests to purchase merchandise. Secondly, they provide their own merchandise. This occurs in a number of ways, including: tee-shirts with the name of the park on them; EPCOT clothing or souvenirs with a suitably attired cartoon character on them, such as a "French" Mickey Mouse purchased in the France pavilion or a sporty Goofy purchased in the Wonders of Life pavilion; merchandise deriving from characters specifically associated with the parks, such as Figment (a character in the Journey into Imagination ride in EPCOT); and a petrified Mickey looking out from the top of the Twilight Zone Tower of Terror (a Disney-MGM Studios attraction) emblazoned on clothing. . . .

Over the years, it has become increasingly apparent that more money can be made from feature films through merchandising and licensing than from box office receipts as such. While hugely successful merchandise bonanzas like those associated with *Star Wars, Jurassic Park* and *The Lion King* are by no means typical, they represent the tip of a lucrative iceberg. Like many movies, television series also often form the basis for successful lines of merchandise and indeed it has sometimes been suggested that they are devised with merchandise and licensing potential very much in mind. There are no guarantees, however. If a movie flops, like *Judge Dredd,* even though based on a popular comic book character and having superficial merchandise potential, the products will either not be developed or will not move out of stores. Also, the merchandising of even fairly successful films like *Flintstones* and *Casper* can be disappointing (Pereira, 1996). . . .

But it would be a mistake, of course, to view merchandising purely in terms of the movies and cartoon characters. The new themed restaurant chains all follow the lead of Hard Rock Cafe of developing extensive lines of merchandise, including the ubiquitous tee-shirt which simultaneously informs where wearers have been on their holidays and acts as literally a walking advertisement for the chain. You do not necessarily have to eat in the establishment in order to purchase the items. Very often, if not invariably, you can enter the shop area without needing to eat the food. . . .

EMOTIONAL LABOR

Ritzer (1993) was somewhat silent about the nature of work under McDonaldization, but it is clear from his view that since it incorporates Scientific management and Fordism, the work tends to be dehumanizing and alienating. More recently, Ritzer (1998) has written about "McJobs," that is, jobs specifically connected to the McDonaldization of society.

McJobs have a number of new characteristics including "many distinctive aspects of the control of these workers" (1998: 63). In particular, Ritzer draws attention to the scripting of interaction in service work. Not only does this process result in "new depths in . . . deskilling" (1998: 64) but also it entails control of the self through emotional labor, which has been defined as the "act of expressing socially desired emotions during service transactions" (Ashforth and Humphrey, 1993: 88–9). . . .

This is revealed in the insistence that workers exhibit cheerfulness and friendliness towards customers as part of the service encounter.

Emotional labor is in many ways exemplified by the Disney theme parks. The behavior of Disney theme park employees is controlled in a number of ways and control through scripted interactions and encouraging emotional labor is one of the key elements (Bryman, 1995: 107–13). The friendliness and helpfulness of Disney theme park employees is renowned and is one of the things that visitors often comment on as something that they liked (Sorkin, 1992: 228). Moreover, anyone with even a passing knowledge of the parks *expects* this kind of behavior. The ever-smiling Disney theme park employee has become a stereotype of modern culture. Their demeanor coupled with the distinctive Disney language is designed among other things to convey the impression that the employees are having fun too and therefore not engaging in real work.

REFERENCES

Ashforth, B. E. and R. H. Humphrey. 1993. "Emotional Labor in Service Roles: The Influence of Identity." *Academy of Management Journal* 18: 88–115.

Beardsworth, A. and A. Bryman. 1999. "Late Modernity and the Dynamics of Quasification: The Case of the Themed Restaurant." *The Sociological Review* 47: 1.

Bryman, A. 1995. *Disney and His Worlds.* London: Routledge.

Canemaker, J. 1991. *Felix: The Twisted Tale of the World's Most Famous Cat.* New York: Pantheon.

Crawford, M. 1992. "The World in a Shopping Mall." Pp. 3–30 in M. Sorkin (ed.), *Variations on a Theme Park: The New American City and the End of Public Space.* New York: Noonday.

deCordova, R. 1994. "The Mickey in Macy's Window: Childhood, Consumerism, and Disney Animation." Pp. 203–13 in E. Smoodin (ed.), *Disney Discourse.* New York: Routledge.

Eco, U. 1986. *Travels in Hyperreality.* London: Pan.

Forgacs, D. 1992. "Disney Animation and the Business of Childhood." *Screen* 33: 361–74.

Gottdiener, M. 1997. *The Theming of America: Dreams, Visions, and Commercial Spaces.* Boulder, CO: Westview.

Gould, S. J. 1979. "Mickey Mouse Meets Konrad Lorenz." *Natural History* 88: 30–36.

Grossman, C. L. 1993. "Vegas Deals New Hand of Family Fun." *USA Today* (international edition). 11 August: 5A.

Hopkins, J. S. P. 1990. "West Edmonton Mall: Landscape of Myths and Elsewhereness." *Canadian Geographer* 34: 2–17.

King, M. J. 1983. "McDonald's and Disney." Pp. 106–19 in M. Fishwick (ed.), *Ronald Revisited: The World of Ronald McDonald.* Bowling Green, OH: Bowling Green University Popular Press.

Klein, N. M. 1993. *Seven Minutes: The Life and Death of the American Animated Cartoon.* London: Verso.

Merritt, R. and J. B. Kaufman. 1992. *Walt in Wonderland: The Silent Films of Walt Disney.* Perdenone: Edizioni Bibliotecha dell'Imagine.

Pereira, J. 1996. "Toy Sellers Wish that Pocahontas Were a Lion." *Wall Street Journal* 24 July: B1.

Ritzer, G. 1993. *The McDonaldization of Society.* Thousand Oaks, CA: Pine Forge.

———. 1998. *The McDonaldization Thesis.* London: Sage.

Ritzer, G. and A. Liska. 1997. " 'McDisneyization' and 'Post-Tourism': Complementary Perspectives on Contemporary Tourism." Pp. 96–109 in C. Rojek and J. Urry (eds.), *Touring Cultures: Transformations of Travel and Theory.* London: Routledge.

Sorkin, M. 1992. "See You in Disneyland." Pp. 205–32 in M. Sorkin (ed.), *Variations on a Theme: The New American City and the End of Public Space.* New York: Noonday.

Tumbusch, T. 1989. *Tomart's Illustrated Disneyana Catalog and Price Guide.* Radnor, PA: Wallace-Homestead.

45

Bowling Alone

ROBERT D. PUTNAM

The concern for community may be seen as one of the strongest themes faced in contemporary America. Putnam heralds this concern by articulating the decline of those social institutions and activities—the church, the family, labor unions, civic and political engagement, fraternal organizations, service clubs, parent-teacher associations—through which Americans traditionally fused together and created the fabric of community life. He whimsically focuses the title of this selection around an ironic shift in social behavior that he thinks reflects this trend toward the decline of traditional community: more Americans are bowling today than ever before, but bowling in organized leagues has plummeted. Putnam ponders this decline in "social capital" and its consequent erosion in good neighborliness and social trust. How likely are Americans to know their neighbors, to participate in community and civic events, and to belong to social clubs? Is Putnam right or wrong in his fears about the lack of community involvement of Americans? If he is right, how might this trend continue to affect the individual and group life of Americans?

Many students of the new democracies that have emerged over the past decade and a half have emphasized the importance of a strong and active civil society to the consolidation of democracy. Especially with regard to the postcommunist countries, scholars and democratic activists alike have lamented the absence or obliteration of traditions of independent civic engagement and a widespread tendency toward passive reliance on the state. To those concerned with the weakness of civil societies in the developing or postcommunist world, the advanced Western democracies and above all the Untied States have typically been taken as models to be emulated. There is striking evidence, however, that the vibrancy of American civil society has notably declined over the past several decades.

Ever since the publication of Alexis de Tocqueville's *Democracy in America,* the United States has played a central role in systematic studies of the links between

Reprinted from the *Journal of Democracy,* pp. 65–70, by permission of The Johns Hopkins University Press and National Endowment for Democracy.

democracy and civil society. Although this is in part because trends in American life are often regarded as harbingers of social modernization, it is also because America has traditionally been considered unusually "civic" (a reputation that, as we shall later see, has not been entirely unjustified).

When Tocqueville visited the United States in the 1830s, it was the Americans' propensity for civic association that most impressed him as the key to their unprecedented ability to make democracy work. "Americans of all ages, all stations in life, and all types of disposition," he observed, "are forever forming associations. There are not only commercial and industrial associations in which all take part, but others of a thousand different types—religious, moral, serious, futile, very general and very limited, immensely large and very minute. . . . Nothing, in my view, deserves more attention than the intellectual and moral associations in America.[1]

Recently, American social scientists of a neo-Tocquevillean bent have unearthed a wide range of empirical evidence that the quality of public life and the performance of social institutions (and not only in America) are indeed powerfully influenced by norms and networks of civic engagement. Researchers in such fields as education, urban poverty, unemployment, the control of crime and drug abuse, and even health have discovered that successful outcomes are more likely in civically engaged communities. Similarly, research on the varying economic attainments of different ethnic groups in the United States has demonstrated the importance of social bonds within each group. These results are consistent with research in a wide range of settings that demonstrates the vital importance of social networks for job placement and many other economic outcomes. . . .

The norms and networks of civic engagement also powerfully affect the performance of representative government. That, at least, was the central conclusion of my own 20-year, quasi-experimental study of subnational governments in different regions of Italy.[2] Although all these regional governments seemed identical on paper, their levels of effectiveness varied dramatically. Systematic inquiry showed that the quality of governance was determined by longstanding traditions of civic engagement (or its absence). Voter turnout, newspaper readership, membership in choral societies and football clubs—these were the hallmarks of a successful region. In fact, historical analysis suggested that these networks of organized reciprocity and civic solidarity, far from being an epiphenomenon of socioeconomic modernization, were a precondition for it.

No doubt the mechanisms through which civic engagement and social connectedness produce such results—better schools, faster economic development, lower crime, and more effective government—are multiple and complex. While these briefly recounted findings require further confirmation and perhaps qualification, the parallels across hundreds of empirical studies in a dozen disparate disciplines and subfields are striking. Social scientists in several fields have recently suggested a common framework for understanding these phenomena, a framework that rests on the concept of *social capital*.[3] By analogy with notions of physical capital and human capital—tools and training that enhance individual productivity—"social capital" refers to features of social organization such as networks, norms, and social trust that facilitate coordination and cooperation for mutual benefit.

For a variety of reasons, life is easier in a community blessed with a substantial stock of social capital. In the first place, networks of civic engagement foster sturdy norms of generalized reciprocity and encourage the emergence of social trust. Such networks facilitate coordination and communication, amplify reputations, and thus allow dilemmas of collective action to be resolved. When economic and political negotiation is embedded in dense networks of social interaction, incentives for opportunism are reduced. At the same time, networks of civic engagement embody past success at collaboration, which can serve as a cultural template for future collaboration. Finally, dense networks of interaction probably broaden the participants' sense of self, developing the "I" into the "we," or (in the language of rational-choice theorists) enhancing the participants' "taste" for collective benefits.

I do not intend here to survey (much less contribute to) the development of the theory of social capital. Instead, I use the central premise of that rapidly growing body of work—that social connections and civic engagement pervasively influence our public life, as well as our private prospects—as the starting point for an empirical survey of trends in social capital in contemporary America. I concentrate here entirely on the American case, although the developments I portray may in some measure characterize many contemporary societies.

WHATEVER HAPPENED TO CIVIC ENGAGEMENT?

We begin with familiar evidence on changing patterns of political participation, not least because it is immediately relevant to issues of democracy in the narrow sense. Consider the well-known decline in turnout in national elections over the last three decades. From a relative high point in the early 1960s, voter turnout had by 1990 declined by nearly a quarter; tens of millions of Americans had forsaken their parents' habitual readiness to engage in the simplest act of citizenship. Broadly similar trends also characterize participation in state and local elections.

It is not just the voting booth that has been increasingly deserted by Americans. A series of identical questions posed by the Roper Organization to national samples ten times each year over the last two decades reveals that since 1973 the number of Americans who report that "in the past year" they have "attended a public meeting on town or school affairs" has fallen by more than a third (from 22 percent in 1973 to 13 percent in 1993). Similar (or even greater) relative declines are evident in responses to questions about attending a political rally or speech, serving on a committee of some local organization, and working for a political party. By almost every measure, Americans' direct engagement in politics and government has fallen steadily and sharply over the last generation, despite the fact that average levels of education—the best individual-level predictor of political participation—have risen sharply throughout this period. Every year over the last decade or two, millions more have withdrawn from the affairs of their communities.

Not coincidentally, Americans have also disengaged psychologically from politics and government over this era. The proportion of Americans who reply that they "trust the government in Washington" only "some of the time" or "almost never" has risen steadily from 30 percent in 1966 to 75 percent in 1992.

These trends are well known, of course, and taken by themselves would seem amenable to a strictly political explanation. Perhaps the long litany of political tragedies and scandals since the 1960s (assassinations, Vietnam, Watergate, Irangate, and so on) has triggered an understandable disgust for politics and government among Americans, and that in turn has motivated their withdrawal. I do not doubt that this common interpretation has some merit, but its limitations become plain when we examine trends in civic engagement of a wider sort.

Our survey of organizational membership among Americans can usefully begin with a glance at the aggregate results of the General Social Survey, a scientifically conducted, national-sample survey that has been repeated 14 times over the last two decades. Church-related groups constitute the most common type of organization joined by Americans; they are especially popular with women. Other types of organizations frequently joined by women include school-service groups (mostly parent-teacher associations), sports groups, professional societies, and literary societies. Among men, sports clubs, labor unions, professional societies, fraternal groups, veterans' groups, and service clubs are all relatively popular.

Religious affiliation is by far the most common associational membership among Americans. Indeed, by many measures America continues to be (even more than in Tocqueville's time) an astonishingly "churched" society. For example, the United States has more houses of worship per capita than any other nation on Earth. Yet religious sentiment in America seems to be becoming somewhat less tied to institutions and more self-defined.

How have these complex crosscurrents played out over the last three or four decades in terms of Americans' engagement with organized religion? The general pattern is clear: The 1960s witnessed a significant drop in reported weekly churchgoing—from roughly 48 percent in the late 1950s to roughly 41 percent in the early 1970s. Since then, it has stagnated or (according to some surveys) declined still further. Meanwhile, data from the General Social Survey show a modest decline in membership in all "church-related groups" over the last 20 years. It would seem, then, that net participation by Americans, both in religious services and in church-related groups, has declined modestly (by perhaps a sixth) since the 1960s.

For many years, labor unions provided one of the most common organizational affiliations among American workers. Yet union membership has been falling for nearly four decades, with the steepest decline occurring between 1975 and 1985. Since the mid-1950s, when union membership peaked, the unionized portion of the nonagricultural work force in America has dropped by more than half, falling from 32.5 percent in 1953 to 15.8 percent in 1992. By now, virtually all of the explosive growth in union membership that was associated with the

New Deal has been erased. The solidarity of union halls is now mostly a fading memory of aging men.[4]

The parent-teacher association (PTA) has been an especially important form of civic engagement in twentieth-century America because parental involvement in the educational process represents a particularly productive form of social capital. It is, therefore, dismaying to discover that participation in parent-teacher organizations has dropped drastically over the last generation, from more than 12 million in 1964 to barely 5 million in 1982 before recovering to approximately 7 million now.

Next, we turn to evidence on membership in (and volunteering for) civic and fraternal organizations. These data show some striking patterns. First, membership in traditional women's groups has declined more or less steadily since the mid-1960s. For example, membership in the national Federation of Women's Clubs is down by more than half (59 percent) since 1964, while membership in the League of Women Voters (LWV) is off 42 percent since 1969.[5]

Similar reductions are apparent in the numbers of volunteers for mainline civic organizations, such as the Boy Scouts (off by 26 percent since 1970) and the Red Cross (off by 61 percent since 1970). But what about the possibility that volunteers have simply switched their loyalties to other organizations? Evidence on "regular" (as opposed to occasional or "drop-by") volunteering is available from the Labor Department's Current Population Surveys of 1974 and 1989. These estimates suggest that serious volunteering declined by roughly one-sixth over these 15 years, from 24 percent of adults in 1974 to 20 percent in 1989. The multitudes of Red Cross aides and Boy Scout troop leaders now missing in action have apparently not been offset by equal numbers of new recruits elsewhere.

Fraternal organizations have also witnessed a substantial drop in membership during the 1980s and 1990s. Membership is down significantly in such groups as the Lions (off 12 percent since 1983), the Elks (off 18 percent since 1979), the Shriners (off 27 percent since 1979), the Jaycees (off 44 percent since 1979), and the Masons (down 39 percent since 1959). In sum, after expanding steadily throughout most of this century, many major civic organizations have experienced a sudden, substantial, and nearly simultaneous decline in membership over the last decade or two.

The most whimsical yet discomfiting bit of evidence of social disengagement in contemporary America that I have discovered is this: more Americans are bowling today than ever before, but bowling in organized leagues has plummeted in the last decade or so. Between 1980 and 1993 the total number of bowlers in America increased by 10 percent, while league bowling decreased by 40 percent. (Lest this be thought a wholly trivial example, I should note that nearly 80 million Americans went bowling at least once during 1993, *nearly a third more than voted in the 1994 congressional elections* and roughly the same number as claim to attend church regularly. Even after the 1980s' plunge in league bowling, nearly 3 percent of American adults regularly bowl in leagues.) The rise of solo bowling threatens the livelihood of bowling-lane proprietors because those who bowl as members of leagues consume three times as much beer and pizza as solo bowlers, and the money in bowling is in the beer and pizza, not the balls and shoes. The broader social significance, however, lies in the social interaction and even occasionally civic conversations over beer and pizza that solo bowlers forgo. Whether or not bowling beats balloting in the eyes of most Americans, bowling teams illustrate yet another vanishing form of social capital.

NOTES

1. Alexis de Tocqueville, *Democracy in America,* ed. J. P. Maier, trans. George Lawrence (Garden City, N.Y.: Anchor Books, 1969), 513–17.

2. Robert D. Putnam, *Making Democracy Work: Civic Traditions in Modern Italy* (Princeton: Princeton University Press, 1993).

3. James S. Coleman deserves primary credit for developing the "social capital" theoretical framework. See his "Social Capital in the Creation of Human Capital," *American Journal of Sociology* (Supplement) 94 (1988): S95–S120, as well as his *The Foundations of Social Theory* (Cambridge: Harvard University Press, 1990), 300–21. See also Mark Granovetter, "Economic Action and Social Structure: The Problem of Embeddedness," *American Journal of Sociology* 91 (1985): 481–510; Glenn C. Loury, "Why Should We Care About Group Inequality?" *Social Philosophy and Policy* 5 (1987): 249–71; and Robert D. Putnam, "The Prosperous Community: Social Capital and Public Life," *American Prospect* 13 (1993): 35–42. To my knowledge, the first scholar to use the term "social capital" in its current sense was Jane Jacobs, in *The Death and Life of Great American Cities* (New York: Random House, 1961), 138.

4. Any simplistically political interpretation of the collapse of American unionism would need to confront the fact that the steepest decline began more than six years before the Reagan administration's attack on PATCO. Data from the General Social Survey show a roughly 40-percent decline in reported union membership between 1975 and 1991.

5. Data for the LWV are available over a longer time span and show an interesting pattern: a sharp slump during the Depression, a strong and sustained rise after World War II that more than tripled membership between 1945 and 1969, and then the post-1969 decline, which has already erased virtually all the postwar gains and continues still. This same historical pattern applies to those men's fraternal organizations for which comparable data are available—steady increases for the first seven decades of the century, interrupted only by the Great Depression, followed by a collapse in the 1970s and 1980s that has already wiped out most of the postwar expansion and continues apace.

46

The Net That Binds

ANDREW SHAPIRO

In the vacuum created by the decline Putnam has noted in traditional community forms, new forms of community have arisen. One vibrant movement has been through the Internet, where we have seen intense interest directed to the formation of cyber-communities. These virtual communities represent interest groups that form across physical space, transcending the boundaries and limitations of physical proximity. Cyber communities offer the appeal of relationships and bonds based more strongly on commonalities of interests and goals than by the serendipity of geographic location. Is a community that is based on cyberspace the same as previous communities that were based on face-to-face interaction? Are there qualitative differences that should concern us or is the recent increase in the global community a harbinger of a more tightly integrated society?

One of the curious things about living through a time of whirlwind change is that it is often difficult to understand exactly what is changing. In recent years, new technology has given us the ability to transform basic aspects of our lives: the way we converse and learn; the way we work, play and shop; even the way we participate in political and social life. Dissidents around the world use the Internet to evade censorship and get their message out. Cyber-gossips send dispatches to thousands via e-mail. Musicians bypass record companies and put their songs on the Web for fans to download directly. Day traders roil the stock market, buying securities online with the click of a mouse and selling minutes later when the price jumps.

There is a common thread underlying such developments. It is not just a change in how we compute or communicate. Rather, it is a potentially radical shift in who is in control— of information, experience and resources. The Internet is allowing individuals to make decisions that once were made by governments, corporations and the media. To an unprecedented degree, we can decide what news and entertainment we're exposed to and whom we socialize with. We

Reprinted with permission from the June 21, 1999, issue of *The Nation*.

can earn a living in new ways; we can take more control of how goods are distributed; and we can even exercise a new degree of political power. The potential for personal growth and social progress seems limitless. Yet what makes this shift in power— this control revolution—so much more authentic than those revolutions described by techno-utopian futurists is its volatility and lack of preordained outcome.

Contrary to the claims of cyber-romantics, democratic empowerment via technology is not inevitable. Institutional forces are resisting, and will continue to resist, giving up control to individuals. And some people may wield their new power carelessly, denying themselves its benefits and imperiling democratic values. Nowhere are the mixed blessings of the new individual control more evident than in the relationship of the Internet to communities—not just "virtual communities" of dispersed individuals interacting online but real, geographically based communities.

MASTERS OF OUR OWN DOMAINS

The Internet's impact on community has everything to do with a digital phenomenon known as personalization, which is simply the ability to shape one's experience more precisely— whether it's social encounters, news, work or learning. Traditionally, friendships and acquaintances have been structured by physical proximity; we meet people because they are our neighbors, classmates, co-workers or colleagues in some local organization. Much of our information intake—newspapers and radio, for example—also reflects locality, and we share these media experiences and others (like national television) with those who live around us. The global reach and interactivity of the Internet, however, is challenging this. Individuals can spend more time communicating and sharing experiences with others regardless of where they live. As Internet pioneer J.C.R. Licklider wrote back in the sixties, "Life will be happier for the on-line individual because the people with whom one interacts most strongly will be selected more by commonality of interests and goals than by accidents of proximity."

Virtual communities are perfect for hobbyists and others with quirky or specialized interests—whether they're fans of swing music, chemistry professors or asthma sufferers. Indeed, these associations suggest the possibility of whole new forms of social life and participation. Because individuals are judged online by what they say, virtual communities would appear to soften social barriers erected by age, race, gender and other fixed characteristics. They can be particularly valuable for people who might be reticent about face-to-face social interaction, like gay and lesbian teenagers, political dissidents and the disabled. ("Long live the Internet," one autistic wrote in an online discussion, where "people can see the real me, not just how I interact superficially with other people.")

The Internet also gives individuals a new ability to personalize their news, entertainment and other information. And studies of Internet use show that users are doing so. Rather than having editors and producers choose what they read, hear and watch—as with newspapers or television—they are using the interactivity of the Net to gather just the material they find interesting. This may, among other things, be a winning strategy for dealing with the torrent of information that is increasingly pushed at us.

There is, in fact, plenty to like about personalization. But if we're not careful, customizing our lives to the hilt could undermine the strength and cohesion of local communities, many of which are already woefully weak. For all the uncertainty about what "community" really means and what makes one work, shared experience is an indisputably essential ingredient; without it there can be no chance for mutual understanding, empathy and social cohesion. And this is precisely what personalization threatens to delete. A lack of common information would deprive individuals of a starting point for democratic dialogue, or even fodder for the proverbial water-cooler talk. For many decades, TV and radio have been fairly criticized for drawing us away from direct interaction in our communities. Yet despite this shortcoming (and many others), these mass media at least provide "a kind of social glue, a common cultural reference point in our polyglot, increasingly multicultural society," as media critic David Shaw puts it.

Online experiences rarely provide this glue. Yes, we can share good times with others online who enjoy the same passions as we do. We can educate ourselves and even organize for political change. But ultimately, online associations tend to splinter into narrower and narrower factions. They also don't have the sticking power of physical communities. One important reason for this is the absence of consequences for offensive behavior online; another is the ease of exit for those who are offended. In physical communities, people are inextricably bound by the simple difficulty of picking up and leaving. On the Net, it's always "where do you want to go today?" Are you bored? Ticked off? Then move on! For many, this makes the virtual life an attractive alternative to the hard and often tiresome work of local community building.

Some might think that the weakness of online affiliations would prevent them from posing any real challenge to physical communities. But the ability to meander from one virtual gathering to the next, exploring and changing habitats on a whim, is exactly the problem. The fluidity of these social networks means that we may form weak bonds with others faraway at the expense of strong ties with those who live near us.

Few people, of course, intend to use the Internet in ways that will cause them to be distracted from local commitments. But technology always has unintended consequences, and social science research is beginning to show how this may be true for the Internet. Researchers who conducted one of the first longitudinal studies of the Internet's social impact, the HomeNet study, were surprised when their data suggested that Internet use increases feelings of isolation, loneliness and depression. Contrary to their starting hypotheses, they observed that regular users communicated less with family members, experienced a decline in their contacts with nearby social acquaintances and felt more stress. Although the authors noted the limitations of their findings, the study's methodology has been widely criticized. Until more conclusive results are available, however, what's important is that we take seriously the hazards outlined in the HomeNet study and attempt to prevent them from becoming worse or taking root in the first place.

And how should we do that? Neo-Luddites would likely recommend rejecting technology and returning to our bucolic roots. A more balanced and realistic response, however, calls for a reconciling of personal desire and communal obligations in a digital world. On the one hand, this means acknowledging the sometimes exhilarating adventure of indulging oneself online. No one can deny the value of being able to form relationships with far-flung others based solely on common interests. At the same time, it means not having illusions about the durability of those bonds or their ability to satisfy fully our deepest needs.

We must recognize, for selfish and societal reasons alike, the importance of focusing on the local. This is where we will find a true sense of belonging; shared experience, even if not ideal, creates a sense of commitment. This is where democracy and social justice must first be achieved; getting our own house in order is always the first priority. The Net must therefore be a vehicle not just for occasional escapism but for enhanced local engagement—online and off.

Glossary

achieved status position that is attained through individual effort, such as one's education or occupation

agent of socialization the person or group that provides information about social roles

aggregates large groups of people who actually have no relationship to one another except that they might happen to be in the same place at the same time

alienation term first used by Marx, that refers to the separation of workers from the product or result of their work, which can result in feelings of powerlessness

anticipatory socialization the process by which we prepare ourselves for future roles through thinking about and rehearsing the actions, emotions, and skills that may be involved in these new roles

ascribed status position that is attained through circumstances of birth and that cannot be changed, such as one's race-ethnicity or gender

backstage Goffman's term for the setting, or frame, in which impression management is not needed; contrast to *frontstage*

bourgeoisie term used by Marx and Marxian scholars to describe capitalists, people who own the factories and mills; contrast to *proletariat*

bureaucracies highly structured and formalized organizations that are governed by laws and rules

capitalism the free-enterprise economic system in which private individuals or corporations develop, own, and control business enterprises; contrast to *socialism*

case study a research technique that involves an in-depth look at one case, such as one person, one group, or one organization

causal model graphic device that illustrates sociological relationships between variables

causal relationship an association between variables in which one influences or causes the other

class consciousness Marx's term for people in a social class who are aware of their common interests and concerns and of the fact that these interests conflict with those of another class group

closed-ended questions survey questions that give respondents only certain possible options from which to choose their answers

collective consciousness Durkheim's term describing the common beliefs, values, and norms of people within society

community a group of people who have frequent face-to-face interactions and common values and interests, relatively enduring ties, and a sense of personal closeness to one another

comparative research research method that involves the comparison of data from a variety of groups or settings, such as nations or historical eras

conformity the process of maintaining or changing behavior to comply with the norms established by a society, subculture, or other group

control variable variable that is added to an analysis to see if it affects the relationship between an independent and a dependent variable

correlation the ways in which two variables may be related to each other in a predictable pattern

counterculture a group that strongly rejects dominant societal values and norms and seeks alternative lifestyles

crime actions that a society explicitly prohibits and that are sanctioned through official means

cultural capital Bourdieu's term describing how people behave, dress, and talk and how these manners and styles differentiate those in one class group from those in another

cultural diffusion process whereby elements of one culture or subculture spread from one society or culture to another

culture a common way of life; the complex pattern of living that humans develop and pass on from generation to generation

data factual information that is used as the basis for making decisions and drawing conclusions; the plural form of the word *datum*

deductive reasoning logical process of reasoning that moves from general theories or ideas to specific hypotheses or expectations; the opposite of *inductive reasoning*

demography the scientific study of populations and their effects

dependent variable variable that is said to be influenced or caused by another variable (the independent variable)

deviance behaviors that violate social norms

discrimination differential treatment accorded to a group of people based solely on ascribed characteristics such as race-ethnicity

domination form of power in which one party controls the behavior of others through sanctions; compare with *influence*

dramaturgical theory theory derived from the work of Goffman that uses the metaphor of a drama to explain how individuals play social roles and thus produce social structure

dysfunctions negative consequences of a structure of society for the whole of society

economy social institution that includes all the norms, organizations, roles, and activities involved in the production, distribution, and consumption of goods and services

education social institution that includes all the norms, organizations, roles, and statuses associated with a society's transmission of knowledge and skills to its members

elites powerful people who are able to influence the political process

elitism view that political power and influence are dominated by a small handful of people who are relatively unified and form a comparatively small, tight-knit social network

empirical based on experience and observation rather than preexisting ideas

endogamy marriage rule requiring people to select partners from within their own tribe, community, social class, or racial-ethnic or other such group

ethnic group group of people who have a common geographical origin and biological heritage and who share cultural elements, such as language; traditions, values, and symbols; religious beliefs; and aspects of everyday life, such as food preferences

ethnocentrism the belief that one's own culture or way of life is superior to that of others

exclusion term describing attempts to entirely remove lower-paid groups from a labor market, as, for example, through restrictive immigration laws

exogamy marriage rule requiring people to select partners from outside their own tribe, community, social class, or racial-ethnic or other such group

experiment research method that uses control groups and experimental groups to assess whether a causal relationship exists between an independent variable and a dependent variable

experimental variable see *independent variable*

extended family family that includes relatives besides parents and children; contrast to *nuclear family*

falsification the logic that underlies the testing of hypotheses; we can never prove that a theory is true, but can only say that it has either been falsified (shown to be untrue) or not yet been finished

feminism an ideology that directly challenges gender stratification and male dominance and promotes the development of a society in which men and women have equality in all areas of life

field experiment experiment that takes place in a real-life setting as opposed to a laboratory

field research research method in which a researcher directly observes behaviors and other phenomena in their natural setting

folklore myths and stories that are passed from one generation to the next within a culture

folkways norms that govern the customary way of doing everyday things

frontstage Goffman's term for the setting, or frame, in which behaviors are designed to impress or influence others and in which impression management is important; contrast to *backstage*

function the part a structure plays in maintaining or altering the society

gemeinschaft Tönnies's term describing relationships that might appear in small, close-knit communities in which people are involved in social networks with relatives and long-time friends and neighbors, much like those that appear in primary groups

gender-based division of labor rules about what tasks members of each sex should perform

gender identity one's gut-level belief that one is a male or a female

gender roles the norms and expectations associated with being male or female

gender segregation of the labor force the gender-based division of labor in the occupational world, or the phenomenon of men and women holding very different jobs; also called *occupational gender segregation*

gender socialization learning to see oneself as a male or female and learning the roles and expectations associated with that sex group

gender stratification the organization of society in a way that results in members of one sex group having more access to wealth, prestige, and power than members of the other sex group

generalized other Mead's term for the conception people have of the expectations and norms that others generally hold; the basis of the "me"

gesellschaft Tönnies's term describing relationships that come about through formal organizations and economic relationships rather than kinship or friendship, similar to those that appear in secondary groups

globalization the process by which all areas of the world are becoming interdependent and linked with one another

group in sociological terms, two or more people who regularly and consciously interact with each other through engaging in some common activity and having some relatively stable social relationship

health maintenance organizations (HMOs) organizations that provide, for a set monthly fee, total care with an emphasis on prevention to avoid costly treatment later

hegemony a hidden but pervasive power involving such extreme domination of social life that we seldom recognize it or question its legitimacy

historical-comparative research research that uses historical data to compare two or more societies

historical research research method that involves the examination of data from the past, often written artifacts and records

hypotheses statements about the expected relationship between two or more variables; often derived from theories

ideal types concepts or descriptions of phenomena that may not exist in a pure form in the real world but that define basic aspects of a given situation

ideologies complex and involved cultural belief systems

impression management Goffman's term describing how individuals may manipulate the impression or view that others have of them and give out cues to guide interactions in a particular direction

income how much money a person receives in a given time, such as $20,000 a year

independent variable variable that is said to cause or influence another variable (the dependent variable); also called the *experimental variable* in experiments

inductive reasoning logical process of reasoning that moves from specific ideas and observations to more general hypotheses and theories; the opposite of *deductive reasoning*

influence form of power whereby providing information and knowledge leads others to take different actions; compare with *domination*

institutional norms norms that prescribe appropriate structures and behaviors or organizations and other aspects of social institutions

interest groups political organizations that concentrate their activities on specific policy issues or concerns

internal validity the extent to which the conclusions of a study are true. Specifically in an experiment, the extent to which the observed changes in a dependent variable are caused by the introduction of an independent variable

intervening variable variable that comes between a dependent variable and an independent variable in a causal relationship

labeling theory theory which suggests that definitions of deviant behavior develop from social interactions and that the key element in becoming deviant is how others respond to people's behavior, rather than how they actually behave

latent functions functions that are less obvious and often unintended and that generally are unnoted by the people involved; contrast to *manifest functions*

longitudinal study study which involves data that have been collected at different times

macrolevel theories and analyses that deal with relatively broad areas of society rather than with individuals

male dominance cultural beliefs that give greater value and prestige to men and their roles and activities

managed care any system of cost containment that closely monitors and controls health care providers' decisions about medical procedures, diagnostic tests, and other services that should be provided to patients

manifest functions functions that are easily seen and obvious; contrast to *latent functions*

mean statistical average; computed by simply adding up all values and dividing by the number of cases

means of production term used by Marx and Marxian scholars to refer to the way in which people produce their living, such as by farming or manufacturing or hunting and gathering

median the midway point in a distribution; the point at which 50 percent of the cases are larger and 50 percent are smaller

medicalization the process whereby nonmedical problems become defined and treated as illnesses and disorders

methodology the rules and procedures that guide research and help make it valid

microlevel theories and analyses that deal with relatively narrow aspects of social life, such as individuals' day-to-day activities and relations with other people

middle-range theories theories that focus on relatively limited areas of the social world, as opposed to grand theories; often incorporate aspects of grand theories but are much more directed and applied toward specific research problems and can thus be more easily tested

modal the most common or frequently occurring category or case

monogamy the marriage of one man and one woman

mores norms that are vital to society, and violation of which is seen as morally offensive

neo-Marxist term used to describe recently developed theories that are in the marxian tradition although they may depart from Marx's thought in certain ways

nonparticipant observation type of field research in which a researcher studies a group through observations without actually participating

nonverbal communication all the ways in which we send messages to others without words, including posture and movements, facial expressions, clothes and hairstyles, and manner of speaking

norms cultural rules defining behavior that is expected or required within a group or situation; includes folkways, mores, and laws

nuclear family family group consisting of a mother, a father, and their children; contrast to *extended family*

open-ended questions survey questions that allow respondents to give whatever responses they desire

panel study a longitudinal study that includes information on the same people over a long period of time

participant observation type of field research in which a researcher studies a group or event while actually participating in it

patriarchy a hierarchical system of social organization in which cultural, political, and economic strcutures are controlled by men

placebo a possible, simulated treatment of the control group, which is designed to appear authentic

pluralism view that the political power structure involves a number of powerful groups and individuals, all of which can potentially influence the decision-making process

population the entire group or set of cases that a researcher is interested in generalizing to; see also *sample*

post-test the measurement of the dependent variable that occurs after the introduction of the stimulus or the independent variable

power the ability of one social element, either a group or a person, to compel another social element to do what it wants

prejudice preconceived hostile attitudes toward a group of people simply on the basis of their group membership

pre-test the measurement of the dependent variable that occurs before the introduction of the stimulus or independent variable

primary groups groups that include only a few people and that are characterized by intimate, face-to-face interaction

probability sample sample that can be generalized to a larger group, typically chosen through some type of random selection process

proletariat term used by Marx and Marxian scholars to describe the workers; contrast to *bourgeoisie*

qualitative data measures of data that cannot be assigned real numbers; contrast to *quantitative data*

quantitative data measures that may be assigned real, or meaningful, numbers—for example, income or age

racial-ethnic group subculture that can be distinguished on the basis of skin color and ethnic heritage

random selection process that gives each member of a population an equal chance of being included in a sample

reliability the extent to which a measure yields the same results when used by different researchers on the same subject at different times

religion the social institution that deals with the area of life people regard as holy or sacred; it involves the statuses, roles, organizations, norms, and beliefs that are related to humans' relationship with the supernatural, including shared beliefs, ethical rules, rituals and ceremonies, and communities of people with common beliefs and standards

replication repetition of an earlier study to see if the same results occur and if they hold in other settings

role conflict situation in which a person holds roles with incompatible norms or obligations

role theory perspective that social structure is created and maintained because people generally act in ways that conform to social roles

role transition moves, or transitions, from one role to another during the life course

sample subset of a larger group or population; see also *probability sample*

sanctions social reactions to an individual's behavior, generally reflecting attempts to control the behavior; rewards and punishments

segregation the division or separation of neighborhoods in ways that lead to the inclusion of some groups and the exclusion of others

self one's view of oneself as a distinct person with a clear identity

self-concept the thoughts and feelings we have about ourselves

self-identity a set of categories used to define the self; the way we think about ourselves

significant others people with whom you interact and who are emotionally important to you

social capital resources or benefits people gain from their social networks

social change the way in which societies and cultures alter over time

social class groups of people who occupy a similar level in the stratification system

social control efforts to help ensure conformity to norms

social institutions the complex sets of statuses, roles, organizations, norms, and beliefs, that meet people's basic needs within a society

socialism an economic system that involves public rather than private ownership of the means of production; contrast to *capitalism*

socialization the way in which we develop, through interactions with others, the ability to relate to other people and to play a part in society

social mobility movement between social class groups

social role expectations, obligations, and norms that are associated with a particular position in a social network

social status positions that individuals occupy within the social structure

social stratification the organization of society in a way that results in some people having more and some people having less; divisions in a society based on social class

social structure relatively stable patterns that underlie social life; the ways in which people and groups are related to each other, and the characteristics of groups that influence our behavior

society a group of people who live within a bounded territory and who share a common way of life

sociological imagination Mills's term describing the ability to discern patterns in social events and view personal experiences in light of these patterns

sociology the science of society; the scientific study of the social world and social institutions

spurious correlation correlation between two variables that only occurs because of the influence of a third variable

status term used in the Weberian tradition to designate one dimension of stratification, that involving communities or social networks of people with similar lifestyles and viewpoints; synonymous with *prestige*

status characteristics the statuses that people hold and the evaluations and beliefs (characteristics) that are attached to these statuses

stigma according to Goffman, any physical or social attribute or sign that so devalues a person's social identity that it disqualifies that person from full social acceptance

stimulus the experimental condition of the independent variable that is controlled or introduced by the researcher in an experiment

structural discrimination discrimination that results from the normal and usual functioning of the society (the social structure) rather than from prejudice or from laws and norms that promote segregation or exclusion

structural functionalism sociological theory that tries to account for the nature of social order and the relationship between different parts of society by noting the ways in which these parts or structures function to maintain the entire society

survey research method of data gathering that involves asking people questions, through either interviews or written questionnaires

symbolic interactionism theory that social interaction involves a constant process of presenting and interpreting symbols through thinking about what another person is trying to communicate through the use of symbols

symbols anything that people use to represent something else; for example, language uses the symbols of words to represent objects and ideas

theories broad systems of ideas that help explain patterns in the social world

typology a classification of a group or phenomenon into discrete categories

unobtrusive research research method in which a researcher obtains data without directly talking to or watching people

upward mobility social mobility that involves movement to a social class position that is higher than one's parents occupied

urbanization the process of societal change that involves the movement of people from rural areas or small towns to metropolitan areas

validity the extent to which a measure actually represents the concept it is said to be measuring; when applied to a research design, indicates that we can trust the conclusions

values general standards about what is important to a group

variables logical groupings of attributes; literally, things that vary or have more than one value

wealth assets resulting from the accumulation of income, such as houses, cars, real estate, and stocks and bonds

white-collar crime nonviolent crimes that generally involve fraud and deception and are committed in the workplace